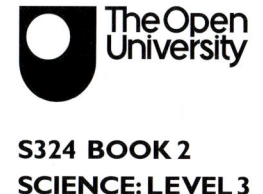

S324 BOOK 2
SCIENCE: LEVEL 3

ENVIRONMENTAL PHYSIOLOGY

EDITED BY ROBIN HARDING

THE S324 COURSE TEAM FOR BOOK 2

COURSE TEAM CHAIR
Robin Harding

COURSE MANAGER
Alastair Ewing

COURSE TEAM ASSISTANTS
Catherine Eden

Dawn Partner

Yvonne Royals

AUTHORS
Patricia Ash

Laurie Haynes

Caroline Pond

David Robinson

ACADEMIC EDITOR
Caroline Pond

CONSULTANT
Richard Jurd

EDITORS
Sheila Dunleavy

Bina Sharma

DESIGN
Jane Sheppard

ILLUSTRATION
Steve Best

PICTURE RESEARCH
Lydia Eaton

Deana Plummer

INDEXER
Jane Henley

BOOK ASSESSOR
Mike Jakobson

This publication forms part of an Open University course S324 *Animal Physiology*. The complete list of texts which make up this course can be found on the back cover. Details of this and other Open University courses can be obtained from the Course Information and Advice Centre, PO Box 724, The Open University, Milton Keynes MK7 6ZS, United Kingdom: tel. +44 (0)1908 653231, e-mail general-enquiries@open.ac.uk

Alternatively, you may visit the Open University website at http://www.open.ac.uk where you can learn more about the wide range of courses and packs offered at all levels by The Open University.

To purchase a selection of Open University course materials visit the webshop at www.ouw.co.uk, or contact Open University Worldwide, Michael Young Building, Walton Hall, Milton Keynes MK7 6AA, United Kingdom for a brochure. tel. +44 (0)1908 858785; fax +44 (0)1908 858787; e-mail ouwenq@open.ac.uk

The Open University
Walton Hall, Milton Keynes
MK7 6AA

First published 2004

Copyright © 2004 The Open University

All rights reserved. No part of this publication may be reproduced, stored in a retrieval system, transmitted or utilized in any form or by any means, electronic, mechanical, photocopying, recording or otherwise, without written permission from the publisher or a licence from the Copyright Licensing Agency Ltd. Details of such licences (for reprographic reproduction) may be obtained from the Copyright Licensing Agency Ltd of 90 Tottenham Court Road, London W1T 4LP.

Open University course materials may also be made available in electronic formats for use by students of the University. All rights, including copyright and related rights and database rights, in electronic course materials and their contents are owned by or licensed to The Open University, or otherwise used by The Open University as permitted by applicable law.

In using electronic course materials and their contents you agree that your use will be solely for the purposes of following an Open University course of study or otherwise as licensed by The Open University or its assigns.

Except as permitted above you undertake not to copy, store in any medium (including electronic storage or use in a website), distribute, transmit or re-transmit, broadcast, modify or show in public such electronic materials in whole or in part without the prior written consent of The Open University or in accordance with the Copyright, Designs and Patents Act 1988.

Edited, designed and typeset by The Open University.

Printed and bound in the United Kingdom by the Alden Group, Oxford.

ISBN 0 7492 58535

1.1

CONTENTS

CHAPTER 1   INTEGRATIVE PHYSIOLOGY 5
1.1 Introduction 5
1.2 Integration across levels of analysis 8
1.3 Integration across disciplines 11
1.4 Integration across species 13
1.5 The importance of phylogeny 14
1.6 Physiology and conservation 16
1.7 Conclusion 18
1.8 Summary of Chapter 1 19
     Learning Outcomes for Chapter 1 19
     Questions for Chapter 1 19

CHAPTER 2   THE EXTERNAL ENVIRONMENT 21
2.1 Introduction 21
2.2 Thermal characteristics of the external environment 22
2.3 Thermal exchanges 25
2.4 Effects of temperature on metabolism 29
2.5 Thermoregulation 38
2.6 Oxygen in the environment 45
2.7 The physics of gas exchange 46
2.8 Effects of hypoxia on metabolism 50
2.9 Effects of temperature and hypoxia at the molecular level 55
2.10 A new approach to environmental physiology: adaptation 58
2.11 Conclusion 62
     Learning Outcomes for Chapter 2 63
     Questions for Chapter 2 64

CHAPTER 3   THE DESERT ENVIRONMENT 69
3.1 Introduction 69
3.2 Environments and populations 73
3.3 Integrating across levels of analysis 85
3.4 Integrating across disciplines 108
3.5 Integrating across species 112
3.6 Phylogeny and cladistic analysis 115
3.7 Conclusion 119
     Learning Outcomes for Chapter 3 120
     Questions for Chapter 3 120

CHAPTER 4   HIBERNATION AND TORPOR   123
4.1   Introduction   123
4.2   The nature and extent of hibernation and torpor in endotherms   126
4.3   Characteristics of hibernation behaviour   130
4.4   Physiological adaptations – molecules and cells   141
4.5   Physiological adaptations – respiration and energy provision   149
4.6   Control systems   158
     Learning Outcomes for Chapter 4   168
     Questions for Chapter 4   169

CHAPTER 5   POLAR BIOLOGY   171
5.1   Introduction   171
5.2   Environmental regulation of physiological processes   174
5.3   Natural feasting and fasting   181
5.4   Thermal insulation   193
5.5   Polar ectotherms   202
5.6   Conclusion   211
     Learning Outcomes for Chapter 5   212
     Questions for Chapter 5   213

CHAPTER 6   ALTITUDE   215
6.1   Introduction: the high altitude environment   215
6.2   Initial physiological and biochemical responses to high altitude   217
6.3   Genotypic adaptations for life at high altitude in vertebrate species   230
6.4   Integrating across disciplines   236
6.5   Integrating across species   241
     Summary of Chapter 6   245
     Learning Outcomes for Chapter 6   246
     Questions for Chapter 6   247

REFERENCES AND FURTHER READING   249

ANSWERS TO QUESTIONS   257

ACKNOWLEDGEMENTS   271

INDEX   275

CHAPTER 1   INTEGRATIVE PHYSIOLOGY

Prepared for the Course Team by David Robinson

1.1 Introduction

Our understanding of animal physiology has changed radically in recent years. In addition to the major advances in laboratory techniques, technological developments have enabled researchers to study physiological variables in free-living animals, rather than just in laboratory experiments, and this has changed the emphasis of many investigations. The study of animals in their undisturbed state has been variously referred to as 'whole-animal physiology', 'physiological ecology' or 'environmental physiology'. However, developments in other disciplines within the Life Sciences have had an impact on physiology too, particularly the new techniques in molecular biology. Now it is becoming possible to study not only the small but significant differences in physiology that exist within and between species, but also to understand what processes and genes underlie these differences and how they might have evolved. So the study of animal physiology now crosses a range of disciplines, including evolutionary biology and genetics.

In Book 1 you saw how physiological processes can be studied at many levels: molecular, cellular, whole organ and whole animal. In this book you will study animal physiology in an integrated way, which emphasizes the links between disciplines and demonstrates that integrative physiology is more than just comparisons between species. This chapter shows how integrative physiology influences research. Many of the examples come from subject areas that are treated in detail in later chapters of the book. There is a continuum from environments and populations to molecules and cells, and physiologists are now looking at traditional subject areas, such as hibernation and temperature regulation, in very different ways from previously. Integrative physiology has become a major subject area and there are courses, conferences and even research departments that use 'Integrative Physiology' in their title.

An example of the integrative approach comes from work on hibernating mammals. Hibernation is considered in detail in Chapter 4.

Small mammals that survive seasonally cold environments often hibernate, entering a state of controlled torpor where the body temperature falls. Hibernation in mammals requires profound changes in body temperature, metabolism, respiration and heart rate, with a relaxation of the strict limits within which these physiological variables are normally regulated. There is a metabolic shift away from the oxidation of carbohydrates and towards the oxidation of stored triacylglycerols as the primary source of energy during torpor.

■ Why is it advantageous to switch from the oxidation of carbohydrates to the use of stored triacylglycerols?

Glucose and ketones are the sole sources of metabolic fuel for the brain, and the concentration of these metabolites in the blood has to be maintained in the absence of feeding, during hibernation. In all tissues except the brain, carbohydrates are spared and the most abundant fuel is fatty acid.

■ How could you demonstrate this switch from the oxidation of carbohydrates to the mobilization and utilization of stored triacylglycerols in a hibernating mammal?

By measuring the respiratory exchange ratio (RER) of a hibernating mammal, it would be possible to get a rough idea of what type of fuel was being used in the production of metabolic energy. RER was described in Book 1, Section 1.4.1.

If fatty acids were the principal fuel, the RER would be expected to be around 0.7 and this value has been observed in hibernators. The value would be 1.0 if the source was pure carbohydrate. How is the switch from carbohydrates to fatty acids achieved?

Fuel selection is controlled by pyruvate dehydrogenase kinase (PDK). PDK exists in several forms, isoenzymes, that are similar in function but differ slightly in amino acid sequence. The isoenzyme PDK4 inhibits pyruvate dehydrogenase, preventing the conversion of pyruvate into acetyl CoA and CO_2, thus minimizing carbohydrate oxidation by preventing the flow of glycolytic products into the tricarboxylic acid cycle.[*] PDK4 has been studied in the 13-lined ground squirrel (*Spermophilus tridecemlineatus*). This is a small, slender species of ground squirrel with longitudinal stripes of dark brown and tan, extending along most of its body. It has a wide geographic range in North America, from Texas to Alberta. Those animals living in cold climates hibernate in underground burrows for as long as August until March. During hibernation, there is a 96% reduction in pyruvate dehydrogenase activity in the heart muscle. Such changes also occur in mammals that go into torpor daily. Gerhard Heldmaier showed that the daily reduction in metabolic rate associated with torpor is accompanied by inactivation of pyruvate dehydrogenase in the heart tissue of Djungarian hamsters (*Phodopus* sp.) (Heldmaier et al., 1999). Pyruvate dehydrogenase is inhibited by reversible phosphorylation catalysed by certain enzymes, of which PDK4 is one.

The level of expression of mRNA that codes for PDK4 is a measure of the activity of the isoenzyme and this can now be readily assayed. Comparison of ten tissues from 13-lined ground squirrels during hibernation shows that PDK4 mRNA expression increases fivefold in skeletal muscle and 15-fold in white adipose tissue (WAT; Book 1, Section 1.4.2) compared with summer levels when the squirrels are active (Buck et al., 2002). Similarly, PDK4 protein increases threefold in heart muscle, fivefold in skeletal muscle, and eightfold in WAT. Monthly sampling over the whole year shows the rise and fall in PDK4 (Figure 1.1). In these squirrels hibernation is interrupted by periods of interbout arousal, when the body temperature rises and the animals are active.

■ During interbout arousal, what happens to the level of expression of mRNA that codes for PDK4?

The level of expression of mRNA, which is much higher in the hibernating animal than in the active one, remains high and possibly increases further during interbout arousal in both skeletal muscle and WAT. (The increase shown in Figure 1.1 cannot be statistically significant.) As PDK4 inhibits carbohydrate oxidation during hibernation, when the animal is not feeding, an increase in PDK4 activity during interbout arousal is consistent with the switch from carbohydrate to fatty acid as a substrate for energy production.

[*] If you need to recall the details of this cycle, look ahead to Figure 2.1 in Chapter 2.

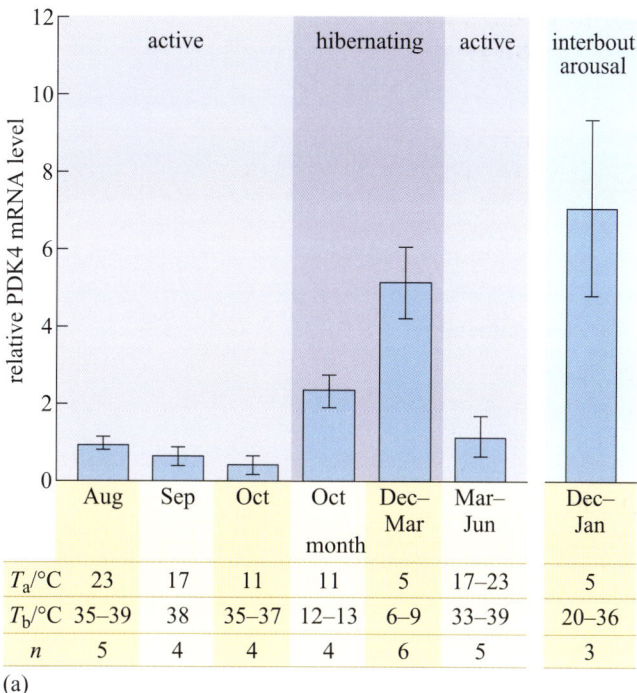
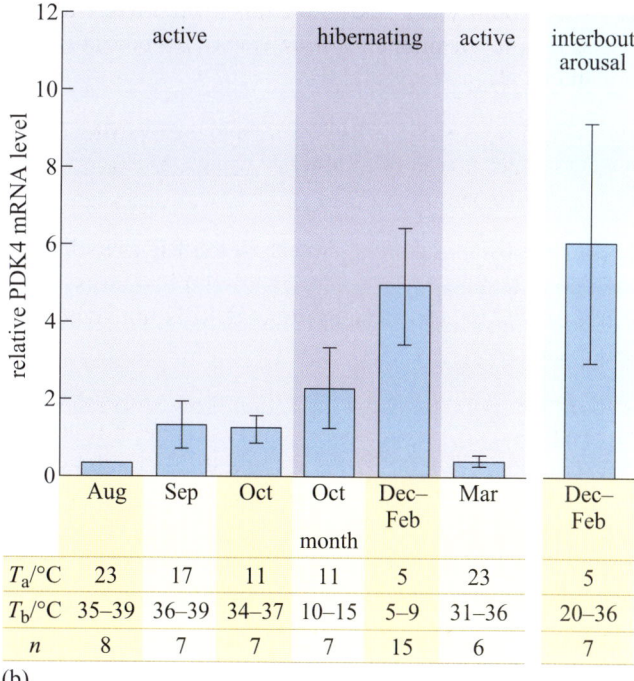

(a)   (b)

Figure 1.1 Comparison of the expression of PDK4 mRNA (± SE) in (a) skeletal muscle and (b) WAT from the 13-lined ground squirrel, over a year. T_a = ambient temperature, which was controlled by the experimenter; T_b = body temperature; n = number of observations.

The amount of PDK4 protein present in the three tissues was also assayed and found to be consistent with the mRNA profiles in Figure 1.1.

The coordinated regulation of PDK4 in three distinct tissues suggests a common signal regulates PDK4 expression and fuel selection during hibernation. There are a number of possible signals. In the 13-lined ground squirrel, high levels of serum insulin are seen during the autumn when PDK4 mRNA levels are low, and it is likely that insulin exerts an inhibitory effect on *PDK4* gene expression. This conclusion is supported by data from diabetic mice. These mice have high levels of mRNA expression for PDK4, but injections of insulin return the PDK4 levels to that of control animals. Another possible signal is the level of long-chain non-esterified (free) fatty acids, such as linoleic acid, which are known to activate the *PDK4* gene, via an indirect route. Elevated free fatty acid levels in the blood have been observed in the period leading up to hibernation, and during hibernation, and may promote *PDK4* gene expression. A rise in fatty acid levels in the blood results from lipolysis of triacylglycerols stored in WAT and the primary lipolytic enzyme in WAT is a hormone-sensitive lipase. However, in the 13-lined ground squirrel, a second lipolytic enzyme PTL (pancreatic triacylglycerol lipase), which mediates lipolysis and production of free fatty acids, is expressed in WAT before and during hibernation. The action of PTL is not inhibited by insulin. Insulin regulates gene expression for PDK4, but insulin regulation can be over-ridden by rising free fatty acid levels, since insulin does not inhibit PTL. PDK4 expression can occur even if there are high levels of insulin in the blood. The significance of this for the ground squirrel is that the switch to free fatty acid metabolism at the onset of hibernation can take place even though insulin levels are high, a state which normally would inhibit PDK4 expression.

The study of hibernating ground squirrels has drawn in elements of ecology, thermobiology, biochemistry and genetics, illustrating the scope of integrative physiology. There are three facets of integration, each of which will be introduced in a separate section here and then highlighted in later chapters of this book. The three facets are:

- integration across levels of analysis;
- integration across disciplines;
- integration across species.

In making comparisons, the most interesting ones from a physiologist's viewpoint are:

- comparing closely related animals that are adapted to *different* environments;
- comparing unrelated species that are adapted to the *same* environment;
- comparing the responses of individuals of the same species to changing conditions.

You will find examples of these comparisons throughout the book, but comparison becomes more powerful when allied to integration.

1.2 Integration across levels of analysis

Animals that have a body temperature that is entirely or substantially dependent upon deriving heat energy from external sources are **ectotherms**. In contrast, **endothermy** is a physiological process that is largely restricted to mammals and birds, where the large quantities of heat generated internally as a by-product of metabolism enable the maintenance of a stable body temperature. Generation of heat through shivering is also a feature of these two groups. However, the evolution of specialized tissue that generates heat, **thermogenic tissue,** is of great physiological interest since although it is not widespread, it has evolved several times (in different forms) in mammals and in fish. Endothermy can be studied at the whole-animal level by measuring body temperature at different environmental ambient temperatures. In free-ranging animals, it is possible to attach instrument packages that transmit data to orbiting satellites. Endothermy can also be analysed at the systems level, by studying the physiological adaptations that conserve heat, and at the biochemical level within the cells of thermogenic tissue. At the molecular level, the expression of the unique proteins within thermogenic tissue might give insights into the evolution of the heat-generating processes.

Most fish have a body temperature that is essentially the same as that of the water that surrounds them. Water is a very good conductor of heat with a thermal conductivity about 23 times as great as that of still air. As a consequence fish lose heat rapidly to the surrounding water and maintaining a body temperature above that of water requires heat to be produced at a high rate. Thus the vast majority of fish can be regarded as thermoconformers – their body temperature matches that of the surrounding water and they make no physiological effort to regulate it (Section 2.5.1).

However, it has been known for a long time that some fish do maintain a body temperature above that of water; skipjack tuna caught during the voyage of *The Beagle* (1831–1836) were recorded as being warmer than the water. Recent measurements have shown that the **body temperature** (T_b) of bluefin tuna is consistently 5 °C above the **ambient temperature** (T_a) of the water. The heat energy that maintains this temperature difference is generated internally, so tuna are described as endotherms. However, skipjack tuna maintain a greater difference between T_b and T_a in cold water temperatures than they do in warmer water. This observation suggests that skipjack are behaving in a similar way to true **homeotherms**, like most mammals, in that they tend to maintain a stable T_b despite changes in T_a.

Information about the physiology of these tuna comes from observations on the free-living animals. The technological development that has made this possible is the archive tag (Block et al., 2001). There are two forms of archive tag, an implanted one and an external pop-up tag. Both tags contain a computer that records and stores data, and each tag type has advantages and disadvantages. The implanted tag can be retrieved only when the fish is caught, but the pop-up tag can communicate with the Argos satellite tracking system, when it is at the sea surface following release from the fish. The pop-up type contains a radio and is pre-programmed with a release time. When it pops up above the water surface, it can transmit data, so that even if the tag is lost before its programmed release date, the information it carries can still be retrieved. (Radio signals cannot be transmitted from below the water surface.) The point from which the tag transmits, locates the animal to which it is attached. Implanted tags rely on a fisherman spotting an external marker on a fish they have caught, removing the implanted tag and sending it back in good condition, to the researchers (for a $1000 reward!). In many cases the location where the fish was caught may not be known precisely.

Archive tags implanted in bluefin tuna have recorded depth information, ambient temperature, body temperature and light intensity every two minutes over a period of up to 400 days. These records have shown that individuals experience a range of environmental temperatures from 2.8 to 30.6 °C while maintaining an internal temperature relatively constant around 25 °C with a maximum thermal excess of 21 °C above ambient.

The body temperatures of tuna are higher than ambient because the muscles produce large amounts of heat, and physiological mechanisms reduce heat loss to the water (see Book 4, Section 3.3). Several large pelagic fishes that are members of the family Scombroidei possess a unique form of endothermy that keeps the temperature of specific parts of the body above that of the surrounding water. The Scombroidei includes tunas, bonitos, butterfly mackerel and billfishes (which includes swordfish and marlins; Figure 1.2).

In these latter two groups, a specialized heating organ heats the brain and eye tissue. The heater organ appears to have evolved from muscle; it warms these nervous tissues and maintains them at a more stable temperature than the rest of the body. The cooling of the body that occurs when the fish swims into colder water, or slows down, would produce fairly rapid changes of temperature in the central nervous system, but the thermogenic tissue dampens down such changes.

Figure 1.2 The blue marlin (*Makaira nigricans*).

Fish that are active swimmers generate more metabolic heat than do sedentary species. The heat is not dissipated as rapidly in a large fish as it would be in a small one, so the swimming muscle could undergo temperature fluctuations linked to fluctuation in metabolic activity. Temperature fluctuations might also be produced in a large fish that moved rapidly between warm surface water and deeper, cooler water.

Studies on swordfish have shown that when the animal dives into colder water, it can maintain a cranial temperature at 14 °C above that of the water surrounding it (Carey, 1990).

The heater organ in these fish is very unusual. Only one other animal tissue is known to function primarily as a source of heat.

■ What is the other example of thermogenic tissue, and where is it found?

Mammalian brown fat is thermogenic and it is found in neonatal mammals, adult hibernators and in smaller quantities in some other mammals. It produces large quantities of heat when the normal proton gradient across the inner membrane of the mitochondria collapses as a result of the activation of uncoupling protein. The uncoupling protein bypasses the ATP-generation process and diverts respiratory energy from ADP phosphorylation to the production of heat.

No other mitochondria have been found in animal tissues that possess this property of reversible uncoupling of ATP generation from proton movement. An uncoupling protein is not present in the mitochondria of the heater organ of fish, indicating that they have a different mechanism of heat generation.

The cells of the heater organ have an extensive sarcoplasmic reticulum (SR) and T-tubule system, features that are typical of skeletal muscle. In muscle, tubular extensions of the plasma membrane extend into the cells, forming the T-tubule system. The lumen of the tubules is continuous with the extracellular fluid. When the plasma membrane is depolarized, the T-tubule system ensures that the signal reaches the cytoplasm rapidly and triggers the release of calcium ions from the sarcoplasmic reticulum into the cytosol. Calcium ions activate enzymes that produce contraction (Book 3). In the heater organ cell, unlike muscle, there are no contractile proteins and there is no detectable ATPase activity like that of muscle (Book 3). An average 63% of the volume of the cytoplasm is taken up by mitochondria, with the bulk of the remaining volume being occupied by the SR. The SR forms a network around each mitochondrion. The plasma membrane of each cell is rich in acetylcholine receptors. In muscle, acetylcholine is the transmitter that acts at the nerve–muscle junction (the endplate), depolarizing the plasma membrane and causing an efflux of calcium ions from the SR. All these features together suggest that calcium ions regulate the metabolism of heater organ cells. Analysis of the role of calcium shows that release of calcium from the SR membrane systems stimulates thermogenesis (O'Brien and Block, 1996). The re-uptake of calcium is an active process, requiring ATP. The ATP is generated within the mitochondria and no uncoupling takes place. Instead, calcium ions go round a cycle (called a 'futile cycle' by some workers), and the active transport of calcium ions requires large amounts of ATP. The increased metabolism in the tissues that is necessary to supply ATP produces so much heat that it warms the adjacent eye and brain (Figure 1.3).

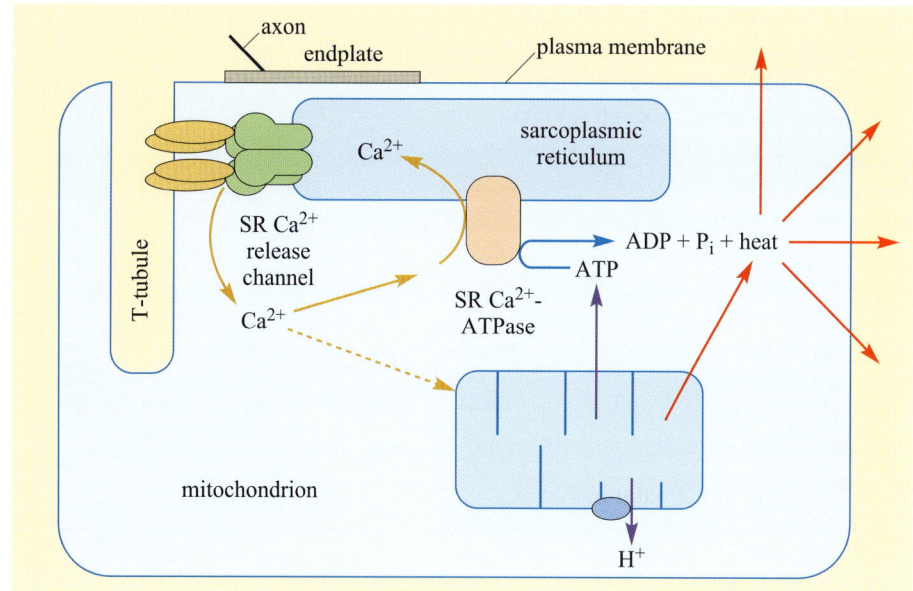

Figure 1.3 A possible model for the mechanism of thermogenesis in cells from the heater organ of the blue marlin. The release of energy as heat is controlled by the level of activity of the Ca^{2+}-ATPase in the sarcoplasmic reticulum. The Ca^{2+}-ATPase, by consuming ATP, increases demand from the mitochondrion, thus increasing the rate of respiration and hence the production of heat. Calcium ions circulate from the sarcoplasmic reticulum through the cytosol and back again.

Interestingly, although the thermogenic mechanism appears similar throughout this group of fish, endothermy appears to have evolved independently at least three times there. This conclusion is based on the results of mapping physiological traits onto a molecular phylogeny, which illustrates the integration across the levels of analysis. A phylogeny for the Scombroidei has been produced using part of the *cytochrome b* gene (Block et al., 1993). The method of analysis used (cladistic analysis) is described in Chapters 2 and 3. The results suggest that partial endothermy evolved in three separate lineages (clades) within the Scombroidei. In two of the lineages there are fish with thermogenic organs (the billfishes and the butterfly mackerel) but they do not share a common ancestor that has a thermogenic organ. Thus the thermogenic organ has evolved twice, a conclusion that is strengthened by the fact that the thermogenic tissues are derived from different eye muscles in each lineage. The third lineage contains the tunas, whose physiological adaptations help to retain general metabolic heat within the body, rather than generating heat locally in a specialized tissue. The analysis also shows that the thermogenic organs of billfish and swordfish are unlikely to have evolved separately, which could have been possible, even within a group.

1.3 Integration across disciplines

Temperature acclimatization is a well-known physiological process, and it is generally studied at the whole-animal level. However, the study of temperature effects on molecular structure, once an entirely different discipline, is now seen to have implications for the whole animal, so, by integrating across disciplines, we can achieve a deeper understanding of physiology.

A very recent addition to the array of techniques available to the physiologist is a new technology that has revolutionized molecular biology. The technique is called **micro-array analysis** and is used to study the total gene expression in an organism. DNA micro-arrays are coated glass slides or silicon 'chips' on which are arranged a large number of spots of single strand DNA (cDNA).[*] mRNA is extracted from tissues and used to produce labelled mRNA sequences that can anneal to the target DNA sequences in the array – the labelled sequences are termed cDNA probes. The amount of annealing can be quantified, giving a measure of the amount of mRNA present, which is a measure of the expression of each gene sequence. A large number of cDNA spots can be put in each array, so many gene sequences can be tested in parallel. (Chapter 4 has more background on this technique.) Previously, physiologists had been limited to studying the effects of single genes, which provides a limited view of the genetic basis of physiological mechanisms. The interaction between gene expression and regulation, which shapes a physiological response, is not easily revealed by identifying single genes and the role that they play. DNA micro-arrays allow the study of the regulation of specific groups of genes that contribute to one physiological function. Potentially, micro-arrays could make it possible to link whole organism function with entire genome expression, because the expression of hundreds of genes can be analysed simultaneously. Physiological experiments on, for example, temperature regulation, can then be carried out, with the analysis of gene expression conducted alongside the whole-animal experiments.

Array technology is not confined to DNA. It is possible to produce protein arrays, enabling studies of the regulation of protein expression to complement those on gene expression. Of course such studies have been primarily on 'model' organisms, such as *E. coli*, yeast, *Drosophila*, mice and humans. Arrays for these organisms are commercially available and are used in medical research. Producing the arrays is very expensive so they are not readily available for non-model organisms. However, for physiologists it is often the non-model organisms that are the most interesting, particularly those that are adapted to extreme conditions or undergo substantial changes during their life history. The adaptations that these organisms possess are almost certainly the result of exaggerated changes in physiology, with similarly large changes in gene expression. Study of these more extreme examples of adaptation can then lead to a better understanding of the physiology of animals under more 'normal' conditions.

Micro-arrays have been used in a study of hypoxia (reduced oxygen concentration) in a fish, the mudsucker (*Gillichthys mirabilis*) that inhabits burrows in the estuarine coast of California. The burrows have low levels of dissolved oxygen, so the mudsuckers are in hypoxic conditions when in their burrows and are tolerant of low oxygen concentration. A group of mudsuckers was placed in hypoxic conditions for up to 6 days and then tissue samples from the liver, heart, brain and muscles were taken, for comparison with tissues from a control group. A micro-array of 5376 cDNA gene elements was constructed from cDNA libraries (collections) that were known to contain hypoxia-regulated genes. RNA was used to produce cDNA probes. The probes were labelled with a fluorescent marker and then allowed to hybridize with the cDNA targets on the micro-array. The resulting pattern of the fluorescence across the array indicated the amount of gene expression. It was noted that the

---

[*] cDNA is single-stranded DNA that is complementary to an mRNA sequence. cDNA can be synthesized from mRNA using an enzyme called reverse transcriptase.

levels of expression of many genes changed progressively over the period of hypoxia and the changes were tissue-specific. 126 distinct cDNAs whose expression was regulated by hypoxia were identified and it was possible to confirm the differential expression of genes. The experiments and the analysis of the results are complicated and the details need not be included here, but the physiological picture is of a wide-scale re-organization of the ATP-generating pathways in the tissues studied, under hypoxic stress. The majority of the genes that are regulated by hypoxia are also present in mammals. Furthermore, many are common to other physiological processes. This example shows the power of the technique but it also shows the difficulties. As the genome of the mudsucker has not been sequenced, there may be unknown genes present that may contribute to the response to hypoxia.

1.4 Integration across species

The power of the comparative approach, in which a range of related species or genera are studied with respect to a single physiological feature over a range of habitats, is well illustrated by the physiology of cold adaptive thermogenesis in small mammals. Li et al. (2001) chose six species of small mammal covering a geographic range from the sub-tropical to the northern temperate zone in China and the semi-desert area of Mongolia. Specimens were collected from the wild and cold acclimated by keeping them at 4 °C for 4 weeks. Then basal metabolic rate was measured, followed by the metabolic response to an injection of noradrenalin. Noradrenalin stimulates **non-shivering thermogenesis** (**NST**), a thermogenic response that does not involve shivering. The investigation found that the adaptive mechanisms to cold of *Tupaia belangeri*, a tree shrew living in the subtropical region, are primarily an increase in BMR and secondly an increase in NST, with the mechanisms of thermogenesis being similar to those in other small tropical mammals. In small mammals native to northern regions, the adaptation to cold is chiefly to increase NST. The greater vole (*Eothenomys miletus*), which also lives in sub-tropical regions, but in the high mountains, reacts in a similar way to the small mammals in the north temperate regions. In other words, it follows the environmental temperature rather than the strict geographical region. The Mongolian gerbil (*Meriones unguiculatus*) is able to tolerate a semi-desert climate where the diurnal temperature range is large. High temperature is the critical stress. The gerbil showed the lowest BMR and NST level, but the greatest increase in BMR (and the second greatest increase in NST) after cold acclimation. Another small mammal that faces large diurnal temperature variation is the plateau pika (*Ochotona curzoniae*). The pika has an unusually high BMR and NST and there was only a small increase in both in the cold acclimation experiment. These experiments show that the physiological responses to cold are not all the same within the group of small mammals studied, but are related directly to the environment that they occupy. The combination of a comparative study across a range of species with a comparison across a range of environments provides a clearer picture of cold-induced thermogenesis and its control.

In Section 3.6 you will read about a study of lark species across a wide ecological range that reveals details of their adaptation to differing desert conditions.

1.5 The importance of phylogeny

When carrying out comparative studies, the physiological characteristics of related groups is understandable only if the underlying phylogeny is known. Closely related species are likely to have diverged from a common ancestor only recently, which means that they will not have had as much time in which to accumulate differences, when compared with more distantly related species. So, for example, in comparing the resting metabolic rates of different species of teleost fish, mapping the data onto an established phylogenetic tree would reveal any evolutionary trends. Unfortunately, though, there is no such phylogeny for these fish. Moreover, although there are significant differences in BMR between six different orders of teleost fish (Clarke and Johnston, 1999), there is also substantial variation between species within an order and, as the phylogenetic relationships are not clear, it is not possible to distinguish any clear evolutionary trends. The lack of a phylogeny is significant for physiologists because there is a correlation between environmental temperature and resting metabolic rate. The BMR of tropical fish is six times that of fish in the oceans around Antarctica and there is are energetic advantages in the decreased maintenance costs of living in cold water. So the question for physiologists is what selection pressure is keeping the resting metabolic rates of tropical fish high. One possible explanation is that a reduced metabolic rate has, as a consequence, a reduced capacity to generate metabolic power. The relationship between the resting power output and the peak power output is known as the relative aerobic scope. It has been calculated that if the relative aerobic scope is independent of temperature, a polar fish of mass 50 g and with a relative aerobic scope of 5 could generate an extra 0.33 watts above rest while a tropical fish with the same aerobic scope could generate 2.05 W. This difference might have great significance and could be the selective advantage for tropical species that has maintained their high resting metabolic rate.

The difference between the two ends of the thermal range of animals is stark when the rates of biological processes are compared. Rates of most biochemical and physiological processes increase by 2–3 times for a 10 °C rise in temperature. The rate change of such processes is represented as:

Q_x where x is the temperature change, so

Q_{10} = 2–3.

For ectothermic polar animals at a temperature of −2 °C, biological processes would proceed at rates between 16 and 90-fold *lower* than those in endothermic or tropical ectothermic animals. In addition, at the very low temperatures of the polar seas T_b would be at, or in some cases below, the freezing point of body fluids and below the range of structural stability of some macromolecules. From a physiological viewpoint the adaptations that are displayed by fish living in the polar seas offer insights into two different classes of adaptation to extreme temperatures. Mechanisms that are necessary for survival in extreme, and potentially lethal thermal conditions are one class of adaptation shown by antarctic fishes. An example of such an adaptation (considered in detail in Section 5.5) is the antifreeze glycoproteins that prevent body fluids freezing at low temperature. The second class is those mechanisms that permit normal biological function at extreme temperatures. Animals living in thermal equilibrium with a very cold environment have proteins that denature at significantly lower temperatures than do the homologous proteins from animals in temperate environments

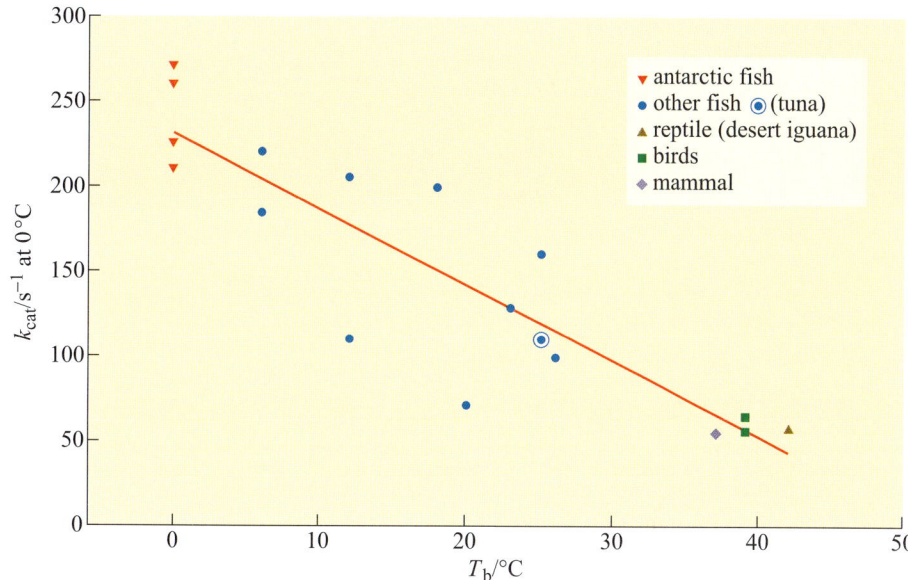

Figure 1.4 The relationship between normal T_b and the catalytic rate constant for the enzyme A4-lactate dehydrogenase from the white skeletal muscle of 18 vertebrate species, estimated at 0 °C (data from Fields and Somero, 1998; Sidell, 2000).

(Somero, 1995). Studies of the enzyme A4-lactate dehydrogenase (LDH), which is the muscle isoenzyme of LDH, show that the rate (k_{cat}) at which this muscle enzyme catalyses the conversion of pyruvate into lactate at 0 °C is inversely related to body temperature (Figure 1.4). The rate of catalysis is higher in antarctic fish than in warmer-bodied fish like the tuna. However, there is an interesting phylogenetic twist in that the antarctic notothenioids have significantly higher rates than the closely related notothenioid fish from South America, whose normal body temperature is only a few degrees higher than that of their polar relatives. Measurements suggest that during the evolution of antarctic fish species, subtle modification of the enzyme structure has taken place, conserving their rates of catalysis by an increase in the flexibility of the tertiary structure of the protein, while at the same time keeping the enzyme-substrate affinity within a narrow range. The protein has become more flexible without losing specificity.

Adaptation to temperature involves compensatory changes in protein structure. For example A4-LDH has a high k_{cat} in low-temperature-adapted antarctic fish (that live at −1.9 °C and die at +4 °C) but a lower k_{cat} in *Dipsosaurus dorsalis*, a lizard (Figure 1.5), whose typical daytime T_b is 32–37 °C in summer (see Figure 1.4). The protein structure of A4-LDH starts to unfold at a lower temperature in the fish than in the lizard. Relative to warm-adapted proteins, cold-adapted ones have a higher k_{cat} and a higher K_M (the substrate concentration at which enzyme activity is half of its possible maximum). The active sites are all the same in cold and temperate fish, but other regions of the molecule show much variation. Two amino-acid substitutions in the molecular structure are sufficient to convert a polar-type LDH into a temperate form. Temperature compensation in proteins is discussed in more detail in Section 2.5.2.

Of course there are other physiological features that display interesting adaptations, for example blood viscosity, oxygen transport and the cardiovascular system. From an evolutionary as well as a physiological perspective, the loss of haemoglobin and myoglobin from some antarctic fish is of great interest. Loss of these proteins, which have such a crucial physiological importance in most animals, has not been lethal probably because cold waters are oxygen rich and

Figure 1.5 The desert iguana (*Dipsosaurus dorsalis*).

permit a low metabolic rate. The big, and unanswered question, is why the apparently deleterious mutation that produces the loss of expression of crucial proteins, has been maintained in the population. Of the 16 species of antarctic icefish known (family Channichthyidae) all have lost haemoglobin, but only six of the species have lost the ability to produce cardiac myoglobin. Study of the phylogeny of these fish has shown that the loss has occurred four times independently during evolution by at least three different mechanisms. Experiments have shown that the myoglobin is functional at low temperatures and the physiologist has to ask why some icefish have this low temperature adaptation and some apparently manage to do without. What is the physiological advantage? There must be one, for four independent events to be conserved by selection.

In this introduction to the physiology of antarctic fishes the importance of phylogenetic studies is apparent. However, this example encompasses a wide range of studies, from protein structure and gene expression to ecology and Earth history. The application of a group of disciplines to a single physiological problem, in this case low temperature, is an example of how powerful a concept integrative physiology has become. A true understanding of physiological adaptations comes from comparison of species whose phylogeny is known. You will read more about antarctic fish and polar biology in Chapter 5.

1.6 Physiology and conservation

The African wild dog (*Lycaon pictus*) is a species that is critically endangered,[*] with perhaps only a few thousand remaining in the wild. Martyn Gorman and his colleagues (Gorman et al., 1998) have carried out a physiological study to investigate energy budgets in these animals. This example integrates different disciplines, ecology and physiology, and the results have an applied value in a third area, conservation.

African wild dogs (Figure 1.6) live in packs of 4–20 adults and dependent young. The dogs hunt cooperatively, preying on ungulates such as gazelle, zebra and wildebeest. A negative relationship between population densities of African wild dogs and spotted hyena (*Crocuta crocuta*) was interpreted as the result of hyenas stealing food from the wild dogs, a habit known as kleptoparasitism. In open habitats, hyenas gather at wild dog kills and drive the dogs off the food. The significant question that Gorman et al.'s study set out to answer was whether there was a significant energetic cost to the dogs in hunting and killing prey that was subsequently eaten by hyenas. They focused on a pack living in the Kruger National Park in South Africa, which consisted of, rather unusually, six adults, 16 yearlings and 27 pups. The dogs hunted for 3.5 hours per day, at dawn and dusk. They rested for the remainder of the day.

To measure daily energy expenditure the team used the **doubly-labelled water technique** with six fully grown male and female dogs. Each dog was captured during the morning when the animals were resting in the shade. They were darted with anaesthetic and injected intravenously with 5 cm³ doubly-labelled water (DLW). DLW contains stable isotopes of both oxygen and hydrogen. Following injection, it is possible to take blood samples at intervals and analyse them for the

---

[*] Critically endangered is the highest category into which a species at risk of extinction can be placed by the UNEP World Conservation Monitoring Centre.

Figure 1.6 African wild dogs (*Lycaon pictus*).

ratio of labelled hydrogen to labelled oxygen. As metabolism proceeds, oxygen is lost as CO_2. Oxygen is also lost as water, but the fact that the hydrogen is labelled as well means that an estimate of the rate of water loss can be calculated and the rate of CO_2 production then deduced. It is the rate of CO_2 production that gives a measure of metabolic rate. After injection of antidote to the anaesthetic the dogs recovered rapidly and ran off to rejoin the pack. Five of the dogs were recaptured at both 48 hours and 96 hours after administration of $^2H_2{}^{18}O$. One dog could be recaptured only at 96 hours. At each recapture 5 cm^3 of blood was taken. The results of determinations of ^{18}O and ^2H content of blood were used for calculation of daily energy expenditure (Table 1.1).

The average daily energy expenditure of the dogs was 15.299 MJ day^{-1}, which was calculated to be equivalent to an intake of 3.5 kg of ungulate meat per day (assuming an energy content of 5.2 MJ kg^{-1} wet mass). Field observations of food intake ranged from 2.5 to 3.5 kg of ungulate meat per day, so the food intake estimated from DLW measurements correlated well with the observed food intake. Table 1.1 shows the high variability of daily energy expenditure between the six dogs, from 7.223 to 20.281 MJ day^{-1}, reflecting the daily variation in involvement in group hunts in the pack studied. In fact this pack had a disproportionately large number of yearlings and pups and it was suggested that the adults may have been hunting more intensely than normal on some days to make up for lack of experience in the yearling dogs.

Table 1.1 Daily energy expenditures of six African wild dogs measured using the doubly-labelled water technique.

Dog	Sex	Body mass/kg	Daily energy expenditure/MJ day^{-1}
1	F	24	7.223
2	F	25	8.729
3	F	23	16.686
4	M	27	18.252
5	M	27	20.281
6	M	25	19.806
Mean daily energy expenditure			15.299

How does a mean daily energy expenditure figure of 15.299 MJ compare with measurements in other dogs? The daily energy expenditure for a moderately active domestic dog is about 6.0 MJ day^{-1}. The food intake of a 25 kg border collie working 6 hours per day is 8.2 MJ day^{-1}. During a 490 km sledge race across arctic Canada, the energy expenditure of Alaskan sledge dogs measured by DLW was 47 MJ day^{-1}. Comparing the African wild dogs with these examples suggests that they are working extremely hard for the 3.5 hours day^{-1} that they are hunting. For African wild dogs, hunting is very costly, and the loss of even a small proportion of their prey to hyenas has a significant effect on foraging time.

■ Assuming a basal metabolic rate for a 25 kg dog of 3.4 MJ day^{-1} and the mean daily energy expenditure in Table 1.1, what is the hourly energy cost of hunting?

$$\text{Energy cost of hunting} = \frac{\text{daily mean energy expenditure} - \text{daily basal metabolic rate}}{\text{number of hours hunting}}$$

$$= \frac{15.299 - 3.4}{3.5} = 3.4 \text{ MJ hour}^{-1}$$

The researchers then calculated that if the dogs lost 25% of their prey to hyenas, they would need to hunt for 12 hours each day to achieve an energy balance, which is physiologically untenable. Indeed, wild dogs hunting for 3.5 hours each day are working close to their physiological limit. Therefore dogs are vulnerable to even low levels of theft of their food. The research suggests that efforts to conserve the African wild dog would be more successful in wooded areas such as the Kruger National Park and relatively unsuccessful in open areas such as the Serengeti where hyenas abound.

The ability to make measurements from the whole, free-living animal was central to the investigation and highlighted the fact that wild dogs live close to the limit of a viable energy budget. Even a small loss of prey to others can threaten their survival in the long term.

1.7 Conclusion

It should be evident to you from the examples given in this chapter that the pace of change in technological innovation, and particularly the biotechnology revolution, has had a massive impact on physiology and the way in which physiological processes are investigated. It is now possible to argue that physiologists are no longer constrained by technology and its limitations. The restraints, apart from the obvious one of money, are the intellectual ones of posing the right questions and designing the right experiments. Of particular significance are the advances in transmitters and satellite monitoring, which allow a range of physiological data to be collected from free-living animals, including marine ones, over long periods of time. The results from such studies have applications to conservation and can be invaluable in developing management plans for species, such as the bluefin tuna, that are exploited by humans.

The advances in molecular biology now provide techniques for the analysis of gene expression that promise a much more detailed understanding of adaptation and the evolutionary steps by which such adaptations came about. The excitement about research in physiology now is the practicality of bringing techniques and ideas from different specializations and focusing them on a single physiological process, as in the example of endothermy mentioned in Section 1.2.

1.8 Summary of Chapter 1

This chapter emphasizes the importance of taking a very wide view of physiology. The changes in research that are driven, in part, by technological development, have opened up new possibilities. Three facets of physiological research are identifiable: integration across levels of analysis, integration across disciplines and integration across species. Each of these integrative approaches has yielded new understanding of physiological processes, for example in hibernation, adaptation to high and low temperatures and endothermy. The application of physiological research in the conservation of animal populations demonstrates that an understanding of energy budgets, obtained from free-living animals, can lead to a realistic management programme for a species.

Learning Outcomes for Chapter 1

When you have completed this chapter you should be able to:

1.1 Define and use, or recognize definitions and applications of, each of the **bold** terms.

1.2 Illustrate each of the three types of integration in physiological studies, using a named example for each.

1.3 Describe the advantages and limitations of electronic recorders attached to free-living animals and apply that knowledge to specific examples.

1.4 Outline the basic principles of the doubly-labelled water technique.

1.5 Appreciate the significance of phylogenetic studies in integrative physiology and illustrate your understanding using examples.

1.6 Outline the mechanism by which heat is generated in a named example of thermogenic tissue.

1.7 Use simple diagrams to explain physiological processes.

1.8 Give examples of the potential contribution made to conservation by physiological studies.

Questions for Chapter 1

(*Answers to questions are at the end of the book.*)

Question 1.1 (LO 1.3)

In preparing an investigation in which you wish to attach a recording device to a seal, what design factors would you need to bear in mind?

Question 1.2 (LO 1.4)

Describe how metabolic rate can be measured using the doubly-labelled water technique.

Question 1.3 (LO 1.5)

How can phylogenetic studies contribute to the understanding of physiology?

Question 1.4 (LOs 1.6 and 1.7)

By means of a simple diagram, illustrate the possible mechanism by which expression of the *PDK4* gene is regulated in the heart and skeletal muscle of the 13-lined ground squirrel.

Question 1.5 (LOs 1.3 and 1.8)

What information about the conservation of tuna could be obtained from implanted archive tags and pop-up tags?

CHAPTER 2 THE EXTERNAL ENVIRONMENT

Prepared for the Course Team by Patricia Ash

2.1 Introduction

Before we can understand the phenotypic and genotypic adaptations of vertebrates for life in their environment we need to examine the physical characteristics of the environment. Vertebrates live both on land and in water so we need to study the thermal characteristics of each, and to understand the processes of heat exchange. We are especially interested in temperature because the rates of all metabolic processes are temperature dependent. As temperature decreases, enzyme activity usually decreases and so do the rates of most other chemical reactions. The highest temperature available for vertebrate metabolism is restricted by the risk of denaturation of proteins and disruption of membrane structure at temperatures greater than about 38 °C. The body temperature of vertebrates represents a balance between heat gained either from metabolism or the environment and heat lost to the environment. Animal physiologists are therefore particularly interested in thermal exchange. Metabolism generates heat, and in endotherms this heat is used to maintain body temperature. Therefore, factors affecting metabolic rate have been much studied. Until recently most of this research was carried out in a laboratory, but techniques have been developed and refined that enable measurements of body temperature and metabolic rate to be carried out on animals in the field.

Nearly all vertebrates depend on oxygen for maintenance of aerobic metabolism, and hence generation of ATP, in tissues. The highest yields of ATP per mole of substrate, are obtained from aerobic catabolism of fuel molecules. Catabolic biochemical pathways convert fuel molecules such as fatty acids and glucose into acetyl coenzyme A (acetyl CoA) (Figure 2.1). Long-chain fatty acids are converted into 2-carbon fragments by oxidation, ending in formation of acetyl CoA. Glucose is converted into pyruvate by the glycolytic pathway in the cytosol; in turn, pyruvate is converted into acetyl CoA. Acetyl CoA enters the tricarboxylic acid (TCA) cycle, combining with the 4-carbon oxaloacetate to form citrate, the 6-carbon intermediate that is progressively decarboxylated and oxidized by the TCA cycle. Deamination of certain amino acids results in formation of TCA cycle intermediates that are fed in at specific stages in the cycle. Each full turn of the TCA cycle results in the loss of two fuel molecule-derived carbon atoms as CO_2.

Hydrogen ions removed from TCA cycle intermediates are taken up by the coenzymes, NAD and FAD. Reduced coenzymes, NAD.2H and FAD.2H, produced during the TCA cycle, are oxidized by means of the electron transport chain in the inner

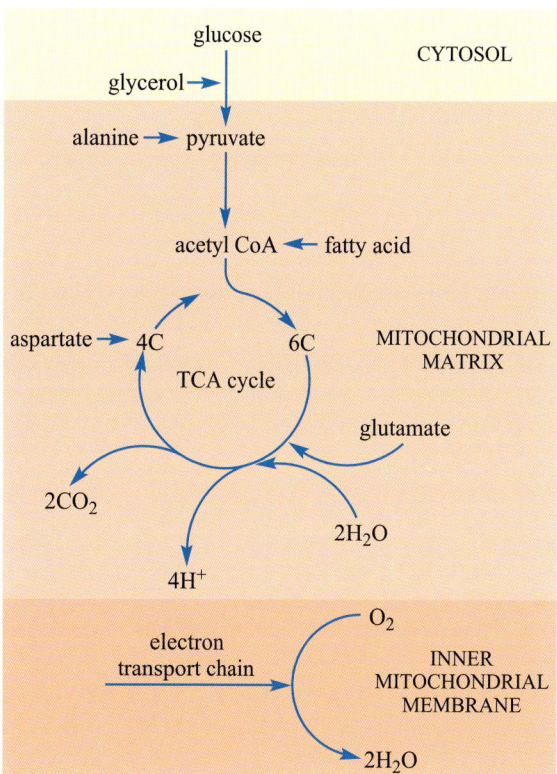

Figure 2.1 Reactions of oxygen and hydrogen atoms during catabolism. Note the entry points for oxidation of glucose, fatty acids and certain amino acids. The final stage of catabolism, the electron transport chain, requires oxygen. Overall, for every molecule of acetyl CoA that enters the cycle, two molecules of water are split and two molecules of carbon dioxide are produced. *Note*: ATP generation is not shown in this figure.

mitochondrial membrane. Most ATP synthesis occurs at this stage, powered ultimately by energy released during electron transfer. As oxygen is the final electron acceptor in the electron transport chain, if oxygen is in short supply, the rate of aerobic catabolism and generation of ATP plummets. Depletion of oxygen in tissues, a situation known as hypoxia, results initially in a switch to anaerobic catabolism of glucose, but the yield of ATP per glucose molecule is low, and high levels of the product – lactate – cause problems. In this chapter we will study the relevant aspects of the physics of gases, and examine the situation at high altitude where oxygen is available, but is in relatively short supply. We will study the implications of Fick's law, which brings together the physical factors that affect the rate of diffusion of gases, and discuss how oxygen is transferred to respiring cells. You should find that the physics explained in Section 2.4 provides useful background for your study of Chapter 6.

Extremes of temperature and hypoxia outside normal ranges elicit special biochemical processes called **stress responses**. Stress responses at the molecular level include those of glycoprotein or protein receptor molecules, either in the cell membrane or cytosol, that detect signals from the environment. Inside the cell, the signal may elicit transcription of specific genes or activation of enzymes, in particular kinases, and often both of these processes. We begin our study of molecular responses to environmental stress in Section 2.8.

We end Chapter 2 with an account of the comparatively new field of evolutionary physiology. Study of evolutionary adaptation requires DNA sequence analysis for determination of evolutionary relationships between species. Accurate phylogenetic trees for animal groups are required for distinguishing between phylogenetic characters and features that are adaptive in particular species. If a mutation in DNA is expressed as an altered protein, and is beneficial in a particular environment, it persists and spreads in a population. The importance of the link between animals' responses to environmental signals at the molecular level and evolutionary adaptation cannot be over-emphasized.

2.2 Thermal characteristics of the external environment

The Sun is the major source of heat energy for the surface of the Earth. The surface temperature of the Earth is the result of the balance between the incoming short-wave radiation from the Sun, and re-radiation of heat energy to space, at a longer wavelength.

■ Why is energy re-radiated by the Earth at a longer wavelength than that of the short-wavelength radiation that is absorbed from the Sun?

The wavelength of radiated energy from a body is related to its temperature – the higher the temperature, the shorter the wavelength. The temperature of the Earth is less than that of the Sun. Therefore, re-radiated energy from the Earth is at a longer wavelength than that absorbed from the Sun.

Figure 2.2 shows the electromagnetic spectrum with energy scales in joules increasing from left to right and wavelength decreasing from left to right. Note that the shorter the wavelength of electromagnetic radiation, the higher the associated energy. If sunlight is split into a spectrum, and the amount of energy at different wavelengths is plotted on a graph, a peak occurs at a wavelength of about 500 nm (Figure 2.3a). The temperature of the Sun near its surface is around 6000 K. For an object at a temperature of about 300 K (= 27 °C, the surface temperature of most land mammals), the peak in emitted energy occurs at a much longer wavelength, about 15 μm (Figure 2.3b). The temperature of the Earth is about 255 K.

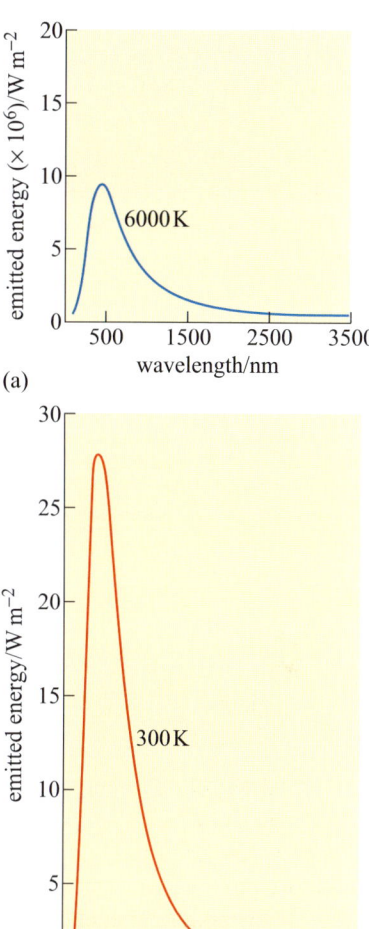

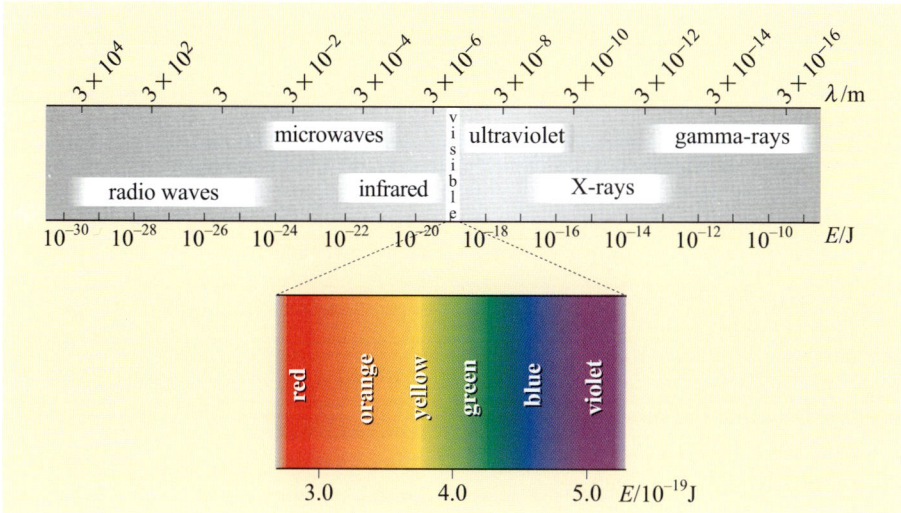

Figure 2.2 The electromagnetic spectrum with photon energy in joules and wavelength, λ, in metres. Note that because the range of photon energies is so large, we have used a power of ten scale; the energy changes by a factor of ten for each division along the scale. Note also that energy increases from left to right.

Figure 2.3 Emission curves for (a) the Sun at 6000 K and (b) a body at 300 K.

Figure 2.4 (overleaf) summarizes the gains and losses of heat derived from incoming solar radiation for the Earth's atmosphere and surface. Approximately 49% of the radiation energy from the Sun that falls on the outer atmosphere reaches the surface directly and is absorbed. Of the other 51%, 31% is scattered by the atmosphere and reflected from the surface and cloud layers, and 20% is absorbed by the atmosphere as long-wavelength radiation. The energy absorbed by the ground/ocean heats the Earth's surface. The warm surface of the Earth radiates long-wave energy (heat) into the atmosphere (R_e), and also loses heat to the atmosphere by convection and evaporation. In turn the surface absorbs infrared radiation from the atmosphere. These processes of energy loss and gain by the Earth's surface are balanced, so that there is no overall long-term heat gain or loss by the Earth's surface.

As air has a low specific heat, a small amount of heat produces a relatively large change in temperature. However, air has a low thermal conductivity, which means that heat is not conducted rapidly, so large thermal gradients can build up. These properties of air contribute to local differences in weather, making the atmosphere complex and diverse. The vegetation adds to the complexity and

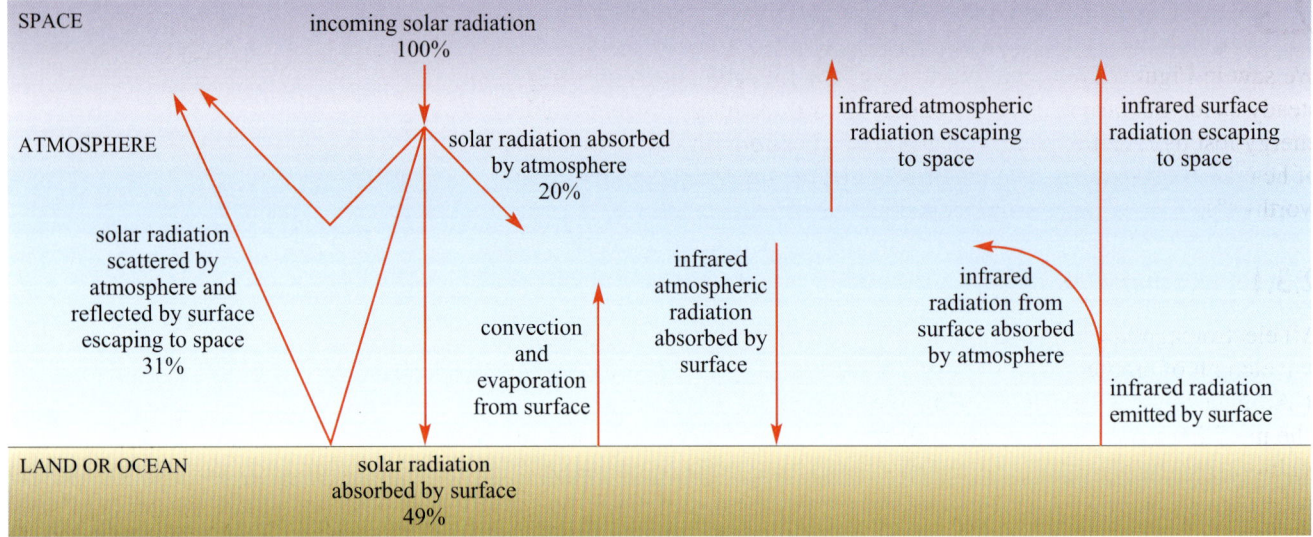

most terrestrial environments are far from uniform. Climatological data for a particular environment give some indication of the type of habitat that it represents, but figures such as air temperatures can only give a general guide. So steep are the gradients within some terrestrial environments that biologists cannot use just climatological data. They are more concerned with the **microclimates** that animals occupy. Microclimates are more influenced by, for example, shading, wind speed, and reflected heat from the ground than general data would indicate. So measurements of the detailed ways in which variables such as temperature affect animals can be very difficult, as we shall appreciate in later chapters.

Air temperature varies not only geographically and seasonally, but there are also marked effects at high altitudes. There is no standard relationship between altitude and temperature as local climatic conditions can affect temperature so much, but generally, temperature falls by 5–10 °C for each 1000 m rise in altitude. Such a decrease is significant to animals that move over large vertical distances, e.g. birds that usually live near sea-level migrating over high mountain ranges.

Water is a very different medium from air, and so the aquatic environment poses other problems for animals. Water has a higher specific heat than air: larger amounts of heat can be absorbed by water, producing only a small change in temperature. Water is also a better conductor of heat than air, so the steep thermal gradients that are characteristic of the atmosphere are not so pronounced as in water.

Figure 2.4 Gains and losses of energy derived from solar radiation by the Earth's surface and atmosphere. The values on the left are rates of energy loss and gain by the Earth's surface and atmosphere directly from solar radiation. 100% represents the rate at which the Earth intercepts solar radiation. The arrows on the right show the routes of energy transfer between the Earth's surface and the atmosphere by radiation, convection and evaporation.

Summary of Section 2.2

The nature of the way that the Sun provides heat energy for the Earth and the physics of heat transfer in air means that there can be large thermal gradients. These gradients, together with the effect of vegetation, create complex thermal environments with many niches that animals can occupy. There are many variables that will influence the precise temperature at a particular location. The effects of these microclimates on animals are important but difficult to measure. Aquatic environments are much more uniform with respect to temperature.

2.3 Thermal exchanges

We saw in Figure 2.4 how the Earth's surface temperature links to a dynamic steady state, the balance between energy gain from solar radiation and long-wave energy lost by radiation, convection and evaporation. The same physical principles of heat loss and gain influence the body temperature of animals and so it is worthwhile reviewing those processes at this point.

2.3.1 Radiation

All electromagnetic radiation travels at the speed of light. We have seen that the wavelength of emitted radiation depends on the temperature of the body that emits it. Animals are relatively cool and therefore emit relatively long-wave radiation (the middle infrared, 7–8 µm), resulting from the motion of atoms and molecules within the body. The rate at which long-wave radiation is emitted is proportional to surface area $\times T^4$ (Stefan's law; T = temperature) so a small increase in surface temperature results in a relatively large increase in rate of emission of radiation. Although emission rates for animals are very low in comparison with rates for hot bodies such as the Sun, radiative heat loss for a mammal can represent a substantial proportion of the energy released from metabolism. The rate of emission also depends on the colour of the surface: physicists define a black body as having 100% absorptance and 100% emissivity of long-wave radiation. However, a black body reflects very little radiation. At physiological temperatures, the surfaces of animals can be regarded as 'black bodies' which means that they absorb almost all long-wavelength radiation such as infrared, and reflect relatively little. However, a black body is also a near perfect emitter of radiation, with the rate of emission depending on the absolute temperature (measured in K), of the surface.

For a black body at temperature K, emission of radiation, P, measured in watts is:

$$P = \varepsilon \sigma A T^4 \qquad (2.1)$$

ε = emissivity, which for a black body is 1

σ = Stefan's constant, 5.7×10^{-8} W m^{-2} K^{-4}

A = surface area of the body/m^2

T = absolute temperature/K.

We will explore the implications of absorption and emission of radiation for the body temperature of a reptile at the end of this section.

Pale surfaces reflect radiation and pale animals are common in hot climates where keeping the body cool is important; dark-coloured animals that bask in the sun warm up more rapidly than pale-coloured animals. However, the relationship between colour and temperature is not simple. Intuitively, it would appear to be advantageous for arctic species to have black fur or feathers for absorbing maximum solar infrared radiation.

However, consider that during the long arctic winter, nights are long and daylight is very weak, so little solar radiation can be absorbed. An apparent advantage of white coats for arctic species is that they are inconspicuous against white snow and ice. Predators such as the polar bear and arctic fox, which have white fur in winter, are therefore less visible to potential prey, and prey species, such as the mountain hare, are less easily seen by their predators.

2.3.2 Conduction

Conduction is the direct transfer of heat between two materials in contact, for example, between an animal and the ground on which it is resting, or between an animal and the air. The two materials may be solid or fluid (fluids include gases). Heat flows from the warmer material, e.g. the animal, to the cooler one, the ground or the air. Animals can select their position within the environment to maximize or minimize heat loss or gain. They can also vary their area of contact with the environment, e.g. flattening the body against a cold substrate such as the floor of a burrow, maximizes the body surface in contact with cool ground, thereby speeding up conductive heat loss. Conversely, insulation by fur or feathers reduces heat loss by conduction by trapping still air next to the skin, since air has a low thermal conductivity.

Physiologists are interested in comparing the insulative properties of fur and feathers of different species, and measure conductive heat transfer, Q_c from Equation 2.2:

$$Q_c = C\Delta T \qquad (2.2)$$

C = thermal conductance/$J\,s^{-1}\,°C^{-1}$

T = thermal gradient/K

Insulative capacity (I) is another useful measurement, directly related to insulation, and is the reciprocal of the total heat flux per unit time in seconds, per unit area per unit of temperature difference:

$$I = 1/J\,s^{-1}\,cm^{-2}\,°C^{-1} \qquad (2.3)$$

measured in units of $°C\,cm^2\,s\,J^{-1}$.

In physiological studies, resistance to heat transfer may be measured in units of $s\,cm^{-1}$, which is the time in seconds taken for heat to be transferred for a distance of 1 cm.

The use of measurements of conductive heat transfer will become more apparent as we study Figure 2.5, which is a plot of insulative capacity against fur thickness.

■ Describe the main features of Figure 2.5 as a bulleted list, including the main trends shown in the data, and any outliers.

- The relationship between fur thickness and insulation is roughly linear.
- In general the larger the animal, the thicker its fur.
- The points for the ringed seal (*Phoca hispida*), are much lower in air than one would expect, considering, for example, that the seal is considerably larger than the lemming.
- Points plotted for the tropical species – tropical deer and monkeys – suggest that tropical mammals have relatively thin fur, e.g. thickness of deer fur is closer to that of the shrew.

Insulation in adult *Phoca hispida* is provided mainly by blubber, which explains the apparent paradox of the points in Figure 2.5 for this species.

For thermoregulatory purposes, an animal is considered to consist of a core and a shell. Generally the core is taken to be the organs in the head and trunk, together with the deeper-seated muscles. The shell consists of superficial tissues and the insulating coat of fur, feathers or clothing. The body core is where most

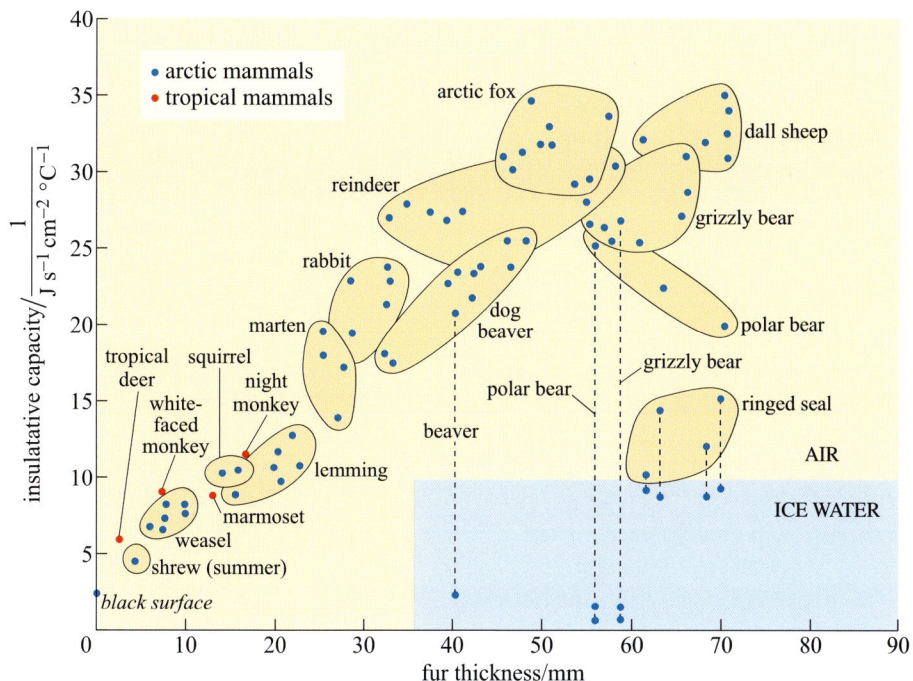

Figure 2.5 Insulation related to winter fur thickness in a series of arctic and tropical mammals. In the aquatic mammals the measurements in °C in air are connected by vertical broken lines to the same measurements in ice water (the pale blue shaded area). Data for individual species or similar sized animals have been grouped together.

thermogenesis (due to metabolism) occurs and the shell provides a buffer between the external environment and the core. Shell temperature may vary widely; core temperature may be quite stable. Heat is lost to and gained from the environment at the body surface, and the role of the shell is to keep heat loss or gain to a minimum.

At low T_a heat loss predominates. The rate of heat loss by conduction and radiation is a function of the thermal gradient, ΔT, between the animal's body surface and the external environment. Terrestrial mammals and birds have a reduced thermal gradient because of the insulating properties of fur or feathers. The chief advantage of fur or feathers is their lightness in relation to the insulation provided by the trapped air. The trapped air insulates well because it is still so the densest coats are the most effective for insulation against cold. The data in Figure 2.5 demonstrate that the thickness of the coat is also important in determining its insulation.[*] Note the points for the arctic fox (*Alopex lagopus*). This species' fur coat is thicker than that of larger species such as reindeer and the fur has the highest range of value for insulation. Note how the fur in the tropical mammals shown in Figure 2.5 does not exceed 15–20 mm. In these mammals fur length tends to decline with body size rather than increase.

2.3.3 Convection

Convection is heat flow by means of fluid flow driven by temperature gradients. For an animal, convection occurs between the boundary layer of fluid over the body surface, and the mass of fluid, air or water, in the environment. The rate of convection depends on the temperature difference between the animal and its

---

[*] Note that Figure 2.5 measures length of hairs; their density is also important especially for tropical species.

environment and the speed of the convection current. As it is difficult in practice to measure convective heat exchange in animals, physiologists generally measure conduction and boundary layer convection together.

Convective heat transfer is usually much faster than conduction, and may derive from rising warm air or water driven by temperature differences, or be speeded up by wind and water currents. It is important to appreciate that a fur or feather coat disturbed by a breeze loses much of its insulating properties because of the disturbance of the layer of still air. In water, convective heat transfer is very fast, and only aquatic mammals and birds can maintain a body temperature higher than that of the water. (A few species of fish, e.g. tuna, can maintain localized regions of their body at higher temperatures but not the whole body.)

2.3.4 Evaporation

Evaporation is an effective way of dissipating heat, because the latent heat of vaporization of water (2400 J g^{-1} at 40 °C) is high, so a large amount of energy is needed to vaporize water. Clearly, only animals living on land and exposed to air can use evaporative cooling. Air consists of a mixture of gases that together exert a total atmospheric pressure at sea-level of 101.3 kPa (kilopascals). Each of the gases that make up air exerts a partial pressure, and all the total pressures add up to give the total atmospheric pressure. The partial pressure of a gas in a mixture occupies the same fraction of the total atmospheric pressure as its percentage concentration in the gas mixture. The partial pressure exerted by water vapour in the air is known as the **vapour pressure**. The rate of evaporation depends on the water vapour pressure gradient, which is the vapour pressure difference between the surface and the atmosphere.* Vapour pressure itself is a measure of **absolute humidity** whereas the familiar concept of **relative humidity** (r.h.), gives the prevailing vapour pressure as a percentage of the maximum amount of water vapour that air at that temperature can hold (i.e. the saturation vapour pressure). So if the vapour pressure of air remains constant and the air is warmed, the r.h. falls, because at higher temperatures, the saturation vapour pressure increases. Figure 2.6 shows the saturation vapour pressure at a variety of air temperatures. For example, at 20 °C, the saturation vapour pressure is 2.4 kPa, which is equivalent to a relative humidity of 100%. When the r.h. falls to 40%, the vapour pressure is only 1.0 kPa. However, consider an air temperature of 30 °C, where the saturation vapour pressure is 4.3 kPa. If the vapour pressure is the same as before (2.4 kPa), the r.h. is now only about 55%. If the vapour pressure is 1.5 kPa (equivalent to 60% r.h. at 20 °C), then at 40 °C, the r.h. is only 20%.

It is worth remembering that the most significant measurement for physiological purposes is not r.h., but the difference between saturation vapour pressure and the prevailing vapour pressure, because this difference is a measure of the ability of the air to take up water by evaporation.

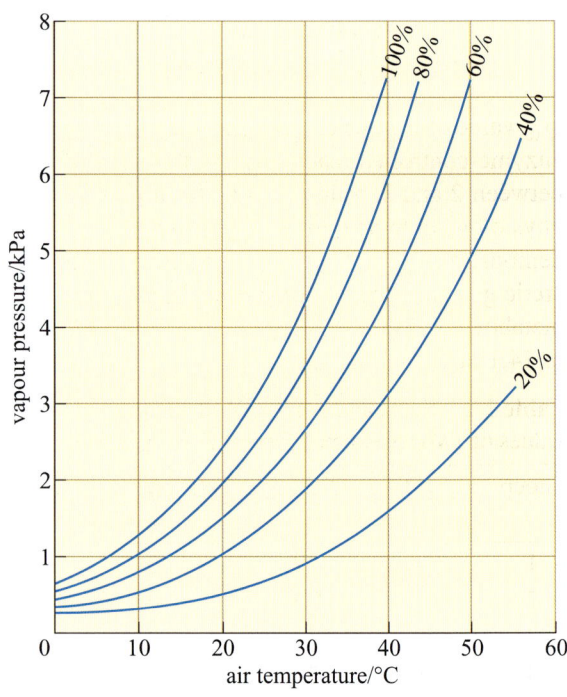

Figure 2.6 The relationship between vapour pressure and air temperature at different relative humidities.

---

* Atmospheric pressure is commonly measured in millibars (mbar) or kPa. To convert mbar to kPa, the SI units of pressure, multiply by 0.1.

Summary of Section 2.3

Thermal exchanges between animals' bodies and the environment comprise heat gains and losses by radiation, conduction, convection and loss by evaporation (latent heat transfer). Rates of absorbance and emission of radiation are affected by an animals' colour; pale objects reflect radiation and absorb less heat than dark objects. Heat loss and gain by conduction and convection are affected by insulation. Animals can minimize or maximize heat loss/gain by behaviour.

2.4 Effects of temperature on metabolism

2.4.1 Metabolic rate and enzyme activity

Our interest in thermal exchanges in animals derives from the effects of temperature on metabolism. The rates of nearly all reactions and physiological processes vary with temperature. A measure of the effect of temperature is Q_{10}, the temperature coefficient, defined as the increase in the rate of a chemical or physical reaction produced by a 10 °C rise in temperature.

In practice, effects of temperature on enzyme-controlled reactions or on metabolic rate are not measured at temperatures that are exactly 10 °C apart. So Q_{10} values are calculated by extrapolation from the increase in rate measured at any two temperatures (Equation 2.4).

$$Q_{10} = \left[\frac{\text{metabolic rate } (T_{b_1}/°C)}{\text{metabolic rate } (T_{b_2}/°C)} \right]^{(10/T_{b_1} - T_{b_2})} \quad (2.4)$$

Q_{10} values for physical processes such as diffusion are usually less than 1.5. For enzyme-controlled reactions, and for basal metabolic rate, values for Q_{10} lie between 2 and 3. Values of Q_{10} for a particular process are not constant over physiological temperature ranges in animals. Q_{10} varies with temperature as demonstrated in Table 2.1 which shows values for Q_{10} measured in a hibernating arctic ground squirrel (*Spermophilus parryii*). During hibernation, T_b and torpor metabolic rate (TMR) vary with T_a. Values for Q_{10} vary from 1.0 at $T_b = 4.7$ °C to 14.1 at $T_b = 17.1$ °C.

Table 2.1 T_a, T_b and average TMR of arctic ground squirrels and corresponding Q_{10} values of TMR between different T_b measurements (Buck and Barnes, 2000).

Group	T_a/°C	T_b/°C	TMR/cm^3 O$_2$ g^{-1} h^{-1}	Range of T_b for Q_{10} calculation/°C	Q_{10}
1	4	4.7	0.012	4.7–8.2	1.0
2	8	8.2	0.012	8.2–12.6	1.5
3	12	12.6	0.014	12.6–17.1	1.8
4	16	17.1	0.018	17.1–20.7	14.1
5	20	20.7	0.047	4.7–20.7	2.4*

*Overall Q_{10} for T_b 4.7–20.7 °C.

Sharp changes in Q_{10} suggest a significant change at a certain temperature in the organelles, cells or tissues involved.

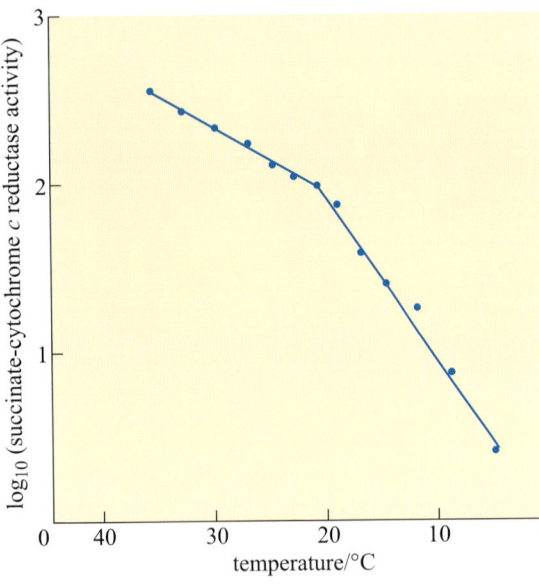

Figure 2.7 The effect of temperature on the activity of succinate-cytochrome c reductase in heart mitochondria from a marsupial mouse. Values for enzyme activity in arbitrary units min^{-1} on the y-axis are plotted as their logarithms. Note that the values on the x-axis are plotted so that temperature decreases from left to right.

Figure 2.7 shows the effects of temperature on the activity of the enzyme succinate-cytochrome c reductase in heart mitochondria from the marsupial mouse (*Sminthopsis crassicaudata*) at temperatures ranging from 5 to 35 °C. Note that enzyme activity is plotted on a log$_{10}$ scale. This means there is an increase of 1 from 0 to 1, a 10-fold increase from 1 to 2 and a 100-fold increase from 2 to 3. The enzyme is an important component of the electron transport chain and its activity provides an indirect measure of the rate of formation of ATP in the electron transport chain.

■ Describe the data in Figure 2.7, and summarize your overall conclusion in one sentence.

As temperature declines from 35 °C to 5 °C, there is a drop in succinate-cytochrome c reductase activity. There is a marked increase in slope at 20 °C, indicating a sharper rate of reduction in enzyme activity as temperature declines from 20 °C to 5 °C.

The data show that enzyme activity is highly temperature dependent.

The dependence of succinate-cytochrome c reductase activity on temperature can be attributed to the principle that rates of chemical reactions vary with temperature. Another factor influencing the activity of enzymes in the electron transport chain is the effect of temperature on the fluid properties of the mitochondrial membrane. The enzymes and electron carriers of the electron transport chain are located in the inner mitochondrial membrane, so change in membrane fluidity could affect their activities, and in turn, the synthesis of ATP. The sharp change in enzyme activity at 20 °C may be related to a change in fluidity of the inner mitochondrial membrane. The supply of ATP to heart cells would change as internal temperature changes. *S. crassicaudata* enters torpor when T_a drops. During torpor, the mechanisms of body temperature control are re-set and body temperature is regulated at a much lower level, just 16 °C. It is possible that there is a direct link between the temperature requirements of the heart mitochondria and the minimum body temperature that the animal's control mechanisms permit.

2.4.2 Tachymetabolism and bradymetabolism

The rates of nearly all metabolic processes of animals are temperature dependent and mechanisms that prevent or minimize fluctuations in body temperature have evolved. The body temperature of an animal is influenced by the external temperature, and also by the metabolic activity of the body's cells. Breakdown of glucose and other fuels to generate ATP is only about 30% efficient, with the remaining 70% energy liberated as heat. This heat energy contributes to the maintenance of a body temperature that is largely independent of the external environment in birds and mammals. In contrast, metabolic rates of amphibians and reptiles are not sufficiently high to generate enough heat to maintain body temperature close to 37 °C in animals that do not have insulating fur or feathers. Loss of body heat is even more significant in amphibians whose thin moist skin promotes evaporative water loss and hence cooling in air. Reptiles and amphibians rely on the external environment as a source of heat energy, but so do mammals and birds in certain circumstances.

Various methods can be used to measure the rate of heat production in animals. Heat released as a result of energy conversions during metabolism can be measured either directly by using calorimetry or indirectly by measuring metabolic rate. Metabolic rate can be measured by direct calorimetry in which rate of heat production is measured (Figure 2.8a) and indirect calorimetry, in which gas production or utilization is measured (Figure 2.8b–d) Heat production, measured by direct calorimetry would have units of joules produced per unit time, e.g. $J\,s^{-1}$, the same unit as the watt. For measuring heat flux directly, the calorimeter containing the rabbit could be immersed completely in water and the heat gained by the water calculated from its rise in temperature. Indirect measures of heat production all involve measuring either carbon dioxide production, or oxygen consumption or both. Measurement of oxygen uptake would have units of volume of oxygen consumed per kg body mass per unit time, e.g. $cm^3\,O_2\,kg^{-1}\,min^{-1}$. Of the two indirect calorimetry measurements, the closed circuit method (Figures 2.8b and 2.8d) is the most widely used. Masks (Figure 2.8c) can only be used where a person or animal is habituated to their use. The rabbit has a high metabolic rate and responds rapidly to changed environmental conditions. In addition, if the animal is placed in a closed container, it is easy to monitor changes in concentration of oxygen that result from its activities. This type of apparatus enables the metabolic rate to be calculated. Experimental set-ups such as those shown in Figure 2.8b and d are known as metabolic chambers and can be built to suit the animal being studied. Williams and colleagues (Williams et al., 2001) constructed a huge metabolic chamber for studies on the oryx. You will look this work in Chapter 3.

■ What source of uncertainty may affect measured values for metabolic rate of an animal inside a metabolic chamber?

The stress of being confined inside a metabolic chamber may affect the metabolic rate of an animal subject.

In practice, researchers take time and care to habituate animals to the metabolic chamber before taking any measurements. Metabolic rates of animals fall within a range of values between a minimum, the **basal metabolic rate**, commonly abbreviated as **BMR**, and the summit metabolic rate.

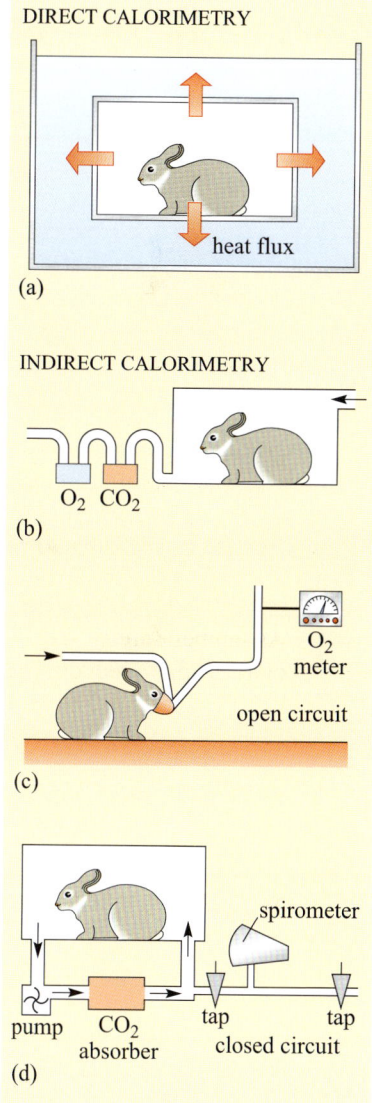

Figure 2.8 Different methods that could be used to measure the rate of heat production in animals. The spirometer in (d) is an instrument for measuring the volume of air entering and leaving the lungs.

■ Under what conditions would you expect to measure BMR?

The animal must be at rest and not feeding or digesting food. As metabolic rates are temperature dependent, BMR must be measured when an animal is making no thermoregulatory effort.

■ Under what conditions would you expect to measure summit metabolic rate?

The summit metabolic rate would be measured when the animal was under the greatest physiological stress, such as intensive exercise, or for a mammal or a bird, extreme cold stress.

It is always difficult to be sure that the measured summit metabolic rate is at a maximum, so there is always a small element of uncertainty in values for summit metabolic rate. Measurement of the metabolic rate of an exercising animal requires a special technique. Figure 2.9 shows an iguana walking on a moving treadmill while wearing a face mask which has a constant flow of air through it. The mask is attached to a system that measures oxygen, carbon dioxide and water vapour partial pressures in the air, providing the data from which metabolic rate can be calculated as rate of uptake of oxygen per unit time and body mass. The apparatus can be scaled up for large animals such as horses, which can be trained to walk, trot and gallop to command. Humans regularly exercise voluntarily on treadmills and pay for the privilege.

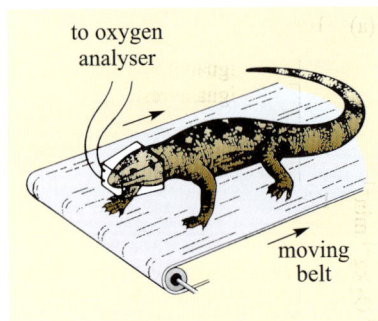

Figure 2.9 A method for measuring the oxygen consumption of a running iguana. For small animals, it is possible to dispense with the mask and just enclose the animal and drive belt in a container, the oxygen content of which can be monitored.

The summit metabolic rate can be measured as the maximum rate of oxygen uptake, $\dot{V}_{O_2\,max}$ during intense exercise. In fact the best measurements of $\dot{V}_{O_2\,max}$ are derived from treadmill measurements. To measure $\dot{V}_{O_2\,max}$ for a human subject, the slope of the treadmill is gradually increased until the subject is exhausted, and the maximum oxygen uptake is the plateau value for oxygen consumption (Figure 2.10).

When measuring BMR for particular species the researcher has to choose a time that the animal is in the post-absorptive phase, i.e. when the animal has completed digestion of its last meal. Digestive processes use energy, resulting in a transient increase in metabolic rate, so measuring basal metabolic rate while an animal is digesting food would give a misleading value.

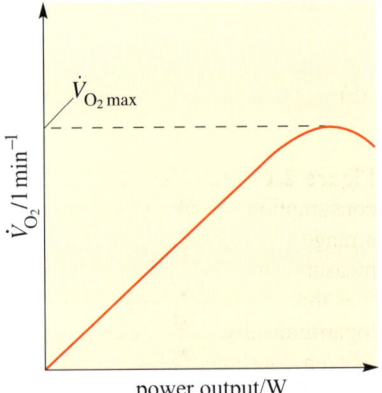

Figure 2.10 Measurement of $\dot{V}_{O_2\,max}$, in a human subject on a treadmill, by plotting oxygen uptake during an increasing workload of exercise.

Figure 2.11a shows data comparing the metabolic rates of a resting and active iguana at a range of T_a from 12 °C to 40 °C. Figure 2.11b is a plot of metabolic rates of a rabbit at a range of ambient temperatures. These animals are about the same size but have very different metabolism.

■ Describe the effect of increasing T_a on the metabolic rate of the resting and active iguana.

As T_a increases, metabolic rates of both resting and active iguana increase. For example, oxygen consumption for the resting iguana at T_a 15 °C is about 0.2 cm^3 O$_2$ kg^{-1} min^{-1} increasing to about 1.2 cm^3 O$_2$ kg^{-1} min^{-1} at T_a 40 °C. The active iguana at T_a 15 °C has an oxygen consumption rate of around 1 cm^3 O$_2$ kg^{-1} min^{-1} compared with about 8 cm^3 O$_2$ kg^{-1} min^{-1} at 40°C. Therefore higher T_a alone also increases rate of oxygen uptake in the iguana.

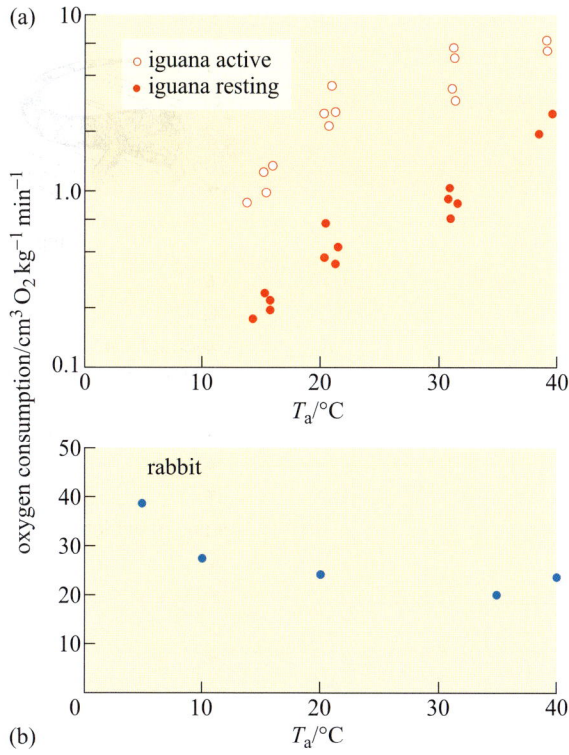

Figure 2.11 (a) Metabolic rates, determined by oxygen consumption, of an active iguana and a resting iguana at a range of ambient temperatures (maximum of four measurements from the animal at each temperature). Note that the values for oxygen consumption are plotted logarithmically. (b) Metabolic rates, determined by oxygen consumption, of a resting rabbit at a range of ambient temperatures.

■ Compare the effect of increased T_a on metabolic rate of the rabbit with that observed in the iguana.

In the rabbit, as T_a increases, metabolic rate decreases from 40 cm³ O_2 kg⁻¹ min⁻¹ at T_a 5 °C, to 20 cm³ O_2 kg⁻¹ min⁻¹ at T_a 35 °C. Metabolic rate for the rabbit is about 100 times higher at T_a 15 °C than that for the resting iguana. Even the highest measured value of metabolic rate for the iguana, 7.5 cm³ O_2 kg⁻¹ min⁻¹ at T_a 40 °C, is lower by a factor of about 3 than the lowest measured metabolic rate of the rabbit, 20 cm³ O_2 kg⁻¹ min⁻¹.

The rabbit, like most other mammals and also birds, uses metabolic heat to maintain a high body temperature. Most mammals and birds have high metabolic rates, with a consequent high demand for energy from food. This strategy is called **tachymetabolism,** meaning high metabolism. The opposite strategy, **bradymetabolism,** is observed in animals with relatively low metabolic rates, which means that relatively little metabolic heat is available for maintaining T_b. Animals that use bradymetabolism include reptiles such as the snake and the iguana (Figure 2.11a).

2.4.2 Field metabolic rate

Measurements of metabolic rate made in the laboratory are useful, but physiologists are particularly interested in how an animal uses energy in the field. The field metabolic rate (FMR) is the metabolic rate of an animal as it goes about its normal activities in its natural environment.

Physiologists use the doubly-labelled water technique to measure the metabolic rate of a free-ranging animal. The technique is widely used: you have already come across it in Chapter 1, in the context of the study on African wild dogs. Radioactive isotopes of oxygen and hydrogen are used as tracers. Oxygen has an atomic mass of 16, but there is a heavier isotope with an atomic mass of 18, written as ^{18}O. Hydrogen has an atomic mass of just 1, but heavier isotopes exist and can be synthesized with atomic masses of 2 and 3, written as ^{2}H and ^{3}H, respectively.

During catabolism in animal cells, substrates such as glucose are broken down into acetyl CoA, which then enters the TCA cycle (Figure 2.1). For every molecule of acetyl CoA that enters the TCA cycle to be broken down, two molecules of water are required and two molecules of carbon dioxide are produced. Half of the oxygen atoms in the CO_2 come from water. In addition, energy released during electron transport pumps protons (H^+) out of the mitochondrial matrix through the inner mitochondrial membrane, into the inter-membrane space. The protons flow down their electrochemical gradient back into the matrix, via the ATPase in the inner mitochondrial membrane, powering ATP synthesis and forming water by combining with oxygen. So oxygen atoms are lost from the animal either in CO_2 or H_2O, whereas hydrogen is only lost in water through evaporation or excretion.

If an animal breathes in labelled oxygen, $^{18}O_2$, or is given water containing oxygen labelled with ^{18}O, then after a period of equilibration, labelled oxygen appears in expired carbon dioxide. Such measurements show that the amount of label in the carbon dioxide is about the same as the amount in the body water and is independent of the source of the original oxygen atoms (as gaseous molecular oxygen or as a component of water). We need to understand why is this is so.

When carbon dioxide dissolves in blood it reacts with water to form carbonic acid (H_2CO_3) which dissociates to form hydrogen ions and hydrogen carbonate ions (HCO_3^-):

$$CO_2 + H_2O \rightleftharpoons H_2CO_3 \rightleftharpoons H^+ + HCO_3^- \qquad (2.5)$$

The first of these reactions is catalysed by carbonic anhydrase and is virtually instantaneous. So oxygen in the dissolved carbon dioxide and oxygen in body water come rapidly into equilibrium. This point is particularly important because it means that if ^{18}O is introduced into an animal it rapidly appears in the pool of internal water. Therefore the amount of label present declines as metabolism proceeds, with label being lost in water and carbon dioxide. Clearly the rate of loss depends on the rate of metabolism, and to measure the rate of loss it is not necessary to analyse expired gas; successive blood samples, when analysed, show the rate of decline. However, the rate is influenced by water loss and carbon dioxide loss, and the rate of carbon dioxide loss is measured as an indicator of the metabolic rate.

- How can the decline in ^{18}O due to carbon dioxide loss can be separated from the decline due to water loss?

If the labelled oxygen is introduced into the animal in the form of water, and the water is also labelled with ^2H, then the decline in labelled hydrogen in the blood can also be measured.

The decline of the labelled hydrogen in the blood is a result of the loss of hydrogen in water, so the difference in turnover of the two labels depends upon the rate at which carbon dioxide is produced and expired. Figure 2.12 illustrates this point.

- What is the measure of metabolic rate when the doubly-labelled water technique is used?

The measure of metabolism is the rate of CO_2 production, which is determined by monitoring the rate of disappearance of labelled oxygen (^{18}O) from the blood. By labelling hydrogen (^2H) it is possible to separate out the loss in water of ^{18}O, from the loss of ^{18}O in CO_2. Hence the rate of CO_2 loss can be calculated.

A number of assumptions are made about this technique, but the results are estimated to be accurate to about 95% compared with 98% for other methods. Although less accurate, the doubly-labelled water technique provides data that could not be collected in any other way since the animals can be free-ranging (although they have to be captured for injection of doubly-labelled water, and again for taking blood samples).

An example of the use of the doubly-labelled water technique is the study carried out by Williams et al. (2002) on wild Rüppell's foxes (*Vulpes rueppelli*) at Mahazat as-Sayd in Saudi Arabia. Mahazat as-Sayd is an arid sandy desert area with no free-standing water available, so animals living there rely entirely on water taken in with their food. One aim of the study was to explore the hypothesis that FMR is reduced in Rüppell's fox in comparison with species that live in more **mesic** (well-watered) environments.

- Suggest why the researchers hypothesized that FMR for Rüppell's fox would be reduced compared with fox species living in more mesic environments.

Lower FMR would be related to lower daily energy requirement of the foxes and also to lower amounts of metabolic heat generated, reducing the need for cooling by evaporative water loss.

FMR is effectively a measure of the daily activity of an animal so the less active the animal the lower its FMR.

Foxes were captured by use of traps baited with food that was enclosed in wire mesh (so food or water intake were not altered). An initial blood sample was taken from each individual. Each fox was injected intraperitoneally (into the abdominal cavity), with a suitable dose of doubly-labelled water $^2H_2{}^{18}O$. At 3 hours post-injection, a 2 cm³ blood sample was taken, the fox was weighed, an ear tag was fitted and the animal released.

Each fox was recaptured 4.8–5.8 days following the $^2H_2{}^{18}O$ injection. The territorial behaviour of Rüppell's foxes means that individuals are always within

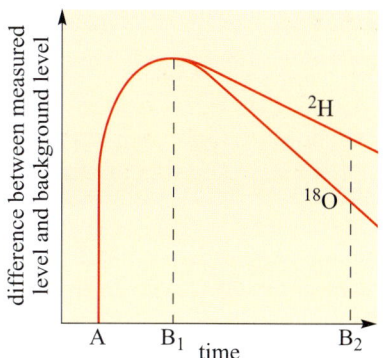

Figure 2.12 An illustration of an experiment to measure carbon dioxide production, and hence metabolic rate, in an animal. A measured dose of water labelled with ^{18}O and ^2H was injected into the animal at time A. At time B_1, the first blood sample was taken, corresponding to the point at which the labelled and unlabelled atoms had come into equilibrium. A second blood sample was taken at time B_2. The difference in turnover of the two isotopes gives a measure of carbon dioxide production.

their own territories, so careful placement of traps ensures successful recapture. A second blood sample was taken from each fox and the animal was weighed again, then released.

Isotope concentrations in the blood samples were determined in the laboratory and the values for the isotope concentrations were used for calculation of energy expenditure, measured as kJ day^{-1}, an expression of whole body FMR.

Table 2.2 summarizes the results of FMR measurements. The researchers used an exponent of 0.869 for body mass when calculating FMR as kJ day^{-1}, based on previous work by Nagy et al. (1999) who derived this exponent for making physiological comparisons between carnivores of varying sizes, including fox species.

Table 2.2 FMR of free-living Rüppell's foxes in Saudi Arabia during October–December 1998 (adapted from Williams et al., 2002).

Animal	Body mass/kg	Litres CO_2 day^{-1}	FMR/kJ day^{-1}	FMR/kJ day^{-1}(mass$^{0.869}$)$^{-1}$	Measurement interval/days
Females (ear tag number)					
174	1.65	22.4	575.4	0.92	5.8
177	1.64	27.4	704.4	1.13	6.0
107	1.47	34.6	888.7	1.57	5.7
Mean ± SE*	1.59 ± 0.10	28.1 ± 6.1	722.8 ± 158	1.20 ± 0.33	5.8 ± 0.15
Males (ear tag number)					
111	1.70	43.8	1126.6	1.75	5.8
08	2.08	52.6	1352.9	1.77	5.3
84	1.98	56.0	1439.9	1.97	4.8
Mean ± SE	1.92 ± 0.2	50.8 ± 6.3	1306.0 ± 162	1.83 ± 0.12	5.3 ± 0.50
Mean for males + females ± SE	1.75	39.5	1014.7	1.52	5.6

*Standard error.

Table 2.3 gives values of FMR for three fox species living in more mesic environments.

Table 2.3 FMR of three fox species.*

Species	Environment	kJ day^{-1}(mass$^{0.869}$)$^{-1}$
swift fox (*Vulpes velox*)[1]	dry grasslands; plains; pasture	2.31
kit fox (*Vulpes macrotis*)[2]	dry grasslands; plains; pasture	2.07
Blanford's fox (*Vulpes cana*)[3]	at altitudes ~2000 m in mountains in Iran, Turkestan and Arabia	1.61

*Data compiled by Williams et al. (2002) from [1]Covell et al. (1996); [2]Girard (1998); [3]Geffen al. (1992a).

- Compare the average FMR for males and females [measured as kJ day^{-1} (mass$^{0.869}$)$^{-1}$] for Rüppell's fox with the values in Table 2.3 for the three species living in more mesic environments. Does your comparison support the hypothesis that Rüppell's fox has a lower FMR than fox species living in more mesic habitats?

The data suggest that Rüppell's fox does have a lower FMR than the three species listed in Table 2.3. Average FMR of male and female Rüppell's foxes is 1.52 kJ day^{-1} (mass$^{0.869}$)$^{-1}$ compared with values of 2.31, 2.07 and 1.61 kJ day^{-1} (mass$^{0.869}$)$^{-1}$ for swift, kit and Blanford's fox, respectively. So the data are consistent with the hypothesis that Rüppell's fox has a lower FMR than fox species living in more mesic habitats.

- Suggest why caution is required when comparing values for FMR measured in several species, when the work on each species is carried out by different research teams working independently.

If the research teams have not standardized their methods, each group is likely to have used variants of the technique. Values for FMR are closely related to the activity of an animal. Research teams may have made their measurements in different seasons. For example, the FMR measurements on the Rüppell's foxes were carried out in the winter. Differences in FMR between species relate directly to their activity, which is likely to show seasonal variation.

Note that animals are mobile, and do not always remain in their 'own' environment as identified by biologists. Until recently, Blanford's fox was regarded as a species found only in mountainous areas, preferring altitudes of around 2000 m. However, a population was found living near the Dead Sea in Israel, the lowest valley in the world, 480 m below sea-level, where rainfall is < 50 mm yr^{-1}, and average daytime T_a in summer exceeds 40 °C (Geffen et al., 1992b). Therefore, Blanford's fox does not always live in more mesic environments than Rüppell's fox.

There are potential pitfalls in making comparisons between FMR values in different species, and drawing conclusions from those measurements. Many mammal species are able to live in a variety of environments and energy budgets of individuals are likely to vary according to food availability, and also seasonal variation in activity. Nevertheless, FMR values provide valuable information about energy expenditure of animals while they are behaving normally in the field.

Summary of Section 2.4

The rates of chemical reactions and in particular enzyme-catalysed reactions in the body are highly temperature dependent. Measurement of the heat production of animals can be used to estimate metabolic rate. Metabolic rate can also be determined by measuring the rate of oxygen consumption or carbon dioxide production.

Measurement of metabolic rate under natural conditions is possible by use of the doubly-labelled water technique. Such studies on Rüppell's foxes, which live in arid conditions, suggest that this species has a lower metabolic rate than related species that live in more mesic environments. However, caution is required when interpreting results obtained by different research groups.

2.5 Thermoregulation

2.5.1 Thermoconformers and thermoregulators

Section 2.4.1 demonstrated the effect of varying temperature on chemical reactions in the cell. Now we need to set that information into the context of the whole animal.

■ Why is it advantageous for an animal to maintain body temperature within limits?

At low temperatures, chemical reactions in the cell proceed at very low rates; as temperature decreases, cell membranes lose their fluidity, which reduces the rate at which the electron transport chain proceeds and hence the rate of synthesis of ATP. At high temperatures, enzymes and other proteins such as haemoglobin denature and cannot function. Therefore it is important for an animal to maintain a body temperature between the extremes of too hot and too cold.

The physiological processes by which body temperature is controlled are known as **thermoregulation.** The ability of an animal to thermoregulate can be measured by comparing the body temperature T_b, with ambient temperature T_a. Figure 2.13 is a plot of T_a against T_a for two hypothetical animals, one that thermoregulates – a **thermoregulator**– and one that does not thermoregulate – a **thermoconformer**.

The equation of the line is:

$$T_b = a + bT_a \qquad (2.6)$$

where a is the intercept of the line with the y-axis and b is the slope. For the thermoregulator, a has a value between 20 °C and 40 °C, and b is zero. For the perfect thermoconformer, a is zero, and as $T_a = T_b$ over the entire range of T_a, b is 1.

Table 2.4 gives values for a and b for a range of animals. Animals above the line, ectotherms, make little use of metabolic heat (bradymetabolism) and rely on external heat sources to raise T_b; those below the line use metabolic heat (tachymetabolism). It may surprise you to learn that not all mammals have such a high T_b as do humans, and that some are not perfect thermoregulators, e.g. the naked mole rat. The aquatic salamander is a perfect thermoconformer having values of a and b of 0 and 1, respectively. The least weasel, on the other hand, is a perfect regulator, with values for a and b of 39.5 and 0, respectively.

Measurements such those shown in Figures 2.11 and 2.13 provide a basis for distinguishing between reptiles and mammals in terms of their patterns of thermoregulation. The term **poikilothermy**, as demonstrated by the iguana, describes a pattern in which large changes in body temperature occur. **Homeothermy** has been described as maintenance of body temperature within ± 2 °C of optimal. This distinction is not so clear-cut, however, as you will appreciate when studying Chapter 3. Many mammals and birds do in fact allow their body temperatures to fluctuate outside these limits, a situation known as 'relaxed homeothermy' (also known as **heterothermy**). Nevertheless, homeothermy is useful as a term to mean 'stable body temperature maintained by tachymetabolism'.

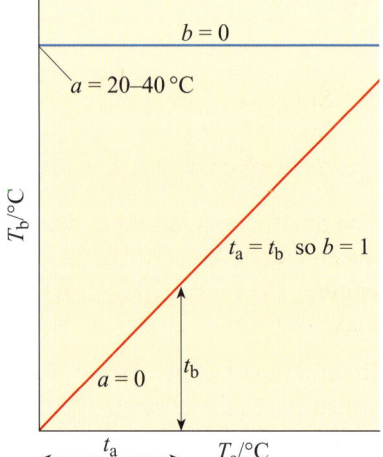

Figure 2.13 Body temperature, T_b, plotted against ambient temperature, T_a, for two hypothetical animals, a thermoregulator (blue line) and a thermoconformer (red line).

Table 2.4 Coefficients for the relationship between T_a and T_b in a range of animals.

	a	b
forest anole lizard	−3.8	1.17
tenebrionid beetle	1.3	1.01
aquatic salamander	0	1.00
Asian honeybee	18.3	0.81
male diamond python	9.6	0.67
bullfrog	12.0	0.60
non-brooding female diamond python	14.5	0.48
desert cicada	24.8	0.44
open habitat anole lizard	16.4	0.43
dragonfly	26.3	0.43
sphinx moth caterpillar	23.7	0.40
naked mole rat	20.6	0.41
bluefin tuna	25.5	0.24
dragonfly	35.1	0.21
poor will	33.6	0.17
brooding female diamond python	28.0	0.12
pocket mouse	35.4	0.081
honey possum	35.1	0.07
queen bumblebee (brooding)	35.0	0.07
cactus mouse	35.8	0.059
Chilean tinamou (related to the partridge)	35.4	0.05
house finch	40.0	0.05
rosy finch	41.0	0.02
Amazonian parrot	40.9	0.01
least weasel	39.5	0
mallee fowl	40.3	−0.04

■ Study Table 2.4 and list those fish, reptiles and insects that are below the line. What does the position of these species below the line suggest about the metabolic rate and thermoregulatory strategy of these animals?

The bluefin tuna, brooding female diamond python, dragonfly and brooding queen bumblebee, are all below the line. Their position suggests that these animals too are capable of tachymetabolism at certain times, and are likely to make use of metabolic heat to maintain T_b.

Bluefin tuna are tachymetabolic when swimming rapidly and generate heat, which raises the temperature of the muscles. Dragonfly flight muscles are tachymetabolic during vigorous flight and generate heat that raises T_b in the thorax by a few °C.

Adult homeotherms maintain a constant deep body temperature by means of elaborate sensory and effector mechanisms integrated by the nervous system. Receptors sensitive to thermal challenges are located both peripherally in the skin, and centrally in the brain and spinal cord, as well as in other organs throughout the body. Information from temperature receptors is passed to the hypothalamus. Appropriate reactions that counter any threat to thermal stability are believed to be initiated by a complex process of integration within the vertebrate hypothalamus. The limits to thermoregulation are set at low ambient temperatures by the heat-generating capacity of the animal's body, and at high temperatures by its heat-dissipating abilities. Within these extremes, there is a balance between the rate of heat production and the rate of heat loss, so that the set level of deep-body temperature is maintained.

Whether any given combination of climatic factors results in 'thermal comfort' or in heat or cold stress depends on the species, the individual's state of acclimatization, its social and developmental status and behavioural characteristics such as habitat selection. Many species occupy a protected micro-environment such as a nest or burrow, or huddle together with other individuals thereby avoiding extremely hot or cold macroenvironments that cannot be tolerated for more than short periods of time.

Figure 2.14 is an idealized diagram that summarizes in outline physiological responses to hot and cold environments in mammals and birds, and explains some of the terms used in thermobiology. More detailed understanding of thermoregulatory physiology can only be gained from study of individual species. Nevertheless, general principles can be applied. Within the zone of least thermoregulatory effort, physiological mechanisms such as increased metabolic rate and cooling by evaporative water loss are not required to maintain constant T_b. When T_a drops to a certain level, known as the lower critical temperature, T_{lc}, increased metabolic rate maintains high stable T_b. As T_a declines, metabolic rate increases further. As T_a increases a point is reached, B, at which evaporative cooling is required to keep T_b constant.

■ Use the information in Figure 2.14 to define upper critical temperature, T_{uc}.

T_{uc} is the ambient temperature at which increased metabolic rate (in addition to evaporative water loss) is required for cooling the body to maintain constant T_b.

In certain mammals, evaporative cooling is enhanced by sweating, secretion of salty fluid from sweat glands. Evaporation of sweat cools the skin. Panting also enhances cooling of T_b by increasing the rate of evaporation of fluid from the

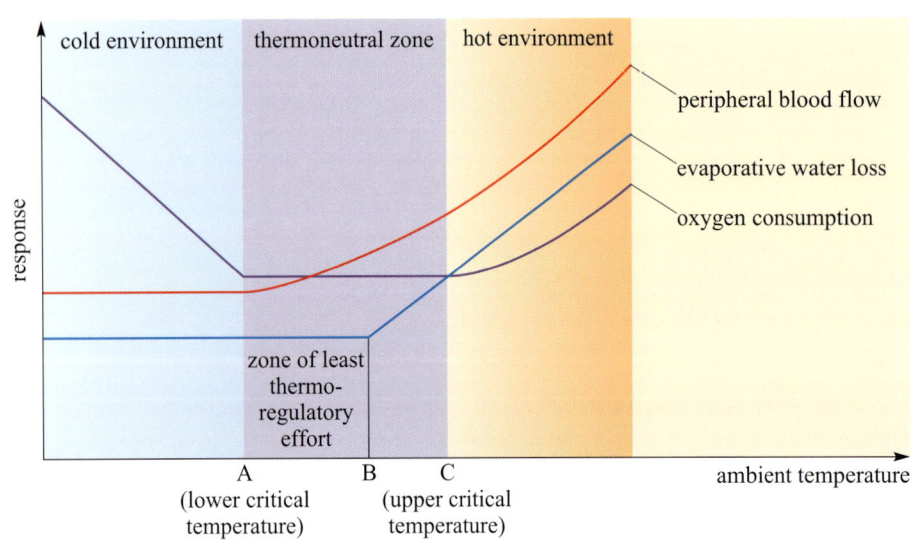

Figure 2.14 A summary of physiological responses of homeotherms to hot and cold environments. The lower critical temperature A (T_{lc}) marks the boundary between cold and hot environments. The zone from A to C (from the lower critical temperature to the upper critical temperature) defines the zone of thermoneutrality. Within that zone, there may be evaporative cooling, which is an active process. Within the thermoneutral zone, the range of T_a over which there is no evaporative cooling is called the zone of least thermoregulatory effort: A to B.

moist surfaces of mouth and tongue. We will be exploring evaporative cooling further in Chapter 3.

- Rüppell's fox is nocturnal. During the day when T_a is extremely high, often 40 °C, the foxes rest inside their underground dens. At night, T_a plummets and the foxes emerge and forage for reptiles and insects. What advantage does the nocturnal life style offer to a desert fox in terms of thermoregulatory effort?

As Rüppell's foxes rest in their dens during the day they avoid the need for physiological thermoregulation, e.g. panting, when T_a outside is high. T_a at night is considerably lower than T_a during the day. Therefore, by emerging at night to forage for food, Rüppell's foxes reduce the need for evaporative water loss, and also reduce energy consumption.

Nocturnal desert foxes demonstrate that behavioural and physiological thermoregulatory strategies are closely intertwined. Behaviour has profound effects on physiology, and physiology affects behaviour.

Even within a single species, both T_{lc} and T_{uc} are affected by an animal's recent thermal history and diet. Animals that have been living at a low T_a for some time show a decreased value for T_{lc} in comparison to individuals of the same species that have been living at high T_a. Such short-term physiological adjustments in response to exposure to a variety of factors in new environments are known as **acclimatization**.*

2.5.2 Measuring the body temperature of wild animals

By now you should be thinking that the familiar distinction between 'cold-blooded' animals such as reptiles, amphibians and fish, and warm-blooded animals, birds and mammals, is not a clear one. Reptiles living in the Sonoran and Mojave deserts maintain high daytime T_b. The desert iguana (*Dipsosaurus dorsalis*) has a daytime T_b of 40–47 °C; the chuckwalla (*Sauromalus ater*) maintains a T_b of 38–40 °C on sunny days, hardly what would be expected for cold-blooded animals. We learned in Section 2.3.1 that during hibernation the marsupial mouse regulates body temperature at 16 °C, a value too low for activity in some 'cold-blooded' animals.

How do physiologists measure T_b of wild animals? A fundamental problem for earlier work on temperature and its regulation was that the act of measuring the body temperature of an animal induced stress that altered the body temperature. Many measurements on reptiles, for example, have been made with a Schultheis thermometer, which is a mercury thermometer with a very fine bulb. When the bulb is inserted into the cloaca of the reptile a value for the deep body temperature can be obtained. However, unless the animal is very large, it is certainly not in an unstressed and natural state during this operation. Physiologists were well aware of this source of error and resorted to ingenious tactics to minimize stress on experimental subjects. When measuring T_b of *Liolaemus* at different altitudes in the Andes Mountains in Chile, Marquet and his colleagues (1989) resorted to stalking the lizards and quickly grabbing them and measuring T_b before the animals had time to become stressed. The data are summarized in Table 2.5.

---

* You will find the term acclimation used interchangeably with acclimatization in some literature. Note that acclimation does have a precise meaning: changes that occur in response to one variable in controlled experiments.

Table 2.5 Activity field temperature (field T_b, mean ± 2 SE), air temperatures (T_a, mean ± 2 SE) and preferred body temperatures (lab T_b, mean ± 2 SE) for high-altitude *Liolaemus* lizards (n = number of individuals measured; N = number of measurements taken). No laboratory data are available for the lower altitude lizards.

Species	Field T_b/°C	Range	T_a/°C	Range	n	Lab T_b/°C	Range	N
Higher altitude								
L. alticolor	29.1 ± 1.2	23.0–34.0	20.8 ± 1.4	14.0–28.5	23	34.5 ± 0.9	25.8–39.5	108
L. jamesi	29.1 ± 2.2	24.0–33.0	19.3 ± 2.6	13.0–25.0	9	36.0 ± 1.4	27.5–40.4	51
Lower altitude								
L. islugensis	28.2 ± 1.3	18.5–37.0	23.4 ± 1.4	12.0–32.0	41			
L. ornatus	30.7 ± 1.3	21.0–35.0	26.6 ± 2.1	17.5–35.0	24			

■ Summarize Marquet et al.'s measured values for T_b of the four *Liolaemus* species using the data included in Table 2.5. Comment on the values for field T_b of the Andean *Liolaemus* species, drawing on the provided values for field T_b of *Dipsosaurus* and *Sauromalus*.

Table 2.5 shows that *L. alticolor* and *L. jamesi* (species living at altitudes of 4000–4500 m), had similar field T_b, around 29 °C, to those of *L. islugensis* and *L. ornatus* (species living at lower altitudes of 3500–4000 m). Compared with T_b values of up to 47 °C and 40 °C for *Dipsosaurus* and *Sauromalus*, respectively, both desert species, a T_b of 29 °C is low.

The relatively low T_b values for Andean *Liolaemus* species reflect the harsh thermal environment at high altitudes where cold winds increase convective heat loss from the body.

Development of radiotelemetry has enabled continuous monitoring of individual and groups of wild animals. For monitoring of body temperature, tiny radiotransmitters with incorporated temperature detectors can be attached or implanted into a wild animal. This technique, radiotelethermometry, is limited by the size of the battery pack required but in recent years, battery packs have become progressively smaller. There is also the problem of the aerial that is needed to transmit the signal; care must be taken to ensure that the animal could not be harmed by say getting the aerial caught on a bush or in a tunnel wall. When these problems are solved, radiotelethermometry provides a useful insight into the thermal biology of an animal in its natural environment. Figure 2.15 shows the results of T_b measurements on a monitor lizard (*Varanus* sp.), by radiotelethermometry, carried out over 4 days.

Figure 2.15 Values of T_b measured over 4 days by radiotelethermometry for a monitor lizard *Varanus*. T_a is also shown (modified from Stebbins and Barwick, 1968).

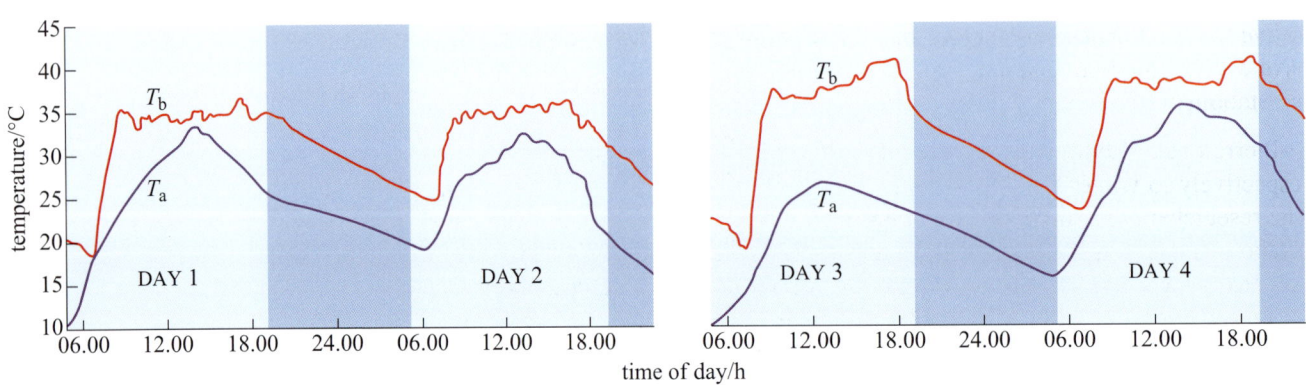

- Describe the pattern of the T_b values over 24 hours from 06.00 on one day to 06.00 the following day. Is *Varanus* an effective thermoregulator?

At 06.00, T_b is relatively low, about 20 °C, but increases rapidly to well above ambient temperature, reaching about 36 °C by 09.00 when T_a is about 20 °C. From 09.00 to about 19.00, the lizard maintains a T_b around 36 °C that fluctuates slightly and is always above T_a. Clearly *Varanus* is an effective thermoregulator during the day. At night, T_b gradually drops, reaching 20 °C by 06.00, but it rarely drops as low as T_a.

Observations on *Varanus* showed that during the early morning, the lizard sunbasks with the length of the body orientated at right angles to the Sun, and with crests erected for maximal heat gain. This phenomenon is sometimes referred to as **heliothermy**. When preferred or **eccritic** T_b of about 36 °C is attained the lizard adopts a posture head-on to the Sun to reduce heat gain. **Shuttling** between sun and shade is also used during daylight hours: *Varanus* rests in the shade when T_b exceeds preferred T_b. Use of solar radiation as a source of heat is not possible at night and the lizard spends the night curled up in its tree hole. Note that even at night T_b does not drop to T_a. The curled up posture of the large lizard which exposes as little as possible of surface area of the animal to the air, thereby reducing radiative and convective heat loss, means that the body has a high **thermal inertia**, i.e. a high resistance to temperature change.

2.5.2 Temperature compensation

In animals such as fish, amphibians and reptiles, in which body temperature fluctuates, the rate of temperature-dependent processes also fluctuates. Fish living in temperate areas may be living in water at 1–4 °C during winter in contrast to summer when water temperature may be 15 °C or more. For processes that are acutely sensitive to temperature, a drop of a few degrees might stop that process altogether. Activity of enzymes involved in catabolism of fuel molecules declines with decreasing temperature. If the decline was at a constant rate, metabolic rate at winter temperature of water for fish might be too low to support normal activity. If cardiac rate is measured in reptiles or amphibians over a range of summer temperatures, and the heart-rate vs temperature curve is extrapolated to winter temperatures, then at some point the heart would stop beating. Yet fish continue activity during winter, and at least some species of amphibians and reptiles survive over winter. Experimental studies have demonstrated that in winter the metabolic rate vs temperature curve for certain fish and amphibian species is shifted upwards on the *y*-axis in comparison to the curve obtained in the summer. Such temperature compensation enables the animal to live to function normally at the colder winter temperatures. Research studies have concentrated on temperature compensation of activity of enzymes involved in catabolism of glucose and fatty acids.

St-Pierre et al. (1998) studied two groups of rainbow trout acclimatized respectively to winter (1 °C) and summer (16 °C) temperatures in outdoor tanks. The researchers measured activity of enzymes involved in glucose and fatty acid catabolism at a range of temperatures in samples of red muscle mitochondrial suspensions prepared from the two groups of fish. Table 2.6 includes the results for cytochrome *c* oxidase (CCO), citrate synthase (CS) and carnitine palmitoyl transferase (CPT, involved in catabolism of fatty acids).

Table 2.6 Effects of thermal acclimatization and assay temperature on measured activities of enzymes in isolated red muscle mitochondria of rainbow trout (*Onchorhyncus mykiss*) (St-Pierre et al., 1998).

Fish	Assay temperature/°C	Cytochrome *c* oxidase*	Citrate synthase*	Carnitine palmitoyl transferase*
WT	1	1.68 ± 0.19	0.64 ± 0.04	$9.6 \pm 1.3 \times 10^{-3}$
ST		ND	ND	ND
WT	8	$2.01 \pm 0.17**$	$0.88 \pm 0.06**$	$1.4 \pm 1.3 \times 10^{-3}$
ST		1.58 ± 0.09	0.69 ± 0.02	$1.1 \pm 0.7 \times 10^{-3}$
WT	15	$2.70 \pm 0.22**$	1.19 ± 0.08	$22.7 \pm 2.3 \times 10^{-3}**$
ST		2.04 ± 0.12	1.13 ± 0.02	$16.0 \pm 0.8 \times 10^{-3}$
WT	22	$3.28 \pm 0.23**$	1.68 ± 0.11	$29.2 \pm 3.1 \times 10^{-3}**$
ST		2.53 ± 0.15	1.51 ± 0.03	$21.8 \pm 0.7 \times 10^{-3}$

Abbreviations: WT, winter-acclimatized trout; ST, summer-acclimatized trout; ND, not determined.
* Enzyme activity/units of product mg^{-1} mitochondrial protein, mean $\pm$SE.
** Significant difference between seasons: $P < 0.05$.

■ Describe the data in Table 2.6.

Activity of CCO from 8 to 22 °C was consistently higher in mitochondria from winter-acclimatized fish than in mitochondria from summer-acclimatized fish ($P < 0.05$). For CS the picture was not so clear, with increased activity in winter-acclimatized fish significant ($P < 0.05$) only at 8 °C. Activity of CS was similar in both winter- and summer-acclimatized fish at 15 and 22 °C. CPT activity was higher in the winter-acclimatized fish at all three temperatures but significant at only 15 and 22 °C. The overall picture is of an upward shift in levels of catabolic enzyme activity in winter-acclimatized fish.

■ Do the results in Table 2.6 support the hypothesis that aerobic metabolism in rainbow trout during the winter is supported by temperature compensation?

Yes. The data in Table 2.6 demonstrate that during cold acclimatization, the capacity of red muscle for aerobic catabolism was increased. Temperature compensation for CCO and CPT was more marked than that observed for CS.

Temperature compensation in rainbow trout does not include enhanced mitochondrial volume density (a measure of volume of mitochondria per unit volume muscle tissue). In contrast, cold acclimatization of goldfish (*Carassius auratus*), crucian carp (*C. carassius*) and striped sea bass (*Morone saxatilis*) is accompanied by an increase in volume density of mitochondria. In these species increased numbers of mitochondria per unit mass of muscle maintains the capacity for aerobic metabolism at low temperatures. However, St-Pierre et al. did measure an increase in mitochondrial protein content in winter-acclimatized rainbow trout, 28.2 ± 1.9 mg g^{-1} fresh muscle, in comparison with summer-acclimatized trout, 16.1 ± 1.7 mg g^{-1} fresh muscle.

The researchers concluded that activities of CCO and CPT per unit mass of mitochondria at low temperatures are increased in the winter-acclimatized fish. Temperature compensation in fish is explored further in Chapter 5.

Summary of Section 2.5

Animals that regulate their body temperature are known as thermoregulators; those that do not are known as thermoconformers. Ectotherms are animals that use external heat to raise T_b in contrast to endotherms that maintain T_b by using metabolic heat. Animals that show large changes in T_b are poikilotherms, those that can maintain T_b within ± 2 °C are homeotherms. Many animals can though adopt a strategy of relaxed homeothermy and allow T_b to fluctuate across a wider range. Figure 2.14 is an important figure that you will need to recall. Measurement of body temperature of wild animals provides important data on thermoregulatory strategies. Temperature compensation is an important mechanism that maintains enzyme activity in poikilotherms during winter when T_b is low.

2.6 Oxygen in the environment

The molecules of the various gases within air move around rapidly. If you imagine a volume of air enclosed within a chamber, the movements of the molecules lead them to collide frequently with the walls of the chamber. These collisions exert pressure on the walls of the container. Atmospheric pressure at sea-level is 101.3 kPa. Each of the gases that make up air exerts a partial pressure, and all the total pressures add up to give the total atmospheric pressure. The partial pressure of a gas in a mixture occupies the same fraction of the total atmospheric pressure, as its percentage concentration in the gas mixture.

■ Dry air at sea-level where atmospheric pressure is 101.3 kPa, contains 20.9% oxygen. What is the partial pressure of oxygen, P_{O_2} at 101.3 kPa?

$$P_{O_2} = \frac{20.9}{100} \times 101.3 \text{ kPa} = 21.17 \text{ kPa}$$

■ At 4340 m altitude, atmospheric pressure is 62.0 kPa. The air contains 20.9% oxygen. What is the partial pressure of oxygen at 4340 m altitude?

$$P_{O_2} = \frac{20.9}{100} \times 62.0 \text{ kPa} = 12.9 \text{ kPa}$$

The sharp drop in P_{O_2} at 4340 m altitude compared to that at sea-level indicates why hypoxia can be a problem for animals at high altitude. Carbon dioxide content of dry air is much lower, only 0.03%. Similar calculations to those provided for oxygen show that at sea-level P_{CO_2} is just 0.03 kPa, and at 4340 m altitude, P_{CO_2} is 6.3×10^{-3} kPa.

We are interested in partial pressures of gases, because gases diffuse down gradients of their partial pressures. So if two gas mixtures are brought together, each component gas within the mixture diffuses down its own partial pressure gradient until the partial pressures of the gases are equalized.

■ The following gas mixtures, 1 and 2, are placed at either side of a membrane that allows gas molecules to cross. In which direction will oxygen, nitrogen and carbon dioxide diffuse?

	Gas mixture 1	Gas mixture 2
P_{O_2}	6.30 kPa	12.8 kPa
P_{CO_2}	7.3×10^{-3} kPa	6.3×10^{-3} kPa
P_{N_2}	48.7 kPa	42.2 kPa

Oxygen diffuses from 2 to 1; carbon dioxide diffuses from 1 to 2; nitrogen diffuses from 1 to 2.

Terrestrial vertebrates, reptiles, birds, mammals and a few fish and some adult amphibians, breathe in air by means of lungs. Inside the lungs, air is in direct contact with water, the fluid that lines the surface of the alveoli. The water contains dissolved gases: oxygen, carbon dioxide and nitrogen. Water evaporates from the alveolar surfaces and the water vapour in turn moves back into the liquid phase. The water molecules in water vapour behave as a gas and therefore exert a partial pressure, commonly known as vapour pressure. Humidity is high in alveolar air, and the air is saturated with water vapour, a situation in which the partial pressure of the water vapour is known as the saturated vapour pressure. The saturated vapour pressure is independent of total pressure and is determined only by temperature.

The effect of having water vapour in the air on P_{O_2} can be appreciated by examining the situation within the alveoli, where air temperature is 37 °C. At this temperature, air saturated with water vapour has a P_{H_2O} of 7.30 kPa, the **saturated vapour pressure**.

■ At sea-level, air contains 20.9% oxygen. When this air is breathed into the lungs, and enters the alveoli, it is immediately saturated with water vapour. Calculate the P_{O_2} of alveolar air at sea-level, where atmospheric pressure is 101.3 kPa.

As the saturated water vapour pressure at 37 °C is 7.30 kPa, at sea-level:

$$\text{alveolar } P_{O_2} = \frac{20.9}{100} \times (101.3 - 7.30) \text{ kPa} = 19.65 \text{ kPa}$$

Measured P_{O_2} in alveoli is about 14.2 kPa, lower than our calculated value because of dilution of inhaled air with residual air left in the lungs after expiration. Total atmospheric pressure is reduced at high altitude, but the percentage composition of the air does not change. However, inside the alveoli, the fixed saturated water vapour pressure becomes an increased fraction of the total air pressure in the alveoli.

■ Calculate the P_{O_2} of alveolar air at altitude 4340 m, where atmospheric pressure is 62.0 kPa. Assume that saturated water vapour pressure in the lungs is 7.30 kPa.

$$P_{O_2} = \frac{20.9}{100} \times (62.0 - 7.30) \text{ kPa} = 11.4 \text{ kPa}$$

So at 4340 m altitude, the total atmospheric pressure is reduced to 61% of that at sea-level but the calculated P_{O_2} in the alveoli is reduced to 58% of its value at sea-level. Despite the sharp reduction in P_{O_2} in alveolar air at high altitude, large populations of animals, including humans, live healthy lives at high altitude.

2.7 The physics of gas exchange

We learned in the previous section that air in the alveoli is saturated with water vapour. With each intake of breath, air in the alveoli is replaced with fresh air, but not completely as there is dead space in the lungs containing air that is not breathed out. The dead space in the lungs accounts for the difference in P_{O_2} between air in the atmosphere and alveolar air. Once in the lungs, oxygen in the air diffuses down its partial pressure gradient into the liquid that covers the alveolar surfaces, and dissolves. Within the liquid, when equilibrium is reached, i.e.

the point at which molecules of a particular gas enter and leave at the same rate, the gas is effectively exerting a partial pressure (tension) in the liquid, equal to its partial pressure when in the gas phase. Note that the tension or partial pressure of a gas in the liquid is not a direct measure of the amount of gas dissolved in it. The quantity of gas in a solution is determined by the solubility of the gas which is a measure of how much gas dissolves in the liquid before equilibrium between solution and dissolution is established. If identical partial pressures of oxygen and carbon dioxide are exposed to water, a much larger *amount* of carbon dioxide dissolves because this gas is much more soluble in water than oxygen. However the partial pressures of the two gases are still equal because partial pressure is a measure of the tendency of the gas to leave the liquid. The solubility of a gas is expressed as its absorption coefficient, which is the volume of gas at **standard temperature and pressure** at sea-level, 0 °C, and 101.3 kPa, that is dissolved in a known volume of solution at equilibrium. So if 0.049 cm^3 oxygen is dissolved in 1 cm^3 water at 0 °C, at 101.3 kPa, then the absorption coefficient for oxygen is 0.049.

Dissolved gases diffuse down gradients of partial pressure and usually in nature, where there is a partial pressure gradient, there is also a concentration gradient. When a gas in simple solution diffuses into another liquid, the gas content of the second liquid rises in direct proportion to the partial pressure of the gas. The alveolar air has a P_{O_2} of about 13–14 kPa, and the end result of gas exchange is that the blood in the pulmonary capillaries leaving the alveoli has the same gas composition as that of the alveolar air. The physical process by which gas exchange is brought about is diffusion, so before we go any further we must examine the factors that affect diffusion.

Intuition tells us that the shorter the diffusion distance and the greater the surface area, the faster dissolved gases can cross a barrier. The difference in gas partial pressure between the two ends of the diffusion path is an important factor. The steeper the partial pressure gradient the faster the diffusion of a gas.

These factors are incorporated into **Fick's equation** (2.7) which enables calculation of the rate of diffusion of a substance, e.g. an ion such as Na$^+$, down its concentration gradient. Fick's equation for free diffusion is:

$$J_{oi} = P_d A (C_o - C_i) \tag{2.7}$$

J_{oi} = rate of diffusion from outside (o) to inside (i) a cell

P_d = diffusion constant, combining a number of properties of the membrane

A = surface area of the membrane

$(C_o - C_i)$ = concentration gradient from outside to inside

Once a substance has crossed the membrane, the rate at which it diffuses through the cytoplasm is inversely proportional to the square of the distance over which diffusion takes place.

When the metabolic rates of animals are calculated it becomes clear that diffusion alone is unlikely to satisfy the oxygen demands of animals where the diffusion distances exceed about a millimetre. In vertebrates, gas exchange occurs via specialized respiratory surfaces in the gills (fish and larval and a few adult amphibians), or lungs (some fish, terrestrial species and those that have become secondarily aquatic such as whales and seals).

Fick's equation can be applied to gas exchange in gills and lungs, and is the basis for the following equation:

$$\dot{M}_{O_2} = \frac{KA\Delta P_{O_2}}{x} \qquad (2.8)$$

$\dot{M}_{O_2}$ = rate of diffusion of oxygen from alveolar air into blood stream/mmol O_2 min^{-1} kg^{-1}. The dot above M signifies that $\dot{M}_{O_2}$ is a symbol for rate of diffusion.

K = Krogh's diffusion constant combining properties of the membranes and tissues between alveolar air and blood inside the capillaries/nmol s^{-1} cm^{-1} kPa^{-1}

A = surface area for respiratory gas exchange (alveoli, gills or tissue capillary bed)/cm^2

ΔP_{O_2} = partial pressure gradient for oxygen from alveolar air to blood in capillary/kPa

x = diffusion distance over respiratory gas exchanger/cm

In mammals, like most terrestrial vertebrates the alveoli of the lungs are the site for exchange of respiratory gases. The mammalian lung contains millions of alveoli, which together have a huge surface area, in the human lung about 70–150 m^2. Fick's equation (Equation 2.7) demonstrates that the larger the surface area for diffusion, the faster diffusion occurs. Pulmonary capillaries bring venous blood to the alveoli. The layer of tissue that separates erythrocytes in the pulmonary capillaries from the alveolar air is very thin. The alveolus is lined by an alveolar membrane which lies adjacent to the capillary endothelium (Figure 2.16a, b, c).

■ Estimate the total blood-to-gas distance in human alveoli using the provided values for the width of each layer separating capillary blood and alveolar air in Figure 2.16c.

Figure 2.16 The microscopic structure of the respiratory portion of the human lung as seen in a thin section of alveoli. (a) Structure of respiratory (alveolar) epithelium and associated tissues. Note the presence of smooth muscle; contraction can have a marked effect on the dimensions of the airways. (b) and (c) Progressively higher magnifications of the alveolo-capillary membrane, which separates alveolar air from capillary blood.

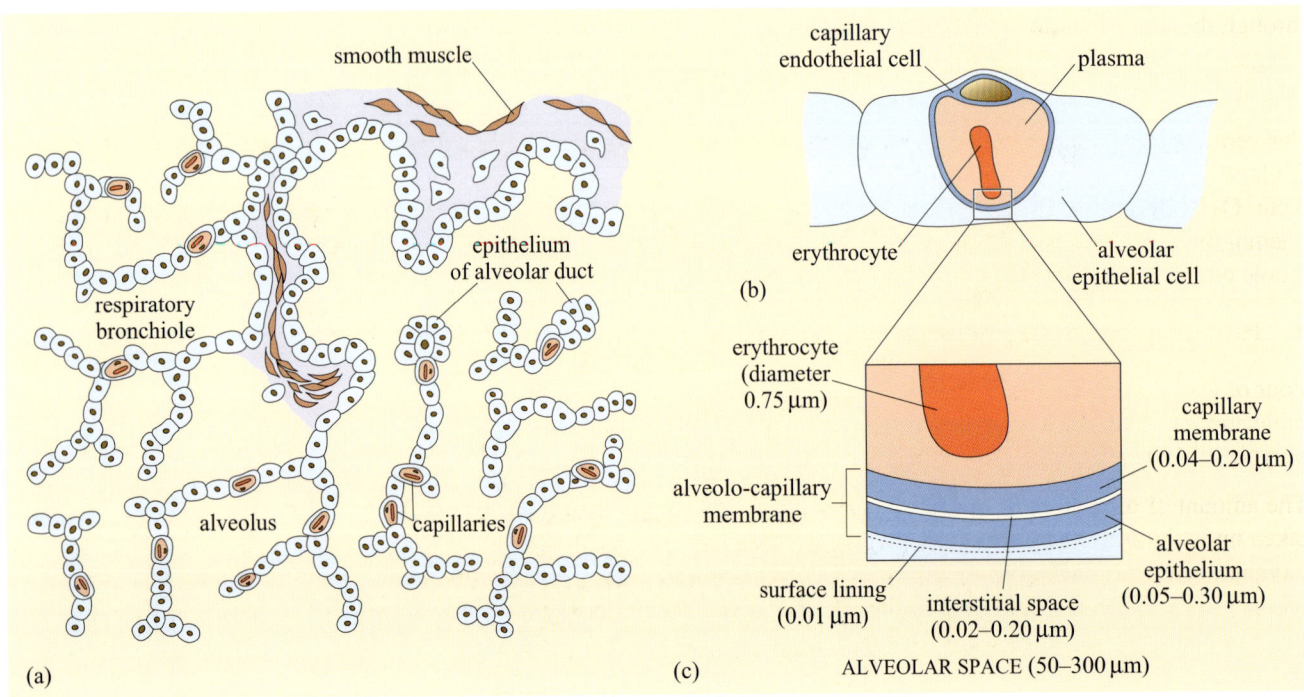

Adding together the widths of capillary membrane, interstitial space, alveolar epithelium and surface lining, gives values in the range of 0.12–0.71 μm.

Measured values for the total blood-to-gas distance in human lung range from 0.2 to 0.6 μm.

■ In some amphibian lungs the total blood-to-gas distance is 4 μm. Use the principles implied by Fick's equation to explain how this distance would affect the rate of diffusion of oxygen into the blood in comparison with the human lung.

Fick's equation has diffusion distance, x, as the denominator so the greater diffusion distance the lower the rate of diffusion. As diffusion distance in amphibian lung is 4 μm compared to a maximum of 0.6 μm in human lung, diffusion rate of oxygen into the blood would be at least 4 μm/0.6 μm = 6.7 times slower in amphibian lung than in human lung.

In the lung the gradient of partial pressure is maintained by the rapid passage of blood in pulmonary capillaries. Although the pulmonary capillary bed is extensive, it only contains about 70–100 cm^3 blood at any instant. Approximately this volume of blood is expelled into the pulmonary artery with each contraction of the right ventricle of the heart. Therefore, at each heartbeat, the pulmonary blood is replaced. Each erythrocyte normally remains in a pulmonary capillary for less than a second. At sea-level, the alveolar air has P_{O_2} of 13.3 kPa and P_{CO_2} of 5.3 kPa. The blood entering the lungs has P_{O_2} of about 7 kPa and P_{CO_2} of about 6.3 kPa, but values vary according to current metabolic activity. The gradients of partial pressure between air and venous blood at sea-level are steep and this drives rapid diffusion. The change in colour of haemoglobin as it becomes oxygenated can be used as a measure of the rate of oxygenation. It takes about 0.4 s for complete oxygen loading of haemoglobin. The 0.3–1.0 s that erythrocytes spend in the pulmonary capillaries provide enough time for oxygen loading. The speed of oxygen loading and the rapid passage of blood through the lungs maintains a continual steep gradient of partial pressure for oxygen and carbon dioxide between blood and alveolar air, thereby maximizing rate of diffusion of respiratory gases, as predicted by Fick's equation.

For vertebrates, the liquid in which respiratory gases are dissolved is the blood. If blood plasma is equilibrated with a gas mixture having a P_{O_2} of 13.3 kPa, then 3 cm^3 O$_2$ l^{-1} dissolves. However, in whole blood, the oxygen binds to the haemoglobin molecules in the erythrocytes, increasing the oxygen content of whole blood to 200 cm^3 l^{-1}.

■ How many oxygen molecules can bind to one molecule of haemoglobin?

Four oxygen molecules can bind to one haemoglobin molecule. Each oxygen molecule binds to one of the four haem groups, each of which is inside the polypeptide chain of one of the four sub-units of the protein.

The amount of oxygen dissolved in plasma is very small compared to the amount taken up by binding to haemoglobin, with no increase in partial pressure of oxygen in plasma. The binding of oxygen molecules to haemoglobin molecules has to proceed to completion before the P_{O_2} in blood can get close to 13.3 kPa.

2.8 Effects of hypoxia on metabolism

Blood in the pulmonary capillaries comes into complete equilibrium with the alveolar gas, so the P_{O_2} of arterial blood is the same as that of the alveolar gas. At sea-level, this quantity is sufficient to supply oxygen to respiring tissues for both rest and sustainable exercise. As we saw in Section 2.3, P_{O_2} decreases as altitude increases. Sudden exposure to high altitudes > 3000 m results in tissue hypoxia, causing unpleasant and sometimes serious symptoms. In time, molecular, biochemical and physiological mechanisms mitigate the effects of hypoxia at altitude, as we shall see in Chapter 6. Many people and diverse species of animals live successfully at altitudes > 3500 m.

Air-breathing vertebrates such as seals, penguins and turtles dive to obtain their food and for other functions. As these animals breath-hold while submerged, their tissues are subjected to hypoxia. Small mammals that spend a long time in burrows may also encounter hypoxia due to poor air circulation in a confined space.

It has been established that hypoxia results in the reversible reduction of metabolic rate and body temperature in many vertebrates, especially small mammalian species (Barros et al., 2001).

■ Why would a reduction in metabolic rate be a useful physiological response for an animal subjected to hypoxia?

A reduction in metabolic rate reduces the demand of tissues for oxygen. In an animal subjected to hypoxia, demand for energy is likely to outstrip the ability to supply tissues with oxygen. If metabolic rate were not reduced the animal would have to resort to anaerobic metabolism, which produces less ATP per mol of glucose than aerobic metabolism. Prolonged anaerobic metabolism is not feasible as build-up of lactate in muscles results in rapid fatigue. So it is advantageous for animals subjected to hypoxia to respond by reducing metabolic rate.

Small mammals encounter hypoxia when they are sheltering inside deep burrows: ventilation is poor and some species, e.g. the kangaroo rat (*Dipodomys* sp.), block the entrances with soil, decreasing ventilation even further. The animals respond to hypoxia by a decline in metabolic rate, which is usually accompanied by a reduction in T_b. Evidence suggests that the decline in T_b is controlled. Tattersall and Milsom (2003) investigated the hypothesis that the initial response to hypoxia in a small mammal includes a transient increase in heat loss that speeds up the decline in body temperature and hence in metabolic rate. The golden-mantled ground squirrel (*Spermophilus lateralis*) was chosen for the investigation because this species spends long periods of time in a hypoxic burrow and may therefore have an enhanced response to hypoxia. Metabolic rates at normoxia and hypoxia were determined by placing an animal in a temperature-controlled metabolic chamber and measuring the rate of oxygen uptake. The rate of oxygen uptake was monitored by measuring amounts of oxygen in the outflow of air from the metabolic chamber and subtracting those values from the known oxygen concentration in the inflowing air. Infrared thermography was used to measure the surface temperatures of the animals. This technique involves the detection of infrared radiation (wavelength 8–12 μm) and converting it to a colour image, an infrared thermogram, within the visible range of light. Computer software was used to convert those images to a grey scale, which was calibrated against values for temperature. Precise values for surface temperatures of the ground squirrels during normoxia, hypoxia, and recovery from hypoxia, were therefore obtained from infrared thermograms. The researchers determined surface temperatures of eyes, ears, nose, feet and flank of a ground squirrel while it was inside the metabolic chamber.

For each experiment, a squirrel was placed inside the metabolic chamber for one hour, while the incoming air contained the normal 21% oxygen. The ambient temperature was set at one of three levels: 10 °C, 22 °C and 30 °C. After 20 minutes of recording oxygen uptake and taking whole body thermograms, the oxygen content of incoming air was reduced to 7% for 60 minutes, after which the concentration was returned to 21%. The metabolic rates measured in normoxia, hypoxia and recovery in normoxia are summarized in Table 2.7.

Table 2.7 Metabolic rates ($\dot{V}_{O_2}$ and $\dot{V}_{CO_2}$ in units of $cm^3\,kg^{-1}\,min^{-1}$) during normoxia (time 0), hypoxia (15 and 60 min) and normoxic recovery (10 and 60 min) in golden-mantled ground squirrels.

	T_a/°C	Normoxia	15 min hypoxia	60 min hypoxia	10 min recovery	60 min recovery
$\dot{V}_{O_2}$	10	40.4 ± 4.4	27.1 ± 2.3*	21.2 ± 1.5*	91.2 ± 10.4*	45.3 ± 6.2
	22	32.0 ± 1.6	23.9 ± 2.1*	23.4 ± 1.6*	53.8 ± 6.5*	39.1 ± 3.5
	30	30.5 ± 1.7	29.4 ± 2.3	24.8 ± 2.0	23.2 ± 1.1*	28.3 ± 1.7
$\dot{V}_{CO_2}$	10	36.5 ± 4.0	26.5 ± 2.2*	16.7 ± 1.5*	72.8 ± 7.1*	42.0 ± 5.6
	22	28.3 ± 1.7	26.7 ± 0.6*	19.2 ± 0.6*	43.3 ± 4.8*	35.2 ± 3.5
	30	25.3 ± 1.1	29.1 ± 2.3	19.7 ± 0.9*	14.4 ± 2.2*	21.3 ± 2.3

Values are means ± SE ($n = 7$).
*A significant effect compared to normoxic control at a specific temperature ($P < 0.05$).

■ Describe the pattern of values for $\dot{V}_{O_2}$ for different T_a during normoxia.

During normoxia, the rate of oxygen uptake, $\dot{V}_{O_2}$, was higher, 40.4 $cm^3\,O_2\,kg^{-1}\,min^{-1}$ at T_a 10 °C, than the rates at T_a 22 °C and 30 °C, 32.0 and 30.5 $cm^3\,O_2\,kg^{-1}\,min^{-1}$, respectively.

■ Describe the effect on $\dot{V}_{O_2}$ of exposure to 7% O_2 for 60 minutes.

$\dot{V}_{O_2}$ declined significantly to below normoxic levels to 21.2 $cm^3\,kg^{-1}\,min^{-1}$ at T_a 10 °C and 23.4 $cm^3\,kg^{-1}\,min^{-1}$ at 22 °C but did not decline significantly at T_a 30 °C, reaching 24.8 $cm^3\,kg^{-1}\,min^{-1}$.

■ What was the effect on $\dot{V}_{O_2}$ of the restoration of normoxia?

By 10 minutes after normal oxygen concentrations in incoming air were restored, $\dot{V}_{O_2}$ had increased significantly to 91.2 $cm^3\,kg^{-1}\,min^{-1}$ at 10 °C, and to 53.8 $cm^3\,kg^{-1}\,min^{-1}$ at 22 °C. In contrast $\dot{V}_{O_2}$ at 30 °C declined significantly in comparison to the original normoxic value. 60 minutes after restoration of normoxia, the values for $\dot{V}_{O_2}$ at all levels of T_a were not significantly different from those measured during the initial normoxia.

The first 10 minutes of recovery were accompanied by bouts of shivering, a physiological strategy for increasing T_b. Shivering uses up large amounts of energy, as can be seen from the sharp increases in $\dot{V}_{O_2}$ measured at 10 minutes post-recovery at 10 °C and 22 °C.

The images obtained from the infrared camera demonstrate the patterns of cooling and warming. You can use Figure 2.17, a thermogram of a golden-mantled ground squirrel next to an actual image of the animal for orientation. The squirrel is facing the camera so that head and front legs, the warmer areas, are visible with the cooler flank region taking up the rear part of the image. The results of thermography are shown in Figure 2.18 and Table 2.8.

Table 2.8 Surface temperatures (T_s) of golden-mantled ground squirrels during normoxia (time 0), hypoxia (15 and 60 minutes at 7% O_2) and normoxic recovery (10 and 60 minutes post-hypoxia) in golden-mantled ground squirrels at three different values of T_a.

Region	T_a/°C	Normoxia	15 min hypoxia	60 min hypoxia	10 min recovery	60 min recovery
Eye	10	25.9 ± 0.5	25.2 ± 0.3	23.0 ± 0.5*	23.8 ± 0.7	25.2 ± 0.7
	22	32.2 ± 0.8	31.0 ± 0.5	30.5 ± 0.6	30.3 ± 0.5	30.9 ± 0.7
	30	35.3 ± 0.4	35.0 ± 0.3	34.1 ± 0.3	33.9 ± 0.4*	34.1 ± 0.4
Nose	10	19.0 ± 0.5	21.0 ± 0.9*	18.4 ± 0.5	18.5 ± 0.4	18.8 ± 0.7
	22	28.9 ± 0.5	29.7 ± 0.6*	28.2 ± 0.8	28.4 ± 0.8	28.4 ± 0.7
	30	31.9 ± 0.4	33.7 ± 0.3*	32.6 ± 0.8	32.2 ± 0.6	32.7 ± 0.5
Ear	10	19.6 ± 0.6	19.8 ± 0.2	18.0 ± 0.5*	18.4 ± 0.4	19.5 ± 0.6
	22	28.7 ± 0.6	30.3 ± 0.8*	29.1 ± 0.8	28.1 ± 0.5	28.0 ± 0.7
	30	33.7 ± 0.5	34.2 ± 0.2	33.0 ± 0.5	32.9 ± 0.5	33.6 ± 0.6
Feet	10	17.6 ± 0.7	20.1 ± 0.9*	17.7 ± 0.3	17.7 ± 0.3	18.1 ± 0.6
	22	26.6 ± 0.5	30.6 ± 0.8*	27.6 ± 0.6	27.3 ± 0.8	27.8 ± 0.7
	30	32.1 ± 0.4	33.7 ± 0.4*	33.1 ± 0.5	32.8 ± 0.3	32.7 ± 0.3
Flank	10	18.6 ± 0.4	18.0 ± 0.3	17.3 ± 0.3*	17.7 ± 0.4*	18.7 ± 0.3
	22	27.6 ± 0.6	28.0 ± 0.5	27.0 ± 0.5	26.9 ± 0.6	27.4 ± 0.6
	30	32.7 ± 0.4	33.5 ± 0.3	32.6 ± 0.5	32.7 ± 0.3	33.0 ± 0.3

Values are means ± SE ($n = 7$).

*A significant effect T_s at a specific T_a.

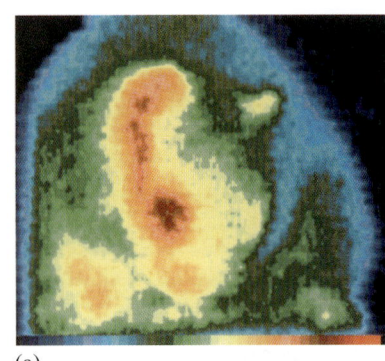
(a)

(b)

Figure 2.17 (a) Infrared thermogram of a golden-mantled ground squirrel. (b) Actual image of a squirrel not taken simultaneously with (a), to demonstrate the angle of the body (but not the head) at which the thermogram was taken.

■ Do the data in Table 2.8 support Tattersall and Milsom's hypothesis that the initial response to hypoxia includes a transient increase in heat loss that speeds up the decline in body temperature? Show how the data support your answer.

Table 2.8 shows that surface temperature, T_s, of the squirrels changed significantly during hypoxia. At T_a 10 °C, T_s of the nose and feet increased significantly within 15 minutes of hypoxia. T_s for eyes, flank and ears had decreased significantly between 15 and 60 minutes of hypoxia. At T_a 22 °C there was a significant increase in T_s of nose, ears and feet within 15 minutes of hypoxia. By 60 minutes after exposure to hypoxia, the T_s of nose, ears and feet had declined to values similar to those measured during normoxia. The data for nose, ears and feet support Tattersall and Milsom's hypothesis that the initial response to hypoxia includes a transient increase in heat loss from the body surface that speeds up the decline in body temperature.

The warming of the nose, ears and feet soon after exposure to hypoxia can be seen in Figure 2.18. The heat loss posture in which the animals lie with their feet raised from their bodies can also be seen clearly. Tattersall and Milsom conclude

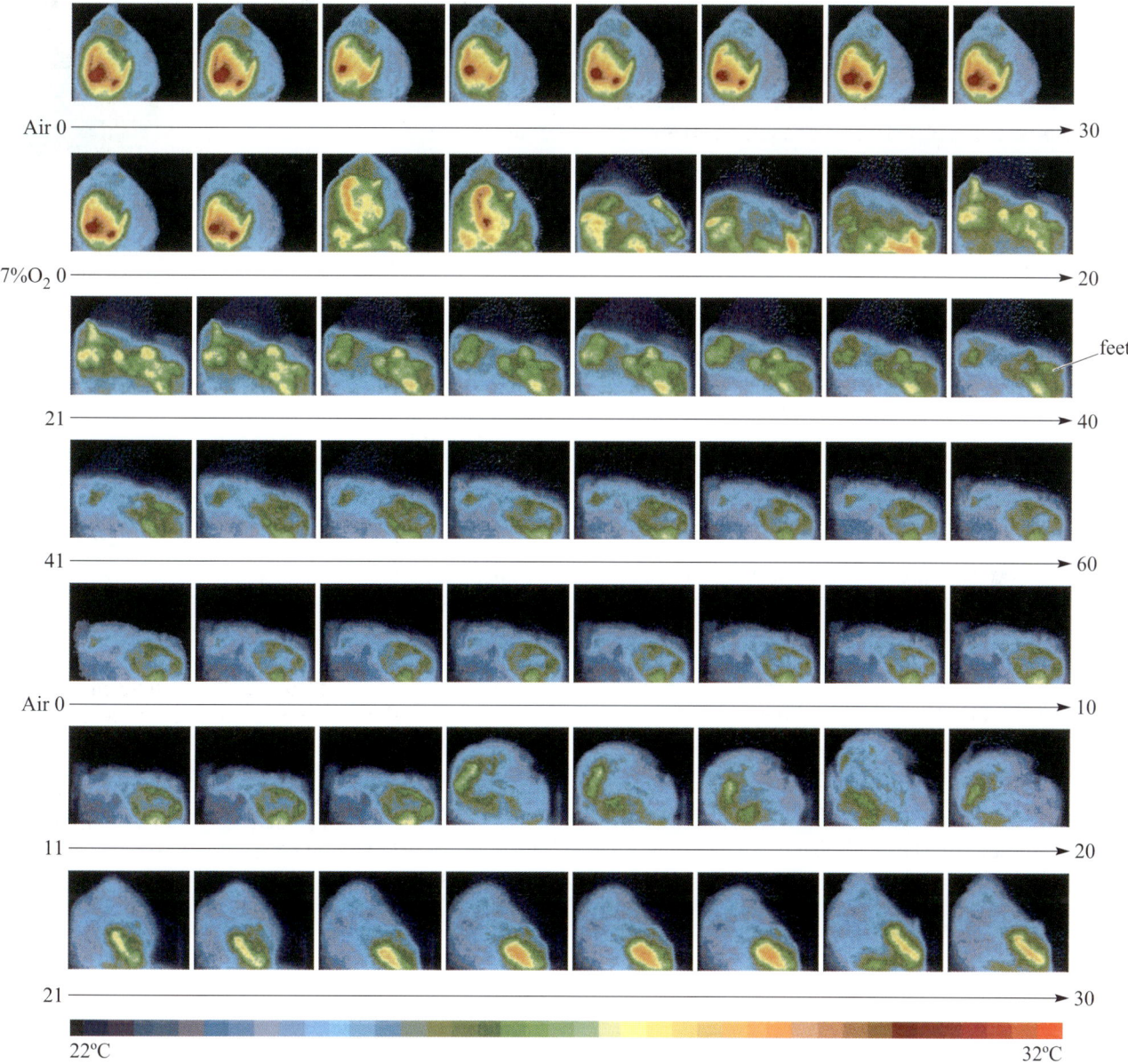

Figure 2.18 Time series of infrared thermograms from one squirrel ($T_a = 22\ °C$) during normoxia (30 minutes), hypoxia (60 minutes) and normoxic recovery (30 minutes). Each row of eight images includes pictures taken over a period of 30 minutes.

that the initial responses to hypoxia in the golden-mantled ground squirrel include the warming of special 'thermal windows' on the body surface, which results in rapid loss of heat, and a reduction in metabolic rate and therefore heat generation. Tattersall and Milsom's observations, together with Barros et al.'s observation of reduction in T_b in response to hypoxia support the view that core heat is transferred to the periphery of the animal from where it is lost by radiation and convection to the air. Tattersall and Milsom's work illustrates how images obtained from infrared cameras can be used in combination with measurements of $\dot{V}_{O_2}$ to track animals' responses to hypoxia.

It is useful at this stage to look at other techniques that are used routinely by researchers studying responses to hypoxia. Terrestrial animals, including humans, encounter hypoxia at high altitudes. Air-breathing animals that dive into water for collecting food or chasing and catching prey need to breath-hold, so they too are subjected to hypoxia.

In Section 2.5.2, we learned how useful radiotelethermometry has proved to be in monitoring body temperature and heart rate in wild animals in their natural environment. However, use of radiotelethermometry for monitoring body temperature, heart rate and behaviour of diving animals is limited by the poor transmission of radio waves through water. Use of data loggers has helped to overcome such problems. A data logger is a device that records data collected by radiotelemetry, for example measurements of T_b and heart rates. Data loggers are available in small sizes that can be attached or implanted in an animal and record data while the animal is submerged in the sea. When the animal is recaptured, the logger is removed and data stored in it can be fed into a computer.

Wilson and Gremillet's (1996) work on bank cormorants (*Phalacrocorax neglectus*) illustrates the practicalities of using data logging. Field work was carried out on Dassen Island, South Africa, where bank cormorants were breeding. A gastric temperature logging unit, 72 mm × 17 mm in size, concealed inside a fish, was fed to each of 19 birds. The unit was retained in the stomach for several hours or days, after which it was regurgitated as a pellet. Fortunately the bank cormorants usually regurgitate indigestible material when they are on the nest, so the researchers recovered the loggers by monitoring nests regularly. Behaviours of the birds were recorded; birds leave the nest to forage between 06.00 and 07.00 h, and fly straight out to sea where they spend their time in the sea, diving to catch fish. Dives are of short duration, ranging from 25 to 68 s, with rest periods of 1–60 s. After an hour or less at sea, the birds return to land. The data logger in a bird's stomach recorded T_b every 16 s. Data from each recovered unit was transferred to a computer. Figure 2.19 is an example of the data obtained from a data logger, showing changes in T_b of a bank cormorant on land and in the sea. Note the sharpest dips (arrowed), which indicate that a prey item such as pilchard or horse mackerel was swallowed, causing a dip in stomach temperature.

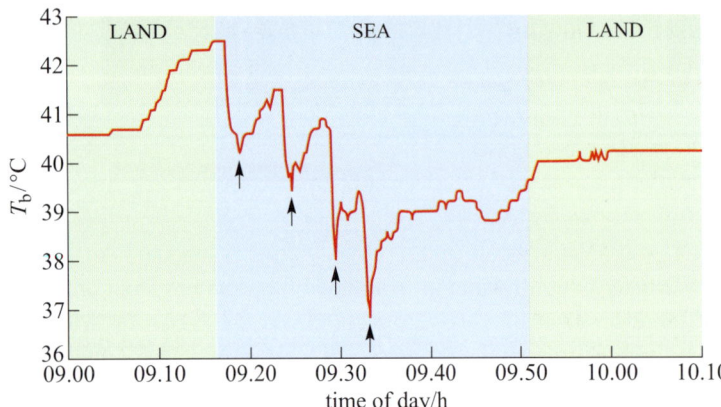

Figure 2.19 Changes in T_b of a bank cormorant recorded by a data logger in the stomach of the bird. The blue tone indicates where the bird was absent from the nest and for most of that time, foraging. Each arrowhead points to a sharp dip in stomach temperature when a cold prey item, a fish, was ingested.

■ Describe the data for T_b in Figure 2.19 and point out links with prey ingestion. What uncertainties are there with regard to measurement of T_b while the bird was at sea?

T_b increased from 40.5 °C to 42.3 °C on land just before the bird took off to go to sea. Once in the sea, T_b dropped sharply. The four arrowed points show that the cormorant ingested four fish during 35 minutes at sea. The lowest T_b recorded was 37 °C following ingestion of a cold fish. Once the bird has swallowed a fish, there is uncertainty in how far the temperature recorded by a data logger in the stomach provides a true measure of T_b because of localized cooling resulting from ingestion of a cold fish. From about 09.45, before return to land, T_b of the bird increased, reaching 40 °C by the time the bird reached land.

The cormorant's dives last only a minute or two, so the bird avoids the consequences of prolonged hypoxia and low T_a. Cormorants rest on land for a time after a foraging bout, during which time they dry out their feathers and raise T_b. The ingested data logger revealed detailed information about the T_b of the cormorants, and feeding behaviour, with minimal disturbance to the birds. Much useful information about diving animals, including data on heart rate as well as T_b, has been collected from data loggers, as discussed in Chapter 7.

Summary of Section 2.8

Experiments on the effects of hypoxia on the metabolic rates of *Spermophilus lateralis* show that an initial response to hypoxia is heat loss from the surface of the animal. This reduces metabolic rate and hence the energy requirements of the squirrel. Data loggers have provided data on temperature and heart rate of cormorants during feeding. The bird raises T_b before leaving land. The ingestion of fish lowers T_b but only by a few degrees. Feeding bouts are limited in duration so the bird avoids problems of hypoxia and low T_b.

2.9 Effects of temperature and hypoxia at the molecular level

So far we have looked at how researchers investigate the effects of temperature and hypoxia in terms of behavioural and physiological responses. Extremes of ambient temperature and hypoxia are examples of environmental stresses.

Advances in molecular biology have made it possible to investigate responses of cells to environmental stresses at the molecular level. Exposure of mammalian cells to relatively small increases in temperature initiates signalling pathways, which either lead to cell death, or enable survival of the cells. Exposure of cells to extremes of high temperature causes a suite of responses termed heat shock, initiated by denaturation of enzymes and other intracellular proteins. The molecular response of animals to temperature extremes begins by the induction of **heat-shock proteins (Hsps)**. We will be studying induction of Hsps in detail in Chapter 3. Our purpose in this chapter is to understand the function of Hsps, so that we can begin to link the molecular responses of cells to environmental stresses, with the physiological responses in the whole animal.

The term heat-shock proteins is applied because levels of Hsps in cells rise rapidly after exposure to abnormally high temperatures, e.g. about 40 °C for mammalian cells. However, note that Hsps are also induced following exposure of cells to other environmental stresses, including hypoxia, toxic chemicals and extremes of low temperature. Hsps comprise at least 12 families of related proteins that are found in cells of all animal and plant species so far investigated, so they are molecules that have been highly conserved in evolution. Hsps are important for species at risk of exposure to high T_a because of their role as chaperone proteins that 'rescue' proteins whose tertiary structure has been disrupted by over-heating. Chaperone proteins bind to denatured regions of a protein and alter the misfolded structure so that the correct three-dimensional structure is regained (Figure 2.20).

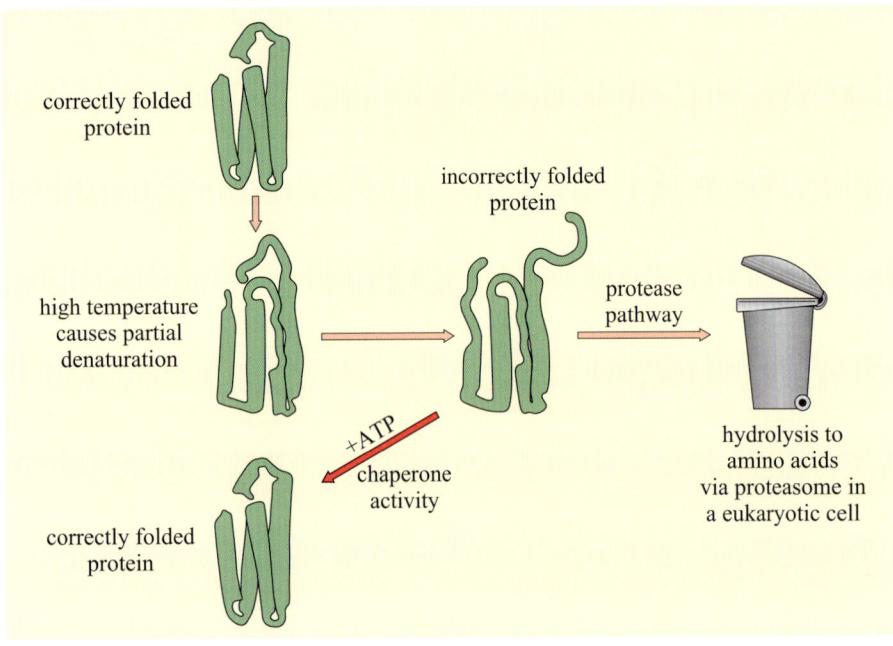

Figure 2.20 Possible mechanism of chaperone-assisted protein refolding. A protein is partly denatured by exposure to high temperature. Interaction with a chaperone protein produces the correct tertiary structure. Proteins that cannot be refolded are recognized as defective and broken down and the amino acids released.

Input of ATP is required for this function. Heat-shock proteins maintain the biologically active conformation of enzymes and other essential proteins in cells during exposure to protein-damaging environmental stresses including extremes of high and low temperatures, ethanol and hypoxia.

Hsps also prevent cells from being killed by the cell death cascade, which is induced in mammalian cells by exposure to temperatures of 43–45 °C. Exposure of mammalian cells to temperatures > 43 °C activates a specific MAP kinase (mitogen-activated protein kinase), known as JNK (c-Jun NH_2 terminal kinase). JNK is a key activator of transcription factors that control cell growth and **apoptosis** (programmed or controlled cell death).

Figure 2.21a shows a simplified summary of the cell death pathway. Activation of JNK leads to escape of cytochrome *c* from mitochondria, probably by phosphorylation, which inhibits the action of apaf-1 (apoptotic protease-activating factor). This inhibition initiates the caspase cascade that leads to apoptosis. Caspases break down proteins by cleaving polypeptide chains between cysteine and aspartate residues. Once caspases are activated, there follows a rapid cascade of activation of various caspases, resulting in rapid breakdown of cell proteins, including nuclear lamins, proteins that maintain structural integrity of the nucleus, and cytosolic proteins. Caspases are cell killers! JNK activation is followed by induction of FAS ligand, one of the 'grim reapers' of the apoptotic pathway. FAS ligand binds to FAS (faint sausage!) protein, initiating activation of caspase-8 leading to activation of more caspases and a cascade of proteolytic activity. The induction of survival pathways is mediated by extracellular signal-regulated kinases (ERKs) and protein kinase B (PKB), which inhibit the effects of JNK.

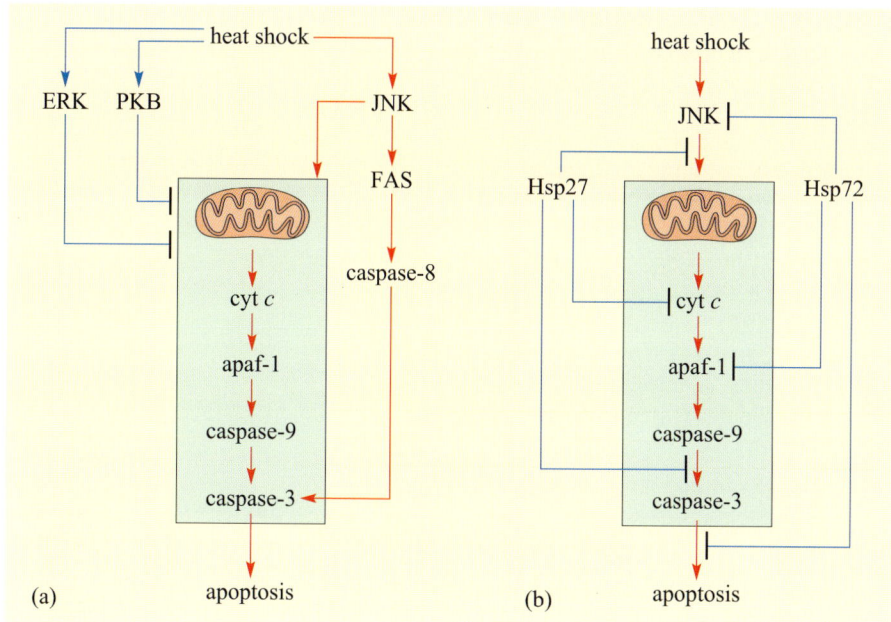

Figure 2.21 (a) Activation of cell death and survival pathway by heat shock. Heat shock activates survival (blue) and death (red) pathways that modulate efflux of cytochrome *c* (cyt *c*) from mitochondria and initiation of caspase cascade. Abbreviations: ERK, extracellular signal-regulated kinase; PKB, protein kinase B; JNK, c-Jun NH_2 terminal kinase; FAS, faint sausage protein; apaf-1: apoptotic protease-activating factor. (b) Suppression of apoptotic pathway by heat-shock proteins Hsp27 and Hsp72.

Although heat shock can initiate the cell death cascade, there are survival pathways that rescue cells from death. We can examine the role of Hsps in modulating these events.

You will not be surprised to learn that there may be interplay between heat-shock induced Hsps and heat-shock-activated signalling pathways. (Figure 2.21b)

Hsps can modulate apoptotic signalling pathways. Hsp27 reduces release of cytochrome *c* from mitochondria during heat shock by preserving the molecular structure of mitochondria. It also inhibits the caspase cascade. Hsp72 binds to apaf-1 thereby preventing activation of caspase-9. In addition, Hsp72 reduces heat-shock-induced activation of JNK and also inhibits the effect of caspase-3 on apoptosis.

■ Explain the effect of reduction of release of cytochrome *c* from mitochondria in the heat-shock response.

Release of cytochrome *c* from mitochondria during heat shock triggers the caspase cascade which ends in apoptosis. A reduction in release of cytochrome *c* would reduce cell apoptosis.

Two major effects of heat shock on living cells have been identified: denaturation of proteins and apoptosis. Hsps function as molecular chaperones and also suppress cell signalling pathways involved in the cell death cascade, and enhance the ability of cells to survive heat shock. We can begin to see a link between the responses of cells to heat shock at the molecular level and the physiological responses of animals to heat shock.

Chapter 3 explores the role of heat-shock proteins further, in the context of whole animal physiology. In Chapter 6, we investigate the effects of hypoxia at the physiological, cellular and molecular levels. Most cell types investigated so far can sense hypoxia and respond by induction of oxygen-regulated genes, in particular *HIF-1α*, which codes for a transcription factor, hypoxia-inducible factor, HIF-1α. Induction of HIF-1α in cells exposed to hypoxia initiates transcription of genes that code for proteins whose action helps to mitigate the effects of hypoxia.

There has been much research interest in identifying the oxygen sensor in mammalian cells, and the signal transduction pathway that leads to activation of HIF-1α. The identity of the oxygen sensor in mammalian cells is not yet known. Various haem-containing molecules have been suggested.

- Suggest the rationale for the hypothesis that a haem-containing protein could be the cellular oxygen sensor in mammalian cells.

Most mammalian proteins that bind oxygen, including haemoglobin and myoglobin, contain iron, in the centre of a haem group. The binding of oxygen to a haem group causes a change in conformation of haemoglobin. A change in conformation of a protein can function as a signal initiating a signalling cascade.

Details of the signalling pathway involved in the induction of *HIF-1α*, and the ensuing physiological effects, are provided in Chapter 6. As a result of new insights gained from use of modern techniques in molecular biology, we have a more detailed understanding of the physiological responses of animals to physical features of their environment including temperature extremes and hypoxia.

Summary of Section 2.9

Heat-shock proteins produced as a result of raised temperature or other stress such as hypoxia can reduce damage to cell proteins. Heat-shock proteins function as chaperone proteins, repairing the conformation of damaged proteins to restore function. They also regulate signalling pathways that lead to cell apoptosis.

2.10 A new approach to environmental physiology: adaptation

We have seen in Chapter 1 that physiology can no longer be separated from other disciplines such as behavioural sciences, evolutionary biology and molecular biology. Understanding of the environmental physiology of animals requires integration of their physiology with behaviour, phylogeny, anatomy, biochemistry and molecular biology. This task may appear daunting and perhaps impossible, but it becomes more manageable when studied in terms of adaptation.

2.10.1 Evolutionary adaptation

Adaptation in evolutionary terms is defined as heritable, that is genotypic, features of animals that we interpret as the result of the process of natural selection, with phenotypic manifestations. For example, we regard the thick fur coats of the polar bear and arctic fox (Figure 2.5), as phenotypic adaptations for life in the freezing polar climate. A thick fur coat provides insulation that helps to prevent loss of heat from the body. Such features are well known and the idea that a thick fur coat is an 'adaptation' for life in cold climate is part of our 'general knowledge'. In contrast, features of large desert mammals can be related to promoting heat loss from the body. The sparse fur coat over most of the camel's body provides little resistance to heat loss. The thicker and longer coat on the dorsal surface of the camel provides shade from the intense solar radiation and keeps the skin cool.

The high levels of heat-shock proteins in cells of desert lizards are regarded as an adaptation for the high values of T_b experienced by those lizards while they are foraging. Lizards that do not live in deserts have negligible amounts of heat-shock protein in their tissues. Physiologists study adaptations at the levels of molecular biology, biochemistry, physiology and anatomy, with the objective of obtaining an integrated picture of adaptation. However, it is not always straightforward to prove that a particular feature is adaptive. The relationship between the structure and function of an anatomical feature was the traditional approach used to support the view that a particular feature is adaptive. Comparisons between species are also used. Many studies have been done in which features of two species are compared. A problem with such studies is that two species may have differences in many traits. Some of these differences will be due to chance, **genetic drift.**

Genetic drift comprises random fluctuations in allele frequencies in different populations of organisms. Such fluctuations are of little significance in large populations, but in small populations random fluctuations in allele frequencies may have a large impact. During meiosis, each haploid gamete receives randomly, one of the alleles for each gene, independently assorted from maternal and paternal alleles. Only a minute proportion of gametes participate in fertilization and few of the resulting zygotes develop into the adults of the next generation. So in a small population, if, by chance, one individual produces no surviving offspring, the frequency of his/her allele could be reduced in the succeeding generation. In an extreme case, some alleles may disappear completely, by chance, not by natural selection, from a small population. If one individual produces more offspring than the others, then his/her alleles become more frequent in the succeeding generation of the population.

If a small population of a species become the 'founders' of a new population, say as a result of a catastrophe that wipes out most individuals, the new population is likely to have different frequencies of alleles than the original one. If the new population eventually evolves into a new species, some of the features seen in the species must have been derived from genetic drift. So differences in a particular trait between two related species are not necessarily the result of natural selection. If we wanted to compare two related species of ground squirrels, *Spermophilus richardsonii*, which lives at relatively low altitudes, with *Spermophilus parryii*, which can found at altitudes as high as 2000 m, how do we know whether the differences between the two are derived from natural selection, or result from genetic drift? If the difference is a result of natural selection, how do we know which species has acquired the 'new' physiological trait? This can only be determined if the physiology of the common ancestor is known, so that comparisons can be made.

In order to make such a comparison, we need to have a phylogenetic tree of the animal group that we are interested in. Such a tree comprises a hierarchy showing the patterns of lineages that have been produced during the evolutionary history of the group. Evidence from fossil anatomy and DNA sequence data, which is becoming available for an increasing number of animal groups, provide the information needed for constructing evolutionary trees. **Cladistics** is a popular and useful system for constructing evolutionary trees. There are three main principles of cladistics. The system accepts that evolution results in changes in characteristics of living organisms with time. It is assumed that groups of organisms that are related to each other evolved from a common ancestor. At each divergence of species, a point called a **node**, there are two branching lines of descendants (Figure 2.22). The branches of an evolutionary tree may be single 'twigs' or be 'bushy' having smaller twigs branching off the main twig. Both bushy and single twigs are known as **clades** and the completed tree is known as a **cladogram**. However, a cladogram is not a description of the complete sequence of steps involved in the evolution of groups of related animals. Within one step of a cladogram, there are more possible evolutionary sequences that could have occurred. The cladogram provides the most parsimonious interpretation of the evolutionary relationships between related groups. Parsimony involves choosing the simplest interpretation of an evolutionary relationship. So the parsimony technique searches for a tree that uses the fewest steps to link groups of species in an evolutionary sequence. The importance of parsimony can be appreciated by imagining that you want to draw a tree for 50 related species. In theory, 2.8×10^{74} trees are possible! Nevertheless, interpretation of a cladogram can be used to formulate hypotheses about which groups are the most closely related.

A branch of the tree that split off and separated from the clade of interest is known as an **outgroup**. Comparison of the features of an outgroup with those of the clade provides clues about whether a feature in the clade of interest is adaptive. If that feature is also present in one or more of the outgroups then it may or may not be adaptive, as we shall see in Figure 2.22, a theoretical case study. We will explore Figure 2.22 to learn how use of phylogenetic trees may help to identify features of animals as being adaptive or not. The case study compares the incidence of thin-walled non-muscular pulmonary small arteries in a *hypothetical* group of species of Ochotonidae, pikas, a family within the order Lagomorpha that includes species that are found both at high altitudes and at sea-level. The hypothesis is that in pikas, thin-walled pulmonary small arteries are an adaptation for life at high altitudes. The thick muscular walls of small pulmonary arteries found in most species of mammals, constrict during hypoxia.

■ Why would constriction of small pulmonary arteries in response to hypoxia be a disadvantage to mammals living at high altitudes?

Constriction of blood vessels reduces blood flow in the lungs which reduces the oxygen loading of blood. High pulmonary blood pressure is also a problem. As P_{O_2} of air is lower at high altitude than at sea-level constriction of pulmonary blood vessels is a serious disadvantage to animals at high altitude.

Our imaginary scenario is that three high-altitude pika species have been found to have thin-walled pulmonary small arteries, in contrast to three sea-level species, which have thick-walled pulmonary small arteries. We want to know whether thin-walled pulmonary small arteries are adaptive for life at high altitude, or were a feature of the ancestral group. Figure 2.22 comprises four phylogenetic trees, and for each we examine possible interpretations of the distribution of two variants of one character, thickness of small pulmonary artery vessel walls.

Even where a character appears to be adaptive, we need to remember that our evidence is correlative. For our theoretical case study, we assumed that for some lagomorph species, life at high altitude correlates with thin-walled small pulmonary arteries. In fact this feature is found in at least one species, *Ochotona*

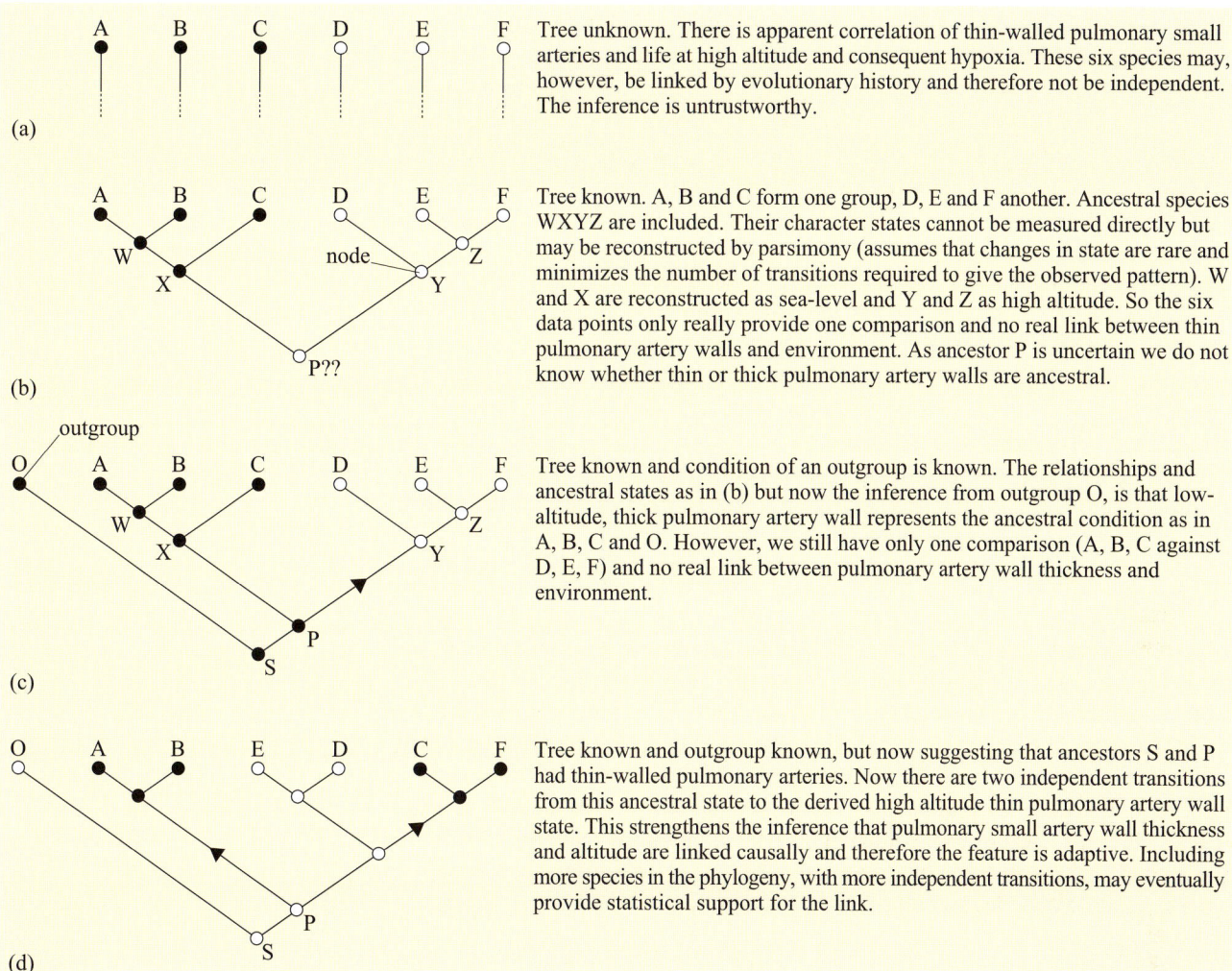

Figure 2.22 The effects of phylogenetic trees on possible interpretations of the distribution of a single character: presence of thick- or thin-walled pulmonary small arteries. Two outgroup comparisons are shown. Solid circles show sea-level taxa with thick-walled pulmonary arteries; open circles show high-altitude taxa with thin-walled pulmonary arteries. Triangles show the sites of changes of state.

curzonii, which we will be studying in Chapter 6. Until information is available on the histology of pulmonary arteries of other species including those that live at high and low altitudes, we cannot be sure that thin-walled pulmonary small arteries are adaptive. More information is required on evolutionary relationships between pika species and to date, not much is available.

By now you will appreciate the point that not all characters are adaptive to individual species. Natural selection can only act on those genes that are already present within a population. Certain traits that were selected a long time ago in the evolutionary history of an ancestral group may persist in descendant species although they are no longer related to selective pressures that exist currently. This effect, known as **phylogenetic inertia**, needs to be eliminated as a possible explanation for features of animals in particular environments before such features can be regarded as adaptive. A study by Tieleman et al. (2003) of physiological adaptations in different lark species along a gradient of aridity takes the possibility of phylogenetic inertia into account. Later in Chapter 3 you will learn how Tieleman et al.'s study of clades in the lark family (Alaulidae), was used to determine whether or not the traits of reduced BMR and evaporative water loss in desert larks are adaptive.

2.10.2 Adaptation: physiological changes in response to environmental cues

Short-term compensatory changes are the result of differential expression of pre-existing genetic traits so that the response is appropriate to existing environmental conditions. These types of changes are sometimes called physiological adaptation, and they are the result of **phenotypic flexibility**.

For example, exposure of people who live at sea-level to high altitude results in a suite of physiological changes initiated by expression of the gene for the transcription factor, hypoxia inducible factor α-1 (HIFα-1) (Section 2.9). This transcription factor switches on a number of genes that promote short-term physiological adaptations for high altitude, including increased rate of production of erythrocytes. Increased numbers of erythrocytes per unit volume of blood increase the oxygen-carrying capacity of blood, important at high altitude where P_{O_2} is low. Short-term physiological adjustments in response to exposure to new environments, acclimatization, are possible because of phenotypic flexibility. The ability of an individual to acclimatize is controlled by genes, because the mechanism for acclimatization is differential gene expression in response to environmental cues.

The term **phenotypic plasticity** is used to describe developmental effects, in which exposure of developing young to particular environmental conditions results in differential expression of genes, the effects being fixed for the remainder of the animal's life. Studies in which littermates of desert rodents were reared with or without access to water illustrate phenotypic plasticity. Littermates with no access to free water develop kidneys that have increased medulla relative to cortex thickness. The medulla contains the loops of Henle that concentrate urine by resorbing water. In contrast littermates reared with free access to water, have a smaller medullary thickness relative to cortex.

Summary of Section 2.10

The use of cladistic analysis provides a phylogenetic perspective for investigating whether apparently adaptive features in related species are due to evolutionary adaptation or are simply characteristics present in ancestral groups. DNA sequence data are providing the evidence to construct cladograms that enable investigation of evolutionary adaptations.

2.11 Conclusion

This chapter has provided an overview of the various disciplines that have been integrated into the study of animal physiology. You may be feeling that animal physiology must be a complex subject area as it encompasses so many disciplines. As you read the following chapters you will begin to appreciate that putting together the various levels of adaptation for understanding the environmental physiology of animals is a difficult but worthwhile process. Animal physiology has moved on from being an experimental science based on relatively simple comparisons between species. The following chapters describe the results of research into aspects of animal physiology integrated with new findings in biochemistry and molecular biology. Integration of research findings at different levels of structure and function enables us to gain an idea of how the whole animal responds to features of the environment in which it is living. New studies on phylogenies provide us with clues as to which features of animals may be truly adaptive. Inevitably there are gaps in our knowledge. Certain species are popular as research subjects either because there are interest groups that provide funding, or because that species is readily available or amenable to being handled. For example, research into the environmental physiology of the camel, is funded by interests related to camel husbandry and racing. The research on the Arabian oryx is funded by interest in the conservation of a species that has great cultural significance for Arab people. Research into the effects of hypoxia at high altitude has become an important branch of medical science because of the increasing number of people who live at high altitudes, or visit such areas for holidays or business. What is exciting about environmental physiology is the opportunity we now have to understand both the initial molecular responses to environmental cues and how that initial response develops into a physiological response. There are examples in the following chapters where the physiological response has been traced back to the initial expression of a transcription factor in response to an environmental signal. Molecular biology provides insights into the evolutionary relationships between species, information that can be used to identify adaptive features of animals.

Learning Outcomes for Chapter 2

When you have completed this chapter you should be able to:

2.1 Define and use, or recognize definitions and applications of, each of the **bold** terms.

2.2 Describe the thermal characteristics of the external environment and explain the importance of radiation, conduction, convection and evaporation for the thermal balance of an animal.

2.3 Describe methods of measuring metabolic rate and temperature of animals and understand the limitations of these techniques.

2.4 Outline how doubly-labelled water studies have helped in the understanding of energy expenditure of foxes living in different habitats.

2.5 Draw a simple diagram that summarizes the physiological responses to hot and cold environments in mammals and birds and apply those principles to case studies.

2.6 Outline some studies that have used the techniques of radiotelemetry to measure temperature and to obtain physiology data in wild animals.

2.7 With reference to studies on fish, outline what is meant by temperature compensation.

2.8 Explain the importance of the concept of partial pressure to the physics of gas exchange.

2.9 Interpret data showing the effects of hypoxia on physiology with reference to studies on ground squirrels and diving air-breathing animals.

2.10 Outline the role of heat-shock proteins in modifying the response of cells to heat-shock stress.

2.11 Outline what you would need to do to investigate whether the physiological differences between related animals that have an adaptive advantage, may have arisen by natural selection.

Questions for Chapter 2

Question 2.1 (LO 2.3)

A group of physiologists, Butler et al. (1993) carried out studies on respiratory and cardiovascular adjustments during exercise of increasing intensity in racehorses. The horses were trained to exercise on a treadmill, and habituated to wearing a face mask connected to a respirometer. The researchers wanted to measure the CO_2 and O_2 content of inhaled and exhaled air during exercise.

Figure 2.23 shows the measured CO_2 content of arterial blood during increasing levels of intensity of exercise in five horses without and with the face mask.

Study the data in Figure 2.23 and then answer the following questions:

(a) Describe the effects of the respiratory mask on arterial blood P_{CO_2}. Note that you need to quote representative values from the data to support your description.

(b) How might the data in Figure 2.23 influence your interpretation of respirometer measurements of CO_2 and O_2 content of inhaled and exhaled air of exercising racehorses at high workloads?

Question 2.2 (LOs 2.3 and 2.4)

These questions relate to the study of Rüppell's foxes by Williams et al. (2002) which we examined in Section 2.4.2. Look back at the data in Table 2.2, and answer the following questions:

(a) Compare the values for field metabolic rate in male and female Rüppell's foxes, both in terms of whole body mass and also measured as kJ day^{-1} (mass$^{0.869}$)$^{-1}$.

(b) Rüppell's foxes are nocturnal and remain in their underground dens during the day emerging only at night when they forage for food (their diet consists of rodents and insects). Suggest an explanation for your answer to Question 2.2a using the data in Figure 2.24 to support or refute your view.

Question 2.3 (LOs 2.1, 2.3 and 2.5)

Table 2.10 includes a series of measurements of ambient temperature, T_a, colonic temperature, T_b, and metabolic rate, derived from observations on a cat. Plot the data on the provided graph paper (Figure 2.25) as follows:

(a) Plot ambient temperature, T_a, from 10 °C to 40 °C on the horizontal x-axis, against oxygen consumption (0-30) on the vertical y-axis, on the left-hand side of the graph. For plotting body temperature, T_b, of the cat, draw a second vertical scale from 38 °C to 40 °C directly above your oxygen consumption graph.

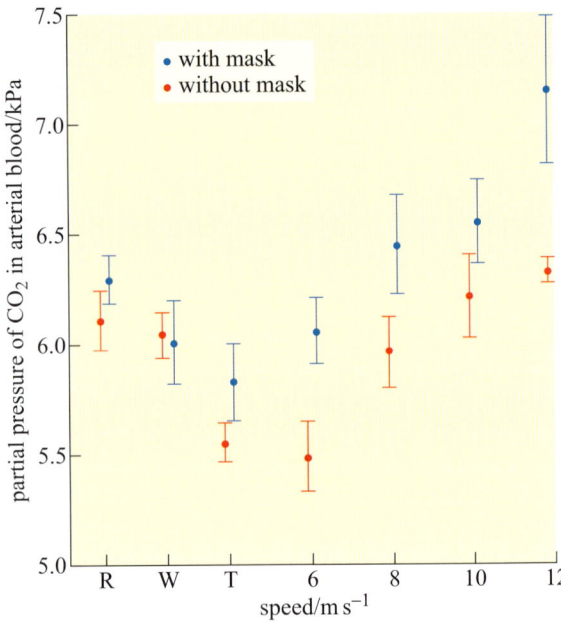

Figure 2.23 The effect of a respiratory mask on P_{CO_2} of arterial blood in racehorses at rest and during different rates of exercise. Mean values ± SE from five horses without and with the face mask during an incremental exercise test with 2 minutes of exercise at each speed and a 5-minute walk in between. R = at rest; W = walking; T = trotting; then cantering at speeds indicated.

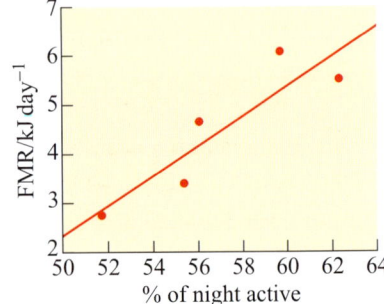

Figure 2.24 Plot of normalized FMR for Rüppell's foxes against % of night spent actively walking or foraging.

Table 2.10 Data obtained from experiments measuring T_b and oxygen consumption of a cat exposed to T_a ranging from 15 °C to 40 °C.

T_a/°C	15.0	17.5	20.0	22.5	25.0	27.5	30.0	32.5	35.0	37.5	40.0
T_b/°C	38.10	38.10	38.15	38.25	38.30	38.35	38.40	38.65	38.90	39.25	39.50
Oxygen consumption/ cm³ O₂ min⁻¹ kg⁻¹	25.0	22.7	20.8	18.0	16.2	13.0	11.2	11.0	11.0	11.1	12.3

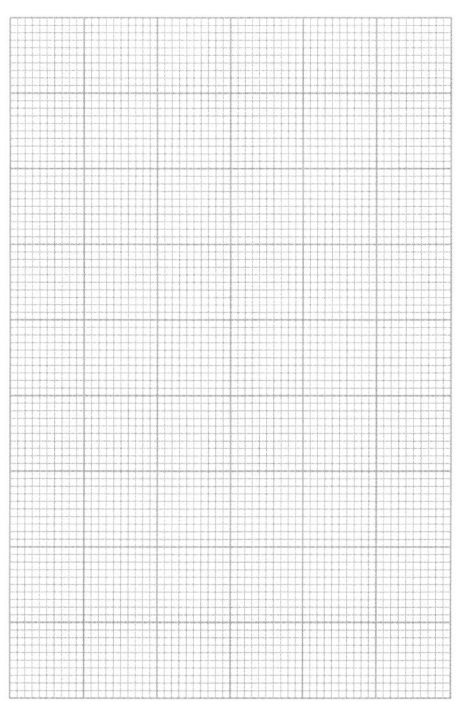

Figure 2.25 Graph paper for use with Question 2.3.

(b) Using your plotted graph as source material, as well as principles taught in Chapter 2, answer the following questions:

(i) Describe the curve you have drawn for T_b against T_a. Does your description of this curve suggest that the cat is a homeotherm or a poikilotherm?

(ii) Calculate the slope of the graph for metabolic rate against T_a (i.e. rate of oxygen consumption against T_a). Your graph should have more than one slope and you should calculate the slope for each.

(iii) Describe the curve you have drawn for metabolic rate, quoting the values for slopes that you have calculated.

(iv) Suggest the range of T_a for the thermoneutral zone.

(v) Deduce a value for the lower critical temperature, T_{lc} for the cat. Deduce a value for the upper critical temperature, T_{uc}, for the cat.

Question 2.4 (LOs 2.6 and 2.9)

A group of researchers, led by Amanda Southwood (1999), studied diving behaviour and monitored heart rates of leatherback sea turtles (*Dermochelys coriacea*) at sea. From late October to early March, female turtles return to their nesting beach to lay eggs, and lay up to 10 clutches on the same beach, returning every 7–14 days for egg-laying. The researchers attached a data logger and a radiotransmitter to the shell of each of six female turtles as they were laying their eggs and burying them. The turtles subsequently returned to the sea, where they remained for 7–14 days before returning to the beach to lay another batch of eggs. The researchers removed the data logger and radiotransmitter from each female while she was laying her eggs. The turtles did not appear to notice that data loggers were being attached or detached. The data loggers provided data for dive depth and heart rate of the turtles.

(a) The researchers attached a radiotransmitter to each turtle even though the physiological and dive depth and duration data were collected and stored in the memory of the data logger. Suggest the purpose of the radiotransmitter. Explain the advantages of using a data logger to record data rather than using radiotelemetry.

(b) Figure 2.26 shows a dive trace with corresponding heart rate for a single dive made by turtle 7610.

　(i)　What was the duration and maximum depth of the dive?

　(ii)　Describe the pattern of heart rate before, during and after the dive.

(c) What concerns might you have about interpreting physiological data from a data logger that was attached to a wild animal?

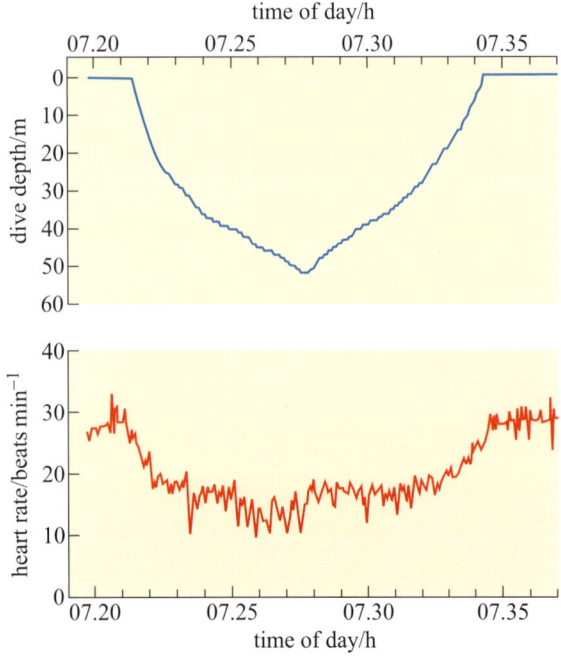

Figure 2.26 Dive trace and corresponding heart rate trace for a single dive made by turtle 7610.

Question 2.5 (LOs 2.1 and 2.2)

Which of the following are examples of acclimation, acclimatization, phenotypic flexibility; phenotypic plasticity; ectothermy, endothermy, poikilothermy, homeothermy, or a combination of several?

(a) A mountaineer walks slowly through Himalayan foothills towards Mount Everest, gradually gaining altitude. He begins his climb of Mount Everest by walking up the slopes gradually, until he reaches vertical rock and ice faces. He then has to use specialized equipment, ropes and ice axes for the slow final stages of the climb. The mountaineer never has to use supplementary oxygen, even when close to or on the summit.

(b) A lizard emerging from a cool burrow early in the morning has a T_b of 15 °C. The lizard finds a sunny spot on a rock and sunbasks until its T_b reaches 30 °C. The lizard then begins to feed on insects. After 2 hours of exposure to solar radiation, the lizard's T_b reaches 40 °C and the animal retreats to a shady spot under a rock to cool off. After 30 minutes in the shade, the lizard has cooled to 34 °C, and emerges into the sun to resume feeding.

(c) A pale dawn breaks over a winter landscape on a university campus. T_a is below freezing, at −5 °C. A fox walks briskly over a cricket pitch; his T_b is 38 °C. Suddenly the fox spots a rabbit and gives chase. He misses the rabbit but the exercise has increased T_b of the fox by 0.5 °C. The fox cools off by panting.

(d) Researchers catch 30 desert larks by mist-netting. In the laboratory, 15 of the birds are kept at 15 °C for 3 weeks; the other 15 are kept at 30 °C for 3 weeks. For both groups, aviary size and design are identical. Day length and timing of dawn and dusk are kept exactly the same as in the birds' natural habitat. Both groups are given water and the same diet *ad libitum.*[*]

---

[*]The term *ad libitum* means 'at will' – in this case the birds were provided with food and water in excess to their needs so that they could eat and drink as much as they wanted to and when they liked.

CHAPTER 3 THE DESERT ENVIRONMENT

Prepared for the Course Team by Patricia Ash

3.1 Introduction

This chapter concerns the integration of behaviour, anatomy, physiology and biochemistry in diverse vertebrates that live in deserts. If you have visited a desert you will have noticed the sparse plant cover, or in certain sandy deserts, the almost complete absence of plant life. The low productivity of deserts derives from their defining feature, which is aridity. Scarcity of water restricts the diversity and amount of plant cover, and in turn the diversity and abundance of animals. However, if you were visiting one of the American deserts after rains, you would be rewarded by the sight of the desert 'in bloom', as vast swathes of annual plants such as the Mojave aster (*Xylorhiza tortifolia*) and sand verbena (*Abronia villosa*) flower simultaneously. You might catch sight of insects such as beetles and locusts, and vertebrates including lizards and occasionally mammalian herbivores, such as gazelle, in African deserts and oryx in the Arabian desert.

Hot deserts located 15°–25° north and south of the Equator have daytime sunshine all year round (Figure 3.1). The persistent descending air and stable high pressures create the hot climate. While daytime temperatures can be as high as 45 °C, night-time temperatures may be close to freezing as heat is lost by radiation into the clear night skies.

Aridity of deserts has three main causes. The deserts in parts of North and South America are arid because they are located on the leeward side of mountain ranges in rain shadow. Rain falls as moisture-laden air rises up the mountains, so that the air is dry by the time it reaches the leeward side. The Gobi and

Figure 3.1 Locations of deserts worldwide showing the broad latitudinal belts north and south of the Equator.

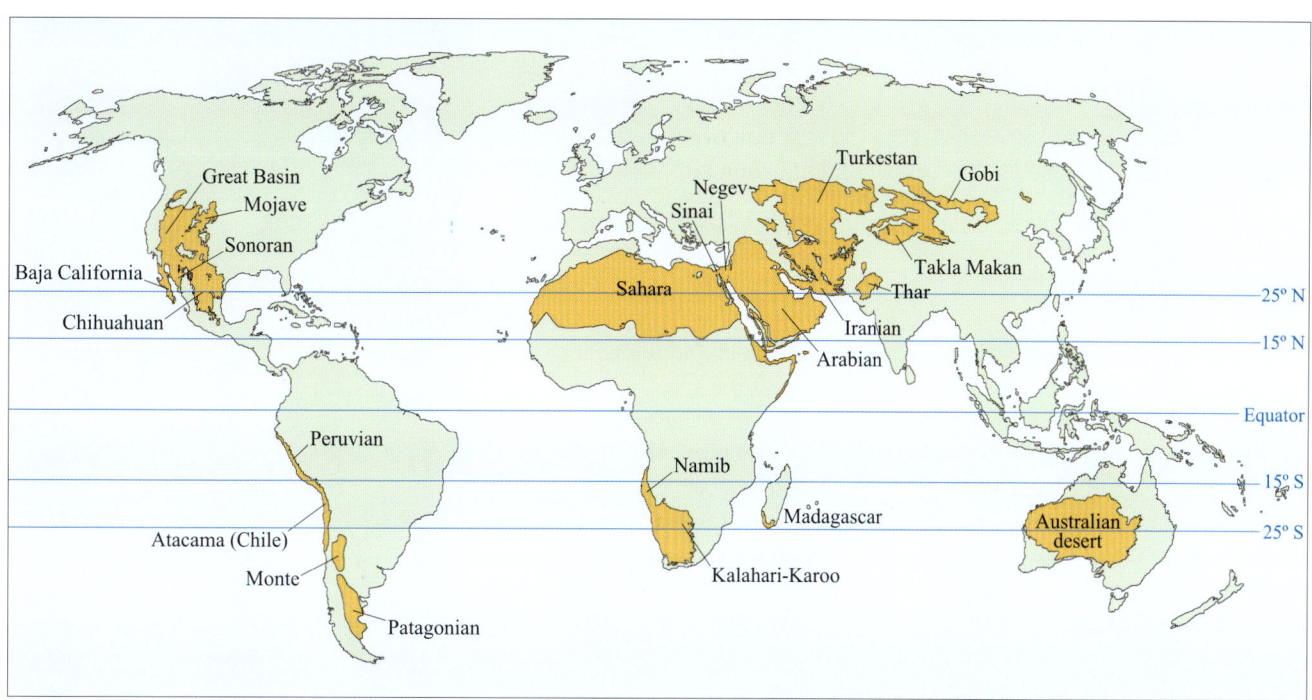

Figure 3.2   The Sahara desert near Erfoud, Morocco.

Figure 3.3   A dromedary camel (*Camelus dromedarius*).

Figure 3.4   Dorcas gazelle (*Gazella dorcas*).

Figure 3.5   Arabian oryx (*Oryx leucoryx*).

Turkestan deserts lie in the centre of a large continent and their lack of rainfall is because they are a long distance from the sea. The Sahara (Figure 3.2) and Arabian deserts are arid because of persistent large high-pressure masses of dry air that prevent penetration of rain-bearing storm systems. A popular concept of deserts is based on the extreme Sahara, where huge areas of sand dunes support little, if any, plant growth. Certain animals such as camels (*Camelus* spp.; Figure 3.3), Dorcas gazelle (*Gazella dorcas*; Figure 3.4) and oryx (*Oryx leucoryx*; Figure 3.5) that survive there by browsing on the sparse plant life and drinking very little or no free water, are regarded as typical desert species.

In fact the picture is much more complex; the environment of each desert is unique, and depends on the interaction between T_a, precipitation, relative humidity and wind. A useful classification is that of Meigs (1953), who defined deserts according to aridity (Table 3.1). The aridity of a desert is determined not just by precipitation but also by the evaporation and transpiration of plants. In order to simplify classification of arid and semi-arid areas, various types of aridity index have been devised. De Martonne's aridity index has been used widely and is calculated from the formula:

$$I_a = \frac{P}{T+10} \tag{3.1}$$

I_a = aridity index

P = mean annual precipitation/mm

T = mean annual temperature/°C

Note that values for average precipitation can be misleading because in arid deserts, especially hyper-arid desert, there are many years that have no rainfall at all. Certain coastal deserts such as the coastal strip of the Arabian Peninsula obtain part of their annual precipitation from thick fog, caused by cold sea breezes that increase humidity sharply. Subdivisions of aridity index are roughly the same as those for average rainfall.

Table 3.1 Definition of deserts according to aridity (Meigs, 1953 and UNEP, 1992).

Rainfall/mm yr^{-1}	Aridity index	Aridity	Examples
<25	<5	Hyper-arid	Namib; Arabian
25–200	5–20	Arid	Mojave
200–500	20–50	Semi-arid	Parts of Sonoran

All deserts have a wide range of T_a, but mean annual temperatures vary from desert to desert. Deserts can be defined as hot, mild, cool or cold (Table 3.2), but the reality is a continuum of desert climates rather than a set of desert climates each with a well-defined rainfall and T_a range.

Table 3.2 Definition of deserts according to mean T_a (adapted from Meigs, 1953).

Climate	% of deserts	Examples	Mean T_a coldest month/°C	Mean T_a warmest month/°C
Hot	43	Central Sahara; central Australian	10–30	>30
Mild	18	Kalahari-Karoo, Chihuahuan	10–20	10–30
Cool	15	Mojave, Namib	0–10	10–30
Cold	24	Gobi	<0	10–30

Deserts may have seasonal climates, with winters being much colder than the summers. The Sonoran desert covers about 260 000 km² and spans the western part of the Mexican state of Sonora, southwest Arizona, southeast California and Baja California. Average rainfall is 120–300 mm yr^{-1}, with the rain falling in two seasons. Storms from the North Pacific bring gentle rain from December to

Figure 3.6 The Lower Colorado River Valley region of the Sonoran desert.

Figure 3.7 *Opuntia* cactus in the Sonoran desert.

March, and surges of wet tropical air bring in rain storms from May to September. Winter temperatures are cool, averaging 13 °C, and summers are extremely hot, reaching 40 °C on average, but peaking at 50 °C. Ambient temperatures can vary by as much as 40 °C in a day. It is also important to appreciate that there are variations in climate, topography and vegetation within a desert. The Lower Colorado River Valley region of the Sonoran desert (Figure 3.6) is the driest hottest area, where annual rainfall may be < 50 mm and summer daytime temperature 50 °C, with surface temperatures reaching 82 °C; according to Meigs' classification this desert would be classed as arid hot. Vegetation there consists of drought-tolerant shrubs and succulent cacti; following winter rainfall there is rapid growth of annual plants, which cover the ground in a mass of blooms. Within the Lower Colorado River Valley there is a large area of sand dunes forming a huge sand sea. In contrast, the Arizona Upland region has an annual rainfall of 70–300 mm, making it an arid or semi-arid desert according to Meigs. There is abundant vegetation including trees such as ironwood, blue paloverde and cat claw acacia. Cacti include 12 species of cholla (*Opuntia* spp.; Figure 3.7) and also saguara cactus. Scrub plants include desert saltbush and creosote bush. Winter annuals, e.g. California poppies, bloom in profusion after rain.

The Mojave desert spans the transition between the Sonoran and Great Basin deserts and extends throughout southeastern California, and parts of Nevada, Arizona and Utah, occupying about 100 000 km². Summers are hot and windy, but during the winter, temperatures can dip to below freezing. The Mojave desert is arid, with only about 130 mm rainfall per year in a winter rainy season, but the rains fail in some years. The plant life in the Mojave desert comprises *Yucca* species, including the joshua tree, big sage brush, bladder sage and creosote bush and at least 200 other endemic species.

Despite the variations in the environment of different deserts, it is correct to say that all desert animals have to cope with water shortage, and animals living in hot deserts cope with extremely high daytime T_a. Physiological problems linked to high T_a are those associated with hyperthermia. Mammals and birds have an optimal core T_b of around 38 °C, and many species cannot tolerate increases > 2 °C or so. The denaturation of crucial proteins, such as enzymes, begins at around 40–42 °C, so hyperthermia also creates physiological problems for ectotherms. Daytime temperatures in desert environments can be much higher than the optimal T_b, e.g. up to 56 °C in Death Valley, California. Homeotherms subjected to heat stress use a suite of physiological mechanisms for cooling the body, which we will explore in later sections. As we have seen in Chapter 2, evaporative cooling is the most effective way for an animal to lower T_b, yet if water is in short supply, dehydration is a serious problem, and the use of evaporative cooling is restricted. Behavioural mechanisms play an important role in cooling the body, both in desert ectotherms such as lizards, snakes and amphibians, and endotherms, such as birds and mammals.

Summary of Section 3.1

Deserts are diverse: sandy deserts in North Africa and Arabia have little or no vegetation but deserts in the USA may have dense vegetation especially after seasonal rain. Nevertheless, all deserts have aridity in common as their salient climatic feature. Classification systems attempt to group deserts in terms of their aridity or mean annual temperatures. Major physiological problems for animals living in deserts include hyperthermia due to intense daytime solar radiation, and also hypothermia at night when desert T_b can be below freezing. The shortage or lack of drinking water in deserts means that evaporative cooling cannot be used freely for physiological thermoregulation.

3.2 Environments and populations

The unique climate and topography of each desert links to the unique and characteristic flora and fauna found there. From the brief description of deserts provided in Section 3.1, you can appreciate that a desert provides a variety of niches for animals and plants. The term 'niche' applied to animals describes its role in a particular environment, and includes a number of characteristics such as habitat range, how the animal feeds, its diet, its environmental requirements and also its predators. So a niche is effectively an animal's particular lifestyle within an ecosystem, and encompasses how it interacts with other organisms and the physical environment within that ecosystem. In desert ecosystems, insectivorous, herbivorous and seed-eating niches are occupied by small animals, including arthropods, lizards, small birds, rodents, squirrels and shrews. Medium and large-sized animals such as hares, gazelle, camels and ostrich occupy grazing and browsing niches. Predators include foxes, e.g. kit fox (*Vulpes macrotis*) and cats, e.g. cougar (*Puma concolor*) in the deserts of the southern USA and Mexico, and Rüppell's fox (*Vulpes rueppelli*) in the Arabian desert. Desert vertebrates make use of a variety of microenvironments and their associated microclimates, small-scale areas in which the climate is different from that of the habitat as a whole. For example, in a desert ecosystem, a cavity beneath a rock, a microenvironment, would have a lower T_a than the surface and hence a different microclimate. A hyper-arid sandy desert such as the Arabian desert, has a relatively low variety of microenvironments and associated microclimates available for vertebrates. Nevertheless, the sand at a few centimetres depth is significantly cooler than at the surface, and provides a relatively cool microenvironment for animals. In contrast, American deserts such as the Sonoran have a diverse range of microenvironments, and contain a richer diversity of vertebrate species.

Although our discussion here is restricted to vertebrates, you should be aware that many invertebrates, particularly insects, inhabit desert environments, and they provide an important food supply for many desert birds and mammals.

3.2.1 How animals interact with the environment is affected by their body size

Willmer et al. (2000) classify desert animals in terms of the range of body sizes and the rate of evaporation (Figure 3.8).

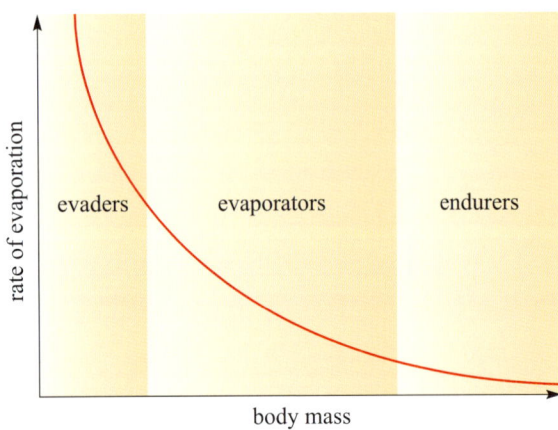

Figure 3.8 Classification of desert animals based on body size and rate of evaporation.

The logic of this classification can be appreciated by the following exercise. If you represent a small animal by a cube, and then make a larger scale model of it twice natural size, the linear dimensions of the larger animal would all be twice as large (Figure 3.9).

However, the surface area of the model would not be increased by a factor of 2, nor would the volume, as can be seen by comparing Figure 3.9a and b. If the linear dimensions double; the surface area increases by a factor of 4 (2^2) and the volume by a factor of 8 (2^3). So the ratio of surface area to volume is lower in a large animal than a smaller one. Since heat is transferred at the surface, a small

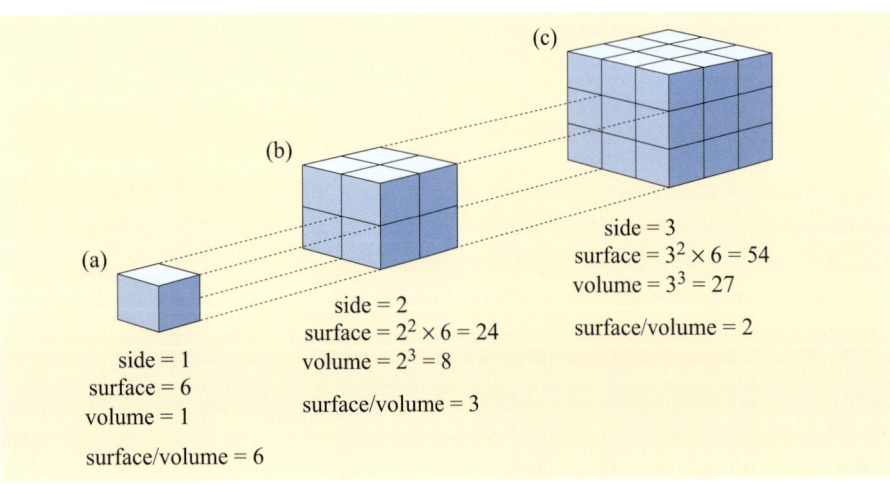

Figure 3.9 The linear dimensions, surface areas and volumes of three different-sized cubes are compared here to show how surface area: volume ratio decreases as the linear dimensions increase.

animal has greater potential for rapidly gaining and losing heat than a larger one because of its relatively large surface area. A smaller animal also has greater relative potential for evaporative water loss through its greater area of skin.

However, animals are not cube-shaped, and as you will learn in Section 3.2.4, certain desert species have features that can increase their surface area relative to their volume.

3.2.2 Behavioural strategies of evaders

Small animals, classified as evaders, include desert amphibians and reptiles, and also mammals, rodents and insectivores. The term 'evaders' refers to the animals' behaviour, which helps to prevent overheating of the body on hot sunny days, and avoids the need for cooling by evaporative water loss, which is not feasible for small animals living in an arid habitat. Evaders make use of microenvironments such as shady rock crevices, underground burrows and shade cast by plants, for behavioural thermoregulation. Evaders also prevent excessive cooling of the body by behaviour, retreating to shelter when T_a plummets at night.

The ultimate evaders are desert frogs such as *Cyclorana* spp. (Figure 3.10) and *Neobatrachus* spp. (Figure 3.11) from Australia, which spend most of the year in aestivation, inside a burrow. Aestivation is a special kind of dormancy, which enables animals to survive lack of water and high T_a during a hot dry season. During the short rainy season, desert frogs accumulate water in the bladder, where it remains during aestivation. A famous example, *Cyclorana platycephala* (Figure 3.10), is known as the water-holding frog; aboriginal people used to dig up the aestivating frogs and squeeze them, in order to collect and drink the water.

Figure 3.10 The water-holding frog (*Cyclorana platycephala*).

Figure 3.11 Painted burrowing frog (*Neobatrachus sudelli*).

During aestivation, the frogs are protected from losing water to the dry soil in the burrow by a cocoon. At the end of the rainy season, the frogs burrow into the soil, and the skin undergoes a type of moulting process in which layers of epidermis are separated from the body but not shed, forming a protective cocoon, covering all parts of the body apart from the nostril openings. The cocoon thickens, becoming heavily keratinized, and prevents loss of water from the frog's body during the 9–10 months of aestivation. At the start of the rainy season, heavy rain with consequent seepage of water into the frogs' burrows, stimulates the frogs to emerge. Breeding and feeding occur during the short wet season.

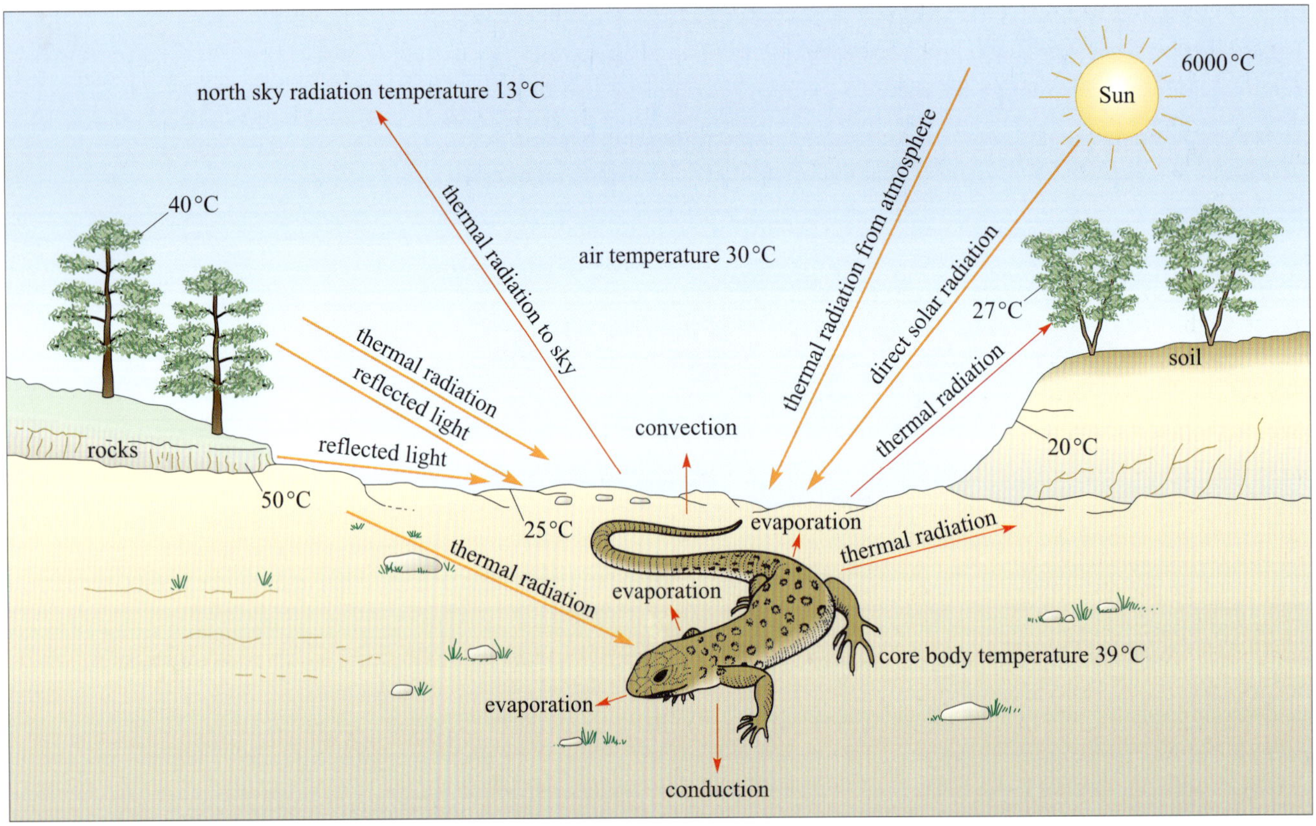

Reptiles with a scaly keratinized skin are not so prone to evaporative water loss as amphibians, and are the vertebrates that you are most likely to see on a visit to a desert. Reptiles are ectotherms and rely on solar radiation for warming the body, and maintaining high T_b during the day. Desert reptiles have no problem in gaining heat for maintaining T_b at a high level on hot sunny days (Figure 3.12).

Figure 3.12 Heat exchange with the environment in a terrestrial reptile on a hot sunny day.

■ What are the sources of energy gain and routes of heat loss for the lizard?

The lizard gains heat energy via thermal radiation from the Sun, the atmosphere and the ground. Heat energy is lost via conduction from the body to the ground, by evaporative water loss, convection and thermal radiation to the sky.

On a hot sunny day, more heat is gained than lost, and it is important for a desert reptile to avoid overheating. It is equally important to reduce loss of body heat when T_a plummets at night or during the winter.

During the day, reptiles may move between warm and cool areas in order to maintain T_b. This movement between warm and cool areas for maintaining eccritic temperature is called **shuttling.** Those species that maintain high stable T_b when environmental conditions allow by adopting heliothermic strategies, are called **thermal specialists**. In contrast, there are some species, known as **thermal generalists**, which allow their T_b to fluctuate and decline, even when they could shuttle between sun and shade to maintain high stable T_b during the day, or use their burrow at night to prevent cooling of T_b to the outside T_a. Bedriagai's skink (*Chalcides bedriagai*; Figure 3.13) is a thermal generalist, preferring to spend a lot of time hiding under rocks rather than basking in the sun.

Figure 3.13 Bedriagai's skink (*Chalcides bedriagai*).

The side-blotched lizard (*Uta stansburiana*; Figure 3.14), found in the Sonoran desert, is a typical thermal specialist. It is a small species, only 4–6 cm long when full grown.

In the morning, *Uta* warms by basking, initially orientating itself at right angles to the Sun's rays and flattening the body against the substratum for maximum exposure to solar radiation. When warmed *Uta* turns the body so that it faces the Sun while resting. *Uta* maintains T_b around 36–38 °C. Active foraging for insects, scorpions and spiders may overheat the body, and for cooling off, especially around noon, *Uta* moves to the shade of rocks and scrubby bushes. Shuttling in this way enables this species to stay active during the day for most of the year except in areas where winter temperatures dip to freezing.

Figure 3.14 The side-blotched lizard (*Uta stansburiana*).

A few desert reptiles are nocturnal; the Moorish gecko (*Tarentola mauretanica*; Figure 3.15), is found in arid regions in North Africa (also in Spain, France and Greece, so it is not restricted to deserts).

Tarentola is most active for a few hours after sunset. During the night, its T_b is as low as 18 °C, and can fluctuate by up to 11 °C. Recall that lizards that tolerate wide fluctuations in T_b, even when they could use features of the environment to maintain a steady T_b, are known as thermal generalists. The Moorish gecko is a thermal generalist at night, when it is active rather than resting in its burrow. During the early morning the Moorish gecko basks in the sunlight and its skin darkens until almost black. At night the gecko is very pale.

Figure 3.15 The Moorish gecko (*Tarentola mauretanica*).

■ What advantages do the changes in skin colour give?

Dark colours absorb and radiate heat better than light colours. At night a light colour should reduce heat loss by radiation, and there is not much heat available to absorb. During the day, dark skin promotes absorption of solar heat. Although radiation to the atmosphere by the dark skin is also promoted, the energy so lost is of little significance compared to the large amount of solar heat absorbed.

The advantage to the gecko of warming up in the morning is uncertain, but it is possible that a physiological process such as digestion of the food eaten during the night requires a higher T_b than the gecko can maintain at night.

Figure 3.16 The Mojave fringe-toed lizard (*Uma scoparia*).

The ability of the gecko to vary skin colour shows that behavioural thermoregulation in reptiles is supplemented by physiological mechanisms, which we will explore further in Section 3.3.3.

Sheltering in the available shade in the desert, or being active at night, are simple strategies for keeping T_b below lethal levels. In sandy desert areas, the sand itself plays an important role in behavioural thermoregulatory strategies. The Mojave fringe-toed lizard (*Uma scoparia*) (Figure 3.16) is restricted to fine, wind-blown sand, e.g. in dunes, dry lake beds and desert scrub in the Mojave desert. Burrows in sand collapse immediately or soon after the animal has moved on, so animals buried in sand rely on air trapped between sand particles for breathing. *Uma* is a 'sand-swimmer' and its dorsoventrally flattened body and shovel-shaped head facilitate movement through the sand, which is especially important when escaping from predators such as snakes and badgers.

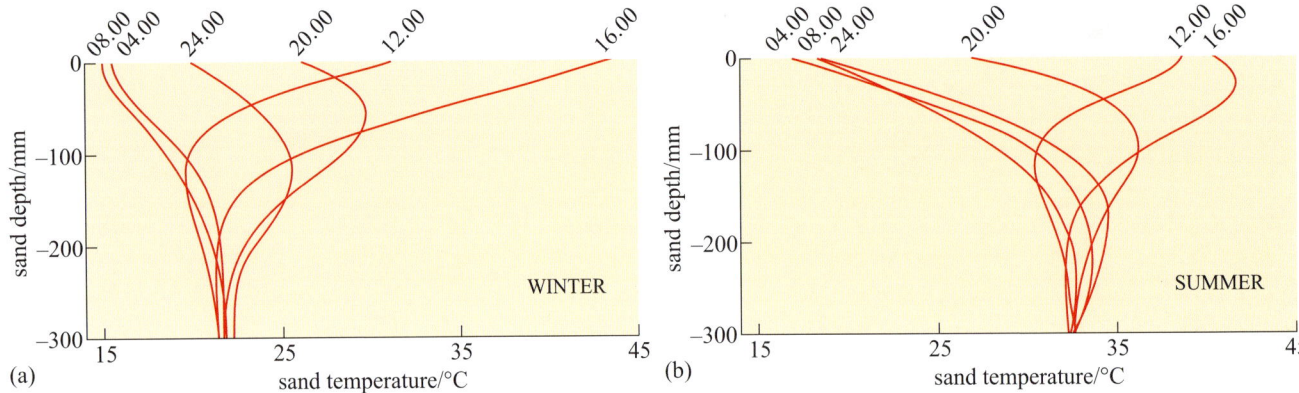

Figure 3.17 Temperatures below the sand surface, measured at 4-hourly intervals on a dune surface in the Namib desert in the (a) winter and (b) summer.

The eyelids are protected from sand by large eyelid fringe scales. The digits have large lamellar fringes, elongated scales, especially long on the hind feet, which enable the lizards to run at speed on the sand surface. *Uma* grows up to about 110 mm in length, and its activity pattern is diurnal, varying according to ambient temperature. In March and April *Uma* is active for short periods because of the low spring temperatures in the Mojave. In summer, from May to September, the lizards are active during mornings and late afternoons, feeding on insects and plants. Sand-swimming lizards are also found in the Namib desert and include the wedge-snouted sand lizard (*Meroles cuneirostris*).

The data in Figure 3.17 were collected from a sand dune slope in the Namib desert.

Although the temperatures of sand at various depths in the Mojave desert would not be precisely the same as those in the Namib, the physical characteristics and thermal environment provided by dry sand are broadly the same in all deserts at similar T_a.

A benign temperature is available below the surface at all times of the day in both seasons, in spite of extremes on the surface. These surface temperature extremes are not very different in summer and winter. The high afternoon surface temperature in winter is due to hot, dry winds (Berg winds) that reach the desert in the winter months.

■ Examine the data in Figure 3.17 and suggest the advantages for a sand-swimming lizard of the following strategies:

  (i) The lizard 'swims' down to 60 mm depth at 12.00 hours in summer, when surface temperatures can reach 40 °C or more.

  (ii) In winter, the sand-swimmer remains in a state of dormancy for a month at 300 mm depth in the sand, when surface temperature can occasionally drop below freezing at night.

(i) At 12.00 hours, when T_a is 40 °C at the surface, by burrowing to a depth of 60 mm the lizard reaches a microenvironment where T_a is significantly lower, about 32°C (Figure 3.17b). The lizard loses body heat by conduction and thereby avoids a dangerous increase in body temperature.

(ii) In winter when ambient temperatures can drop to below freezing, the temperature at 300 mm depth remains constant at around 21 °C (Figure 3.17a). The lizard thereby avoids low T_a at the surface and is not at risk of freezing when T_a drops to < 0 °C.

Chapter 3 The Desert Environment

Figure 3.18 The desert tortoise (*Xerobates agassizii*).

Burrows provide important microenvironments for many desert evaders, and their structure and use vary between species. The desert tortoise (*Xerobates agassizii*; Figure 3.18) lives in deserts in the USA and Mexico, and feeds on annual herbs, cacti and shrubs, obtaining most of its water from the plants.

In the Mojave desert, the tortoises live in sandy areas as well as rocky hillsides, including scrub-type vegetation, joshua tree/yucca and creosote bush/ocatillo habitats. For the tortoises, burrows are important refuges for thermoregulation, summer aestivation and winter hibernation. Tortoise burrows in the Mojave desert are extensive and can be up to 12 m long; the same burrows are used for many generations, and are shared with other species such as burrowing owls and ground squirrels. Each desert tortoise may use up to 12 burrows in its home range and each burrow is used by different tortoises at different times. For short rest periods during the day tortoises dig shallow depressions, known as pallets, which barely cover the carapace.

Susan Bulova (2002) compared temperature and humidity in four unoccupied desert tortoise burrows, and the surface over 24 hours on a summer day in the Mojave desert (Figure 3.19a–c).

■ (a) Compare the fluctuations in T_a (Figure 3.19a), in the burrows and on the surface.

(b) Compare the fluctuations in absolute humidity (a.h.) and relative humidity (r.h.) (Figures 3.19b and c) in the burrows and on the surface.

(a) T_a inside each of the burrows fluctuated by about 2–10 °C, ranging from about 27–37 °C. Surface T_a fluctuated from a low of 20 °C at 05.00 h to 45 °C at 11.00 h. Compared to surface T_a, burrow T_a remained relatively stable, being cooler than surface T_a from 09.00 h–17.00 h and warmer than surface T_a from 17.00 h–09.00 h.

(b) Absolute humidity was generally lower on the surface than inside the burrows, but for two of the burrows measured

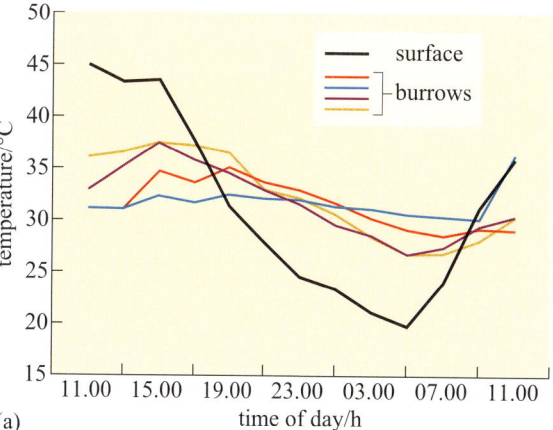

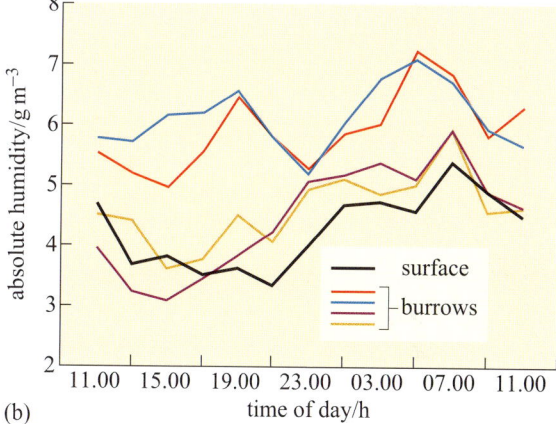

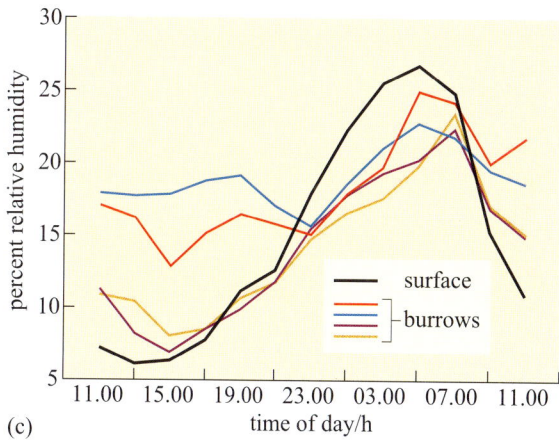

Figure 3.19 Burrow and surface microclimate over 24 h from 11.00 h. Measurements were made inside and on the surface near four burrows, and were taken at 2 h intervals. (a) temperature, (b) absolute humidity and (c) percent relative humidity.

a.h. values were always close to surface a.h. values. Relative humidity was higher in the burrow than on the surface from 09.00 h to 21.00 h but two of the burrows had r.h. values quite close to surface r.h. from 15.00 h to 23.00 h.

Bear in mind that when occupied by a tortoise, a burrow's relative humidity may rise to 40% because of the tortoise's water loss by evaporation from the lungs, exposed skin and eyes. Stable T_a and humidity in the burrow protect the tortoise from extremes of high T_a and from winter frosts. Bulova noticed that tortoises are fussy about the burrow selected for resting. At the end of foraging, tortoises were observed to enter and leave several burrows before settling. Mojave desert tortoises are active between March and June, a time when the winter rains have stimulated the growth of annual plants, providing abundant food for the tortoises. The tortoises begin foraging during the morning but usually by noon they have moved into pallets and burrows to shelter from high T_a. At night, burrows provide shelter from low T_a and also protection from nocturnal predators such as kit foxes and badgers. By the end of June, when surface temperature may reach 60 °C, and annual plants have dried up, the tortoises retreat to their deep burrows and aestivate, a behaviour that helps to conserve body water. During aestivation, up to a quarter of the tortoises' body mass may be water stored in the bladder. Occasionally an aestivating tortoise emerges to drink during summer thunderstorms. In the eastern Mojave desert tortoises are active for most of the summer because there, summer rainstorms provide sufficient new plant growth. For their winter hibernation, tortoises aggregate in the burrows; up to 25 individuals have been found in one burrow. Hibernation lasts from October to the end of February, and during this time T_b of the tortoises is the same temperature as the burrow, around 5–16 °C in winter. Note therefore that hibernation in the desert tortoise is not the same physiological process as it is in hibernating mammals (Chapter 4). Reptiles do not regulate T_b physiologically during hibernation; T_b is the same as burrow T_a. You will find that in some references, reptile 'hibernation' is termed 'brumation'.

You may be surprised to learn that like desert ectotherms, small desert rodents also depend on burrows for thermoregulation. Merriam's kangaroo rat (*Dipodomys merriami*; Figure 3.20), is a typical evader, living in the Sonoran desert, Arizona, and in Death Valley, California, one of the hottest and driest areas in the Western Hemisphere.

Individuals live in a maze of burrows, which they defend. They remain in their burrows during the day, and often plug the entrance with soil. At night kangaroo rats emerge from their burrows for just 2 hours to collect seeds, in particular seeds of the creosote bush, which they push into their cheek pouches, returning at intervals to empty the food into their burrow. In this way, food caches are built up; kangaroo rats always eat inside the burrow, drawing on their food cache. Inside the burrow, the air is cooler and more humid than above the ground, as moisture from respiratory water loss accumulates. Measurements made on similar burrows in the Negev desert, Israel, showed T_a of around 26 °C at 1 metre depth for 24 hours per day when ambient temperature above ground ranged from 16–44 °C. However, not all small desert animals can burrow.

The desert wood rat (*Neotoma lepida*) lives in deserts in the southern USA, including Death Valley, California. Wood rats do not burrow but build elaborate houses around the base of cacti or shrubs, amongst a patch of agaves, or beneath a rock outcrop. Wood rat houses can reach huge sizes and their interior

Figure 3.20 Merriam's kangaroo rat (*Dipodomys merriami*).

is significantly cooler, by about 5 °C, than the outside during the heat of the day. Desert wood rats shelter in their houses during the day, and emerge to forage at night, eating creosote bush, cholla, prickly pear cactus and agave.

3.2.3 Behavioural strategies of evaporators

Willmer (2000) defines evaporators as animals that depend on sufficient water intake to enable them to cool T_b by evaporation. Few of these species can survive in deserts, and those that do either live on the edges of deserts where they can access water, or have behavioural and physiological adaptations that reduce reliance on evaporative cooling. So for evaporators, evasion may be an important part of their thermoregulatory strategy. Evaporators include medium-sized mammals such as jack rabbits, dogs, foxes, and also desert birds such as larks.

The jack rabbit (*Lepus californicus*; Figure 3.21) is a hare, living in the Sonoran and Mojave deserts. Jack rabbits do not burrow, although they are quite small, weighing about 2 kg.

A jack rabbit would need to lose at least 4% of its body mass per hour to thermoregulate by evaporation. There is little or no free water around; water is obtained from the diet, green plants, including cacti in the summer. Knut Schmidt-Nielsen's work (1967) showed that behaviour is important for the jack rabbit's survival. During the hottest part of the day the animal chooses a shaded depression in the ground, often in the lee of a bush, in which it crouches (Figure 3.22).

The bottom of such a depression has a much lower temperature than that of the rest of the surface, the hot desert wind and much of the radiation passing over the animal's head. From its sheltered position, the jack rabbit's large radiator-like ears can be exposed, not directly to the Sun, but to a clear blue sky. The radiation temperature of the north sky at midday is only 13 °C so if the ears, which are richly vascularized, have a temperature of 38 °C, and have a surface area of

Figure 3.21 The jack rabbit (*Lepus californicus*).

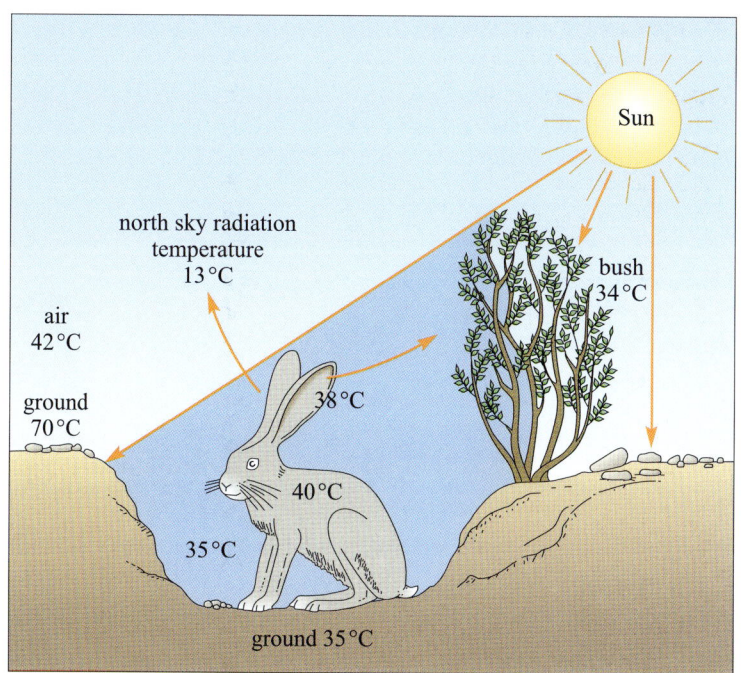

Figure 3.22 The desert jack rabbit in a shaded depression showing a behavioural adaptation to cope with the severe environment.

400 cm², are directed towards the sky, they can radiate about 13 kJ h⁻¹, which is about half of the animal's metabolic heat production. The jack rabbit forages during the night.

The kit fox (*Vulpes macrotis*; Figure 3.23) lives in the Sonoran, Mojave and Great Basin deserts in the southwestern USA. Kit foxes have very large ears, which are thought to provide an increased surface area for cooling the body.

They are carnivores, and hunt at night, preying on kangaroo rats, tortoises and jack rabbits, and occasionally catching ground-nesting birds, reptiles and insects. They reduce evaporative water loss by spending the day in underground dens, emerging at sunset to begin hunting. The physiological importance of dens for desert foxes should not be underestimated. By remaining in the den during the day, a desert fox reduces drastically the need for panting, a mechanism used by foxes and dogs for cooling the body by evaporative water loss (Section 3.3.2).

Figure 3.23 The kit fox (*Vulpes macrotis*).

A few species of small birds live in the most extreme deserts. Dune larks (*Mirafra erythroclamys*; Figure 3.24) are the only birds that live year round in the Namib sand sea, one of the driest regions of the world. Dune larks feed on insects and spiders, which they collect during the day, while walking over the sand surface; they also peck insects from just below the sand surface. In winter the birds feed on seeds blown in from adjacent grass land. The scarcity of water in the Namib sand sea means that dune larks drink rarely and the birds rely on water in their food and on metabolic water. Birds do not sweat, but they use both cutaneous and respiratory evaporative water loss for cooling the body. During the hottest part of the day, from around 12.00 to 15.00 h, dune larks seek shade and stand still. Presumably this behaviour helps the birds to cool T_b and reduces evaporative water loss.

Williams (2001) used taxidermic mounts to determine **operative environmental temperature**, T_e, for the birds. T_e is the temperature that an animal would reach in the environment if it was biologically inactive, i.e. only the physical characteristics of the animal are taken into account. It is defined, in physical terms, as the temperature of a black body of uniform temperature, in an identical situation to that which the animal occupies, with the same values for conduction, convection and radiation. As the definition is purely physical, it is possible to make models of animals and to use them to measure T_e experimentally. Figure 3.25 shows three examples of daily profiles for a model of a dune lark made from a copper cast of a bird covered in plumage.

Figure 3.24 The dune lark (*Mirafra erythroclamys*).

Figure 3.25b shows that during July 1991 (winter in Namibia) mean T_e shade, did not exceed T_{uc} (35.1 °C) for the larks. The results suggest that in winter, the strategy of finding a shady spot during the hottest part of the day lowers T_b sufficiently, so there is no need for physiological cooling, in particular evaporative water loss, for maintaining T_b.

■ Do the results shown in Figures 3.25a and c suggest that dune larks do not need to use evaporative cooling to maintain T_b in the summer? What is the main advantage of resting in the shade for the dune lark? Identify one disadvantage.

Even in the summer, T_e in shade is significantly lower than T_e in sun. In Jan–Feb 1991, at midday, all mounts exposed to the sun reached a T_e of 46–50 °C, higher than T_{uc} of 35.1 °C. T_e in the shade was significantly lower. In summer 1991,

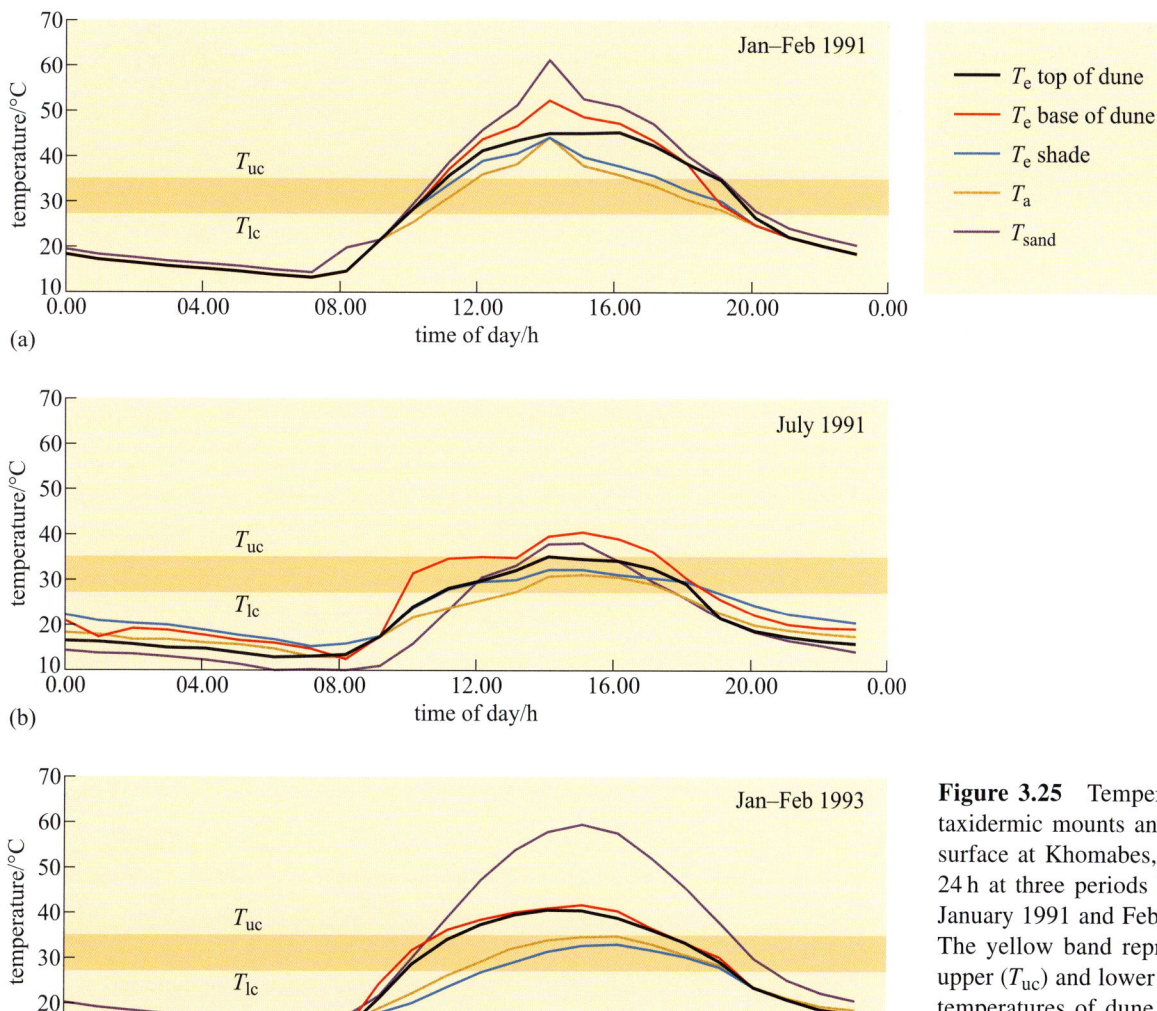

Figure 3.25 Temperature of taxidermic mounts and of the sand surface at Khomabes, Namibia over 24 h at three periods between January 1991 and February 1993. The yellow band represents the upper (T_{uc}) and lower (T_{lc}) critical temperatures of dune larks. Each data point represents a mean for each hour of all days for the period shown.

mounts exposed to full sun reached T_e values of 40–50 °C from about 12.00–20.00 h, whereas those in the shade peaked at 40–45 °C from about 12.00–16.00 h. In summer 1993, T_e for mounts in the shade never exceeded T_{uc}. For the dune lark, the simple strategy of standing in shade during the heat of the day provides significant cooling, even in a particularly hot summer like that of 1991. It is likely that by standing in shade, the need for evaporative cooling must be reduced at least.

However, the disadvantage of this strategy is that while standing in the shade, dune larks cannot forage, so the bird has to find a balance between the need for food and the necessity of avoiding excessively high T_b.

While desert animals classed as 'evaporators' could use evaporative cooling for maintaining T_b at high T_a, the need for this is avoided by simple behavioural strategies. Nocturnal foraging and daytime use of dens, burrows and shade for cooling, reduce the need for physiological cooling by evaporative water loss, thereby conserving water.

3.2.4 Behavioural strategies of endurers

Endurers are defined as large desert mammals such as oryx and camel, and large desert birds, ostrich and emu. The term 'endurers' suggests that these animals are forced to endure the extreme conditions of the desert climate because they cannot shelter from high T_a and intense solar radiation during the day or low T_a at night, as they are too large to hide in burrows or dens. Nevertheless, in spite of their size, endurers do take advantage of aspects of the environment for cooling by means of behavioural strategies. Large mammals tend to be inactive during the hottest part of the day, thereby reducing metabolic heat production. The Arabian oryx (*Oryx leucoryx*; Figure 3.5) lives in the Arabian desert, including areas where free-standing water is rarely if ever available. On hot days oryx dig into the sand with their hooves, exposing the cool sand below the surface, and sit in the depressions. Body heat is lost to the cooler sand by conduction. Where possible, the oryx also spends time sitting in the shade of evergreen trees (*Maerua crassifolia*) during the hottest part of the day. Oryx forage at night during the summer, avoiding exposure to high T_a and intense solar radiation. They feed on grasses and rely on the water content of the plants for their intake of water.

Dorcas gazelle (*Gazella dorcas*; Figure 3.4) live at the borders of the Sahara desert and are the smallest species of gazelle, weighing just 15–20 kg. They have very long limbs in proportion to their body size, and large ears: both features maximize any convective cooling caused by breezes. Dorcas are described as the most desert-adapted of all gazelles, as like the oryx, they are reputed to be able to survive without drinking any water at all. Their feet are splayed, an adaptation for walking and running on sand. Dorcas gazelle graze and browse at night and at dawn and dusk, feeding on leaves, flowers and pods of acacia trees, and using their hooves to dig for bulbs.

Long limbs, tails or necks provide large surface areas from which heat can be dissipated, and behaviour patterns may maximize loss of heat from these areas. The ostrich (*Struthio camelus*) is the largest living bird, weighing up to 150 kg. Ostriches forage during the day. The birds select plants with high water content when grazing, especially during times of water shortage. The naked neck of the ostrich and its long naked legs provide a large surface area for convective and radiative cooling, especially in breezy conditions. The ostrich uses behaviour to enhance the cooling effects of feather erection at a high ambient temperature and incident solar radiation. Sparsely distributed long feathers on the dorsal surface of the bird erect in response to warming of the skin, thereby increasing the thickness of the insulation between solar radiation and skin. The gaps between the feathers allow through air movements, which cool the skin by convection. The birds supplement the physiological response during the hottest part of the day, by orientating themselves towards the Sun and bowing out their wings away from the thorax, forming an 'umbrella' which shades the exposed thorax. The naked skin of the thorax acts as a surface for heat loss by both radiation and convection. At night when ambient temperatures plummet, ostriches conserve heat by folding the wings close to the thorax and tucking the naked legs under the body while they sit on the ground. The dorsal feathers respond to low T_a by flattening and interlocking, which traps an insulating layer of air next to the skin, and keeps most of the skin at 34.5 °C.

You know from Chapter 2 that evaporative water loss is the most effective means of reducing body temperature during heat stress. However, in deserts, very little, if any, free-standing water is available. For all groups of desert vertebrates, behavioural strategies for maintaining T_b play a crucial role in preventing overheating of the body, which reduces the need for evaporative cooling and thereby conserves water. In the following section, we will see how in desert vertebrates, behavioural strategies for controlling body temperature are integrated closely with biochemical and physiological mechanisms.

Summary of Section 3.2

Desert animals are classified in terms of their body size and physiology into three groups: evaders, evaporators and endurers. The logic for this classification is that the smaller the animal, the larger its surface area to volume ratio. Small animals therefore gain and lose heat faster than large animals, warming rapidly when exposed to intense solar radiation, and cooling rapidly at night. Small endothermic evaders, e.g. kangaroo rats, rest in cool microenvironments, e.g. shade or burrows, during the day. Lizards, ectothermic evaders, regulate T_b during the day by shuttling between sun and shelter. They avoid night-time hypothermia by resting in burrows. Nocturnal evaporators, e.g. kit foxes, remain in cool dens during the day. Some endurers, large species such as the oryx, graze nocturnally in summer, sitting in shade during the day. Behavioural strategies for avoiding intense solar radiation link intimately to physiology. Such behaviour prevents large fluctuations in T_b and conserves water by removing the need for evaporative cooling, which is of crucial importance in deserts where water is scarce.

3.3 Integrating across levels of analysis

In mammals and bird, homeostasis, the provision of a stable internal environment, includes keeping certain physiological variables, T_b, cellular and extracellular water and blood glucose at near constant levels. T_b of reptiles varies with T_a, but reptiles can only function over a limited range of T_b. Nevertheless, vertebrate species live successfully in deserts, which are arid, have low productivity and extremes of T_a. Our approach to understanding animals' physiological responses and genotypic adaptations linked to environmental features of deserts, is study and integration of different levels of analysis, covering behaviour, anatomy, physiology and biochemistry. Grouping desert animals into evaders, evaporators and endurers is useful for our purpose. We describe just a few examples here, but the principles identified can help in interpreting data derived from other species.

3.3.1 Integration of anatomy and behaviour with biochemical and physiological strategies in evaders

We know from Section 3.2.2 that small desert rodents remain cool by staying in their burrows for all or part of the day. Kangaroo rats (*Dipodomys* spp.; Figure 3.20) depend on metabolic water as there is little or no water available in their diet of seeds. Kangaroo rats appear to be ill-adapted for life in a desert; like other rodents they neither sweat nor pant. Nevertheless, inside the burrow, they could lose water by evaporation from the lungs, which would be enhanced by T_b being higher than burrow T_a. As the water-carrying capacity of air increases with temperature, warm expired air contains more water than the cooler inhaled air.

However, the temperature of the exhaled air in kangaroo rats is lower than that of T_b, and often close to T_a (Figure 3.26). This is because the nasal passages (**turbinates**) of kangaroo rats are extremely narrow and convoluted and provide a temporal counter-current cooling system, which operates as a heat exchanger (Figure 3.27).

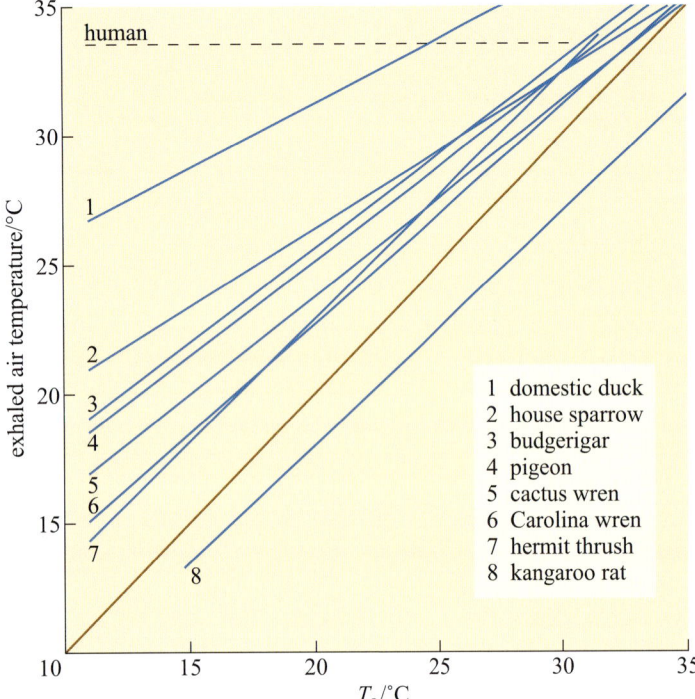

Figure 3.26 Exhaled air temperatures of seven species of bird and the kangaroo rat at various ambient temperatures. (Much the same relationship applies in many small mammals.) The line labelled 'human' represents one individual only. The brown isothermal line shows what the relationship would be if the temperatures of inhaled and expelled air were equal.

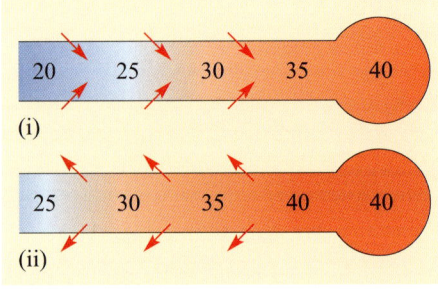

Figure 3.27 A diagram of a temporal counter-current heat exchanger: the nasal heat exchanger. Inspired air draws heat and water from the walls of the respiratory tract (i) and gives both back again at exhalation (ii). The figures represent approximate temperatures varying from those in the deep body tissues (40 °C) to those near the surface. The red arrows show the direction of heat transfer.

Both inspired and expired air pass over the same surface, the nasal mucosa. Air entering the nasal passages takes up both heat and moisture from the mucosa and is therefore both warmed and moistened before entering the lungs. In the short interval between inhalation and exhalation, the thermal gradient between nose and trachea is maintained. When air is exhaled from the lungs, initially its temperature is 37–38 °C and it is humidified by heat and moisture derived from the warm tissues in the nasal passages, trachea and bronchi. As the exhaled air approaches the nasal passages, the temperature and vapour pressure gradients between the mucosa and the adjacent air are reversed and heat is lost from air to the mucosa. During cooling, water condenses on the mucosal surfaces. At the tip of the nose, the air is expired at ambient temperature, still saturated with water vapour, but because its temperature is reduced, it carries much less water. The efficiency of the heat and water exchanger reflects the large surface area to bore ratio of the nasal passages of a small animal like the kangaroo rat. When the kangaroo rat breathes air at 25% relative humidity, the temperature of the expired air ranges from 31 °C at T_a 35 °C to 13 °C at T_a 15 °C. About 54% of the water vapour derived from evaporation from the respiratory surfaces is thereby conserved at 30 °C and 83% at 15 °C. In contrast, humans, with short wide nasal passages, cannot recover more than 16% water vapour at T_a ranging from 12 to 35 °C.

The following material gives some important background on kidney function and to appreciate this you need to revise the concept of osmosis. Osmosis is the movement of water between two solutions which have different solute concentrations, and which are separated by a semi-permeable membrane. Water will move from the side that has lower solute concentration to the side that has the higher solute concentration. Osmolarity is an expression of the osmotic concentration of the solution. You may find texts where solutions with a high osmolarity are referred to as having a high osmotic pressure. You can think of this in terms of the pressure that would have to be applied to the solution on the side with a high solute concentration to prevent the movement of water by osmosis. The greater the solute concentration, the greater the pressure needed to prevent osmosis.

Kangaroo rats and other desert rodents, e.g. the Australian hopping mouse *Notomys*, conserve water by producing extremely hyperosmotic urine, on average 5500 mOsmol l^{-1} in *Dipodomys* and 9000 mOsmol l^{-1} in *Notomys*. Compare the osmolarity of the urine of *Dipodomys* with that of other mammalian species (Table 3.3), and note how small xeric mammals produce more highly concentrated urine than do species living in mesic habitats. Note also that large mammals living in xeric habitats, e.g. camels, do not produce urine as concentrated as that produced by small xeric mammals. Values for the net ratios of osmolarity for urine and plasma (U/P ratios), are provided to demonstrate the concentration of urine relative to that of the blood. Osmolarity is the concentration of solute particles, not the concentration of moles although the two are related. For example, a solution containing 1 mol l^{-1} sodium chloride has an osmolarity of 2 Osmol l^{-1}, because in solution, sodium chloride molecules break down into equal numbers of sodium and chloride ions. In contrast, molarity and osmolarity for a glucose solution are the same because glucose molecules remain intact in solution.

The ability of the kangaroo rat and other desert rodents to produce a hyper-concentrated urine is attributed to their possession of extremely long loops of Henle, which is often quoted as an extreme adaptation for life in parched deserts. But is the ability to produce a concentrated urine an 'extreme adaptation'? Mammalian kidneys are effector organs that maintain the concentration of salts and excretory products, especially urea, in the blood within very narrow limits.

Table 3.3 Urine concentrations and urine/plasma ratios in mammal species from different habitats. For land animals, the values are generally maximal, measured from dehydrated individuals (adapted from Willmer, 2000).

Mammal	Habitat	Urine concentration/mOsmol l^{-1}	U/P ratio
Small mammals			
rat	mesic	2900	9
domestic cat	mesic	3100	10
kangaroo rat	xeric	5500	16
Large mammals			
beaver	freshwater/land	520	1.7
human	mesic	1400	4–5
porpoise	marine	1800	5
eland	xeric	1880	6
camel	xeric	2800	8

The mammalian kidney is a compact organ consisting of an outer dark cortex and an inner pale medulla (Figure 3.28). The kidney tissue is made up of nephrons, which are thin-walled tubules (not to scale in this figure). The nephrons are concentrated in areas known as pyramids. Ducts that collect the urine and transfer it to the ureter are located in the papilla areas.

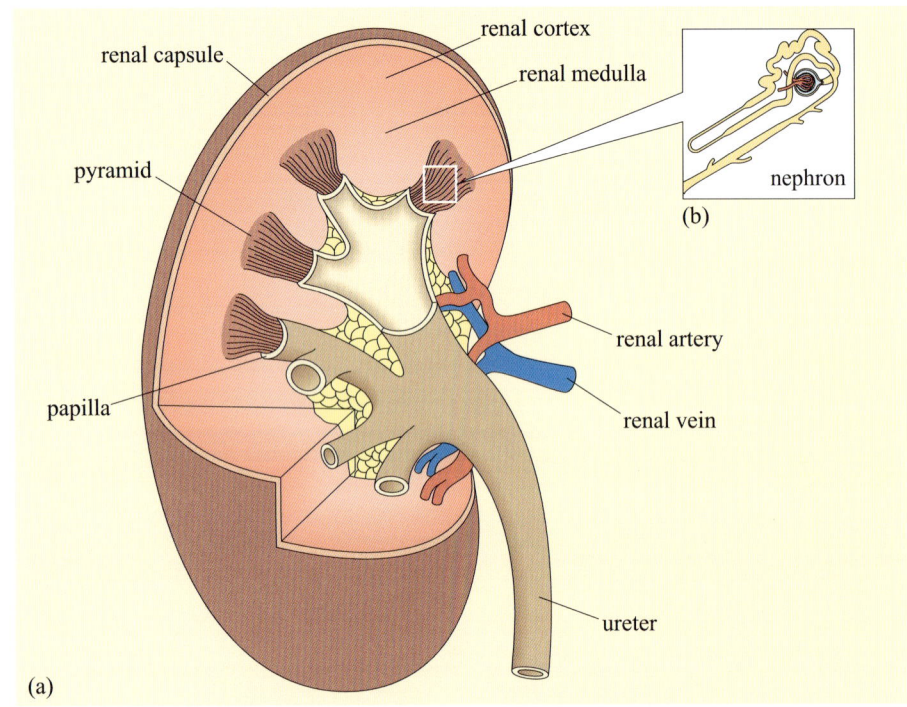

Figure 3.28 (a) Diagrammatic representation of a human kidney showing the gross structure and (b) an enlarged diagram of a nephron.

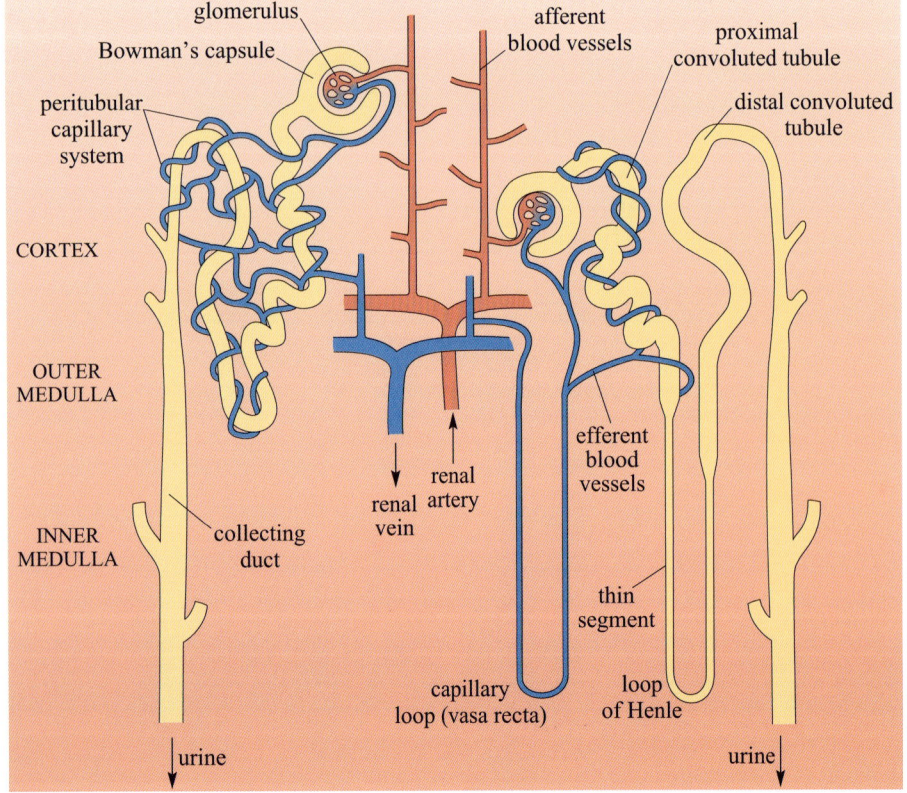

Figure 3.29 Two typical nephrons. The one on the right is 'long-looped' (located close to the border of the medulla and therefore called juxtamedullary and the one on the left is 'short-looped' (or cortical). The long-looped nephron is paralleled by a loop formed by the blood capillary. A capillary network surrounds the short-looped nephron. Most mammalian kidneys contain a mixture of the two types but some species have only one or the other.

Each nephron begins with a cup-shaped structure, the Bowman's capsule. This encloses the glomerulus, a cluster of capillaries.* Bowman's capsule opens into the coiled proximal convoluted tubule, which leads to the loop of Henle (Figure 3.29). There are two types of nephron, distinguished by the length of their loops of Henle.

Cortical nephrons have short-reach loops that just penetrate the boundary between the inner and outer zones of the medulla. Juxtamedullary nephrons have long-reach loops that penetrate deep into the medulla. In humans about 15% of nephrons are juxtamedullary and 85% are cortical. Blood reaching the Bowman's capsule undergoes ultrafiltration. The blood pressure in the glomerular capillaries is high, and it is maintained by the pumping of the heart and the mechanical properties of the blood vessels. Consequently blood in the glomerulus is filtered through the basement membrane of the capsule. Blood cells and proteins remain in the blood, so that the filtrate that enters the nephron tubule has a similar composition to plasma minus its proteins and large lipids. Osmolarity of plasma and filtrate are the same, 300 mOsmol l^{-1}.

As the filtrate travels through the nephron its composition undergoes considerable modification. The movements of sodium, chloride and water are summarized in Figure 3.30.

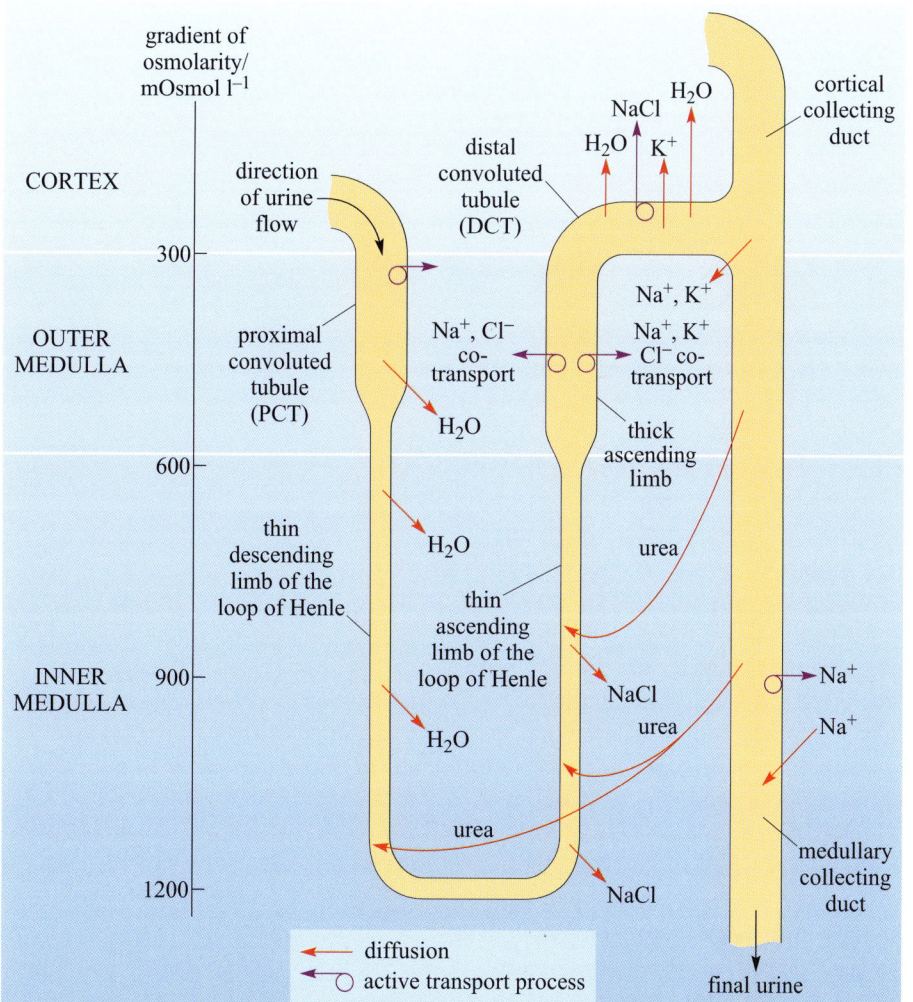

Figure 3.30 Schematic view of part of a mammalian nephron showing the proximal convoluted tubule, the limbs of the loop of Henle, the distal convoluted tubule and collecting ducts. See text for explanation.

*The Bowman's capsule and the glomerulus together are sometimes referred to as the Malpighian body.

The values on the left in Figure 3.30 are the osmolarity of the interstitial tissue. You can see a gradient of osmolarity from 300 mOsmol l^{-1} in the cortex to 1200 mOsmol l^{-1} in the inner medulla. The fluid in the loop of Henle has about the same osmolarity as the fluids in the surrounding tissue. Movements from the tubules of Na$^+$, K$^+$, Cl$^-$, urea and water occur as shown. Relatively small amounts of salt moving from the ascending limb to the interstitial tissues will cause osmotic movement of water out of the descending limb. Due to the principles of counter-current multiplication, a small difference in the concentration between adjacent points on the two limbs leads to a large difference in concentration between the top and bottom of the loop. Since the collecting tubule is very permeable to urea, urea moves into the interstitial tissues. This will increase the osmolarity in the medulla. The thin ascending limb has some permeability for urea so you can view the collecting duct and ascending limb as 'recycling' urea. As urea is moving through the medulla, this increases the osmolarity in this region of the kidney.

The process begins in the proximal convoluted tubule (PCT), where the epithelial cells absorb much of the filtrate passing it back into the blood flowing in the surrounding vessels. Active transport of sodium out of the PCT epithelial cells into the interstitial tissues increases the osmolarity in the tissue. Thus water moves by osmosis out of the PCT. Movement of glucose, amino acids and water is coupled to movement of Na$^+$ out of the tubules. The water permeability of the PCT is high because of the abundance of special membrane channel proteins, aquaporins, in the cell membrane. Permeability of PCT epithelial cells is relatively low for urea, so the 75% reduction in fluid volume in the PCT results in a four-fold increase in urea concentration.

The counter-current system of loop of Henle concentrates the urine. Figure 3.30 shows that the hairpin-like loop of Henle lies between the proximal convoluted tubule and the distal convoluted tubule. Fluid entering the loop flows down the descending limb and then turns the corner, before flowing up the ascending limb. The loop of Henle functions as a counter-current multiplier system as a result of the opposing direction of fluid flow in the descending and ascending limbs. Although the filtered liquid flows into the descending limb of the loop of Henle first, we need to look at processes in the ascending limb so that we can understand what happens in the descending limb. In the ascending limb, sodium and chloride ions are reabsorbed into the medullary interstitial tissues, passively in the lower part of the limb and actively by means of Na$^+$-K$^+$ ATPase pumps in the thick upper part of the ascending limb. The active transport of Na$^+$ out of the tubule cells creates low [Na$^+$] and [Cl$^-$] in the cell cytoplasm; this creates a concentration gradient drawing in Na$^+$ and Cl$^-$ ions from the lumen of the tubule into the tubule epithelial cells via luminal membrane transport molecules in the upper part of the limb. Unlike the descending limb, the ascending limb is relatively impermeable to water, so little water follows the salt. The interstitial fluid of the medulla thereby becomes hyperosmotic compared with the fluid in the ascending limb. The apical membranes of the epithelial cells lining the descending loop of Henle have a very low permeability to ions and urea but a very high permeability to water. Water therefore diffuses out of the fluid in the tubule and into the epithelial cells and then into the interstitial fluid. Water diffuses out of the descending limb into the more concentrated interstitial fluid until the osmolarity between ascending and descending limbs is equal. As the ascending limb is continually pumping sodium and chloride ions, the concentration difference between it and the interstitial fluid is maintained. The osmolarity difference of 200 mOsmol l^{-1} is multiplied to 1400 mOsmol l^{-1} at the bend in the loop.

When the fluid reaches the distal convoluted tubule (DCT) its osmolarity has become reduced to just 100 mOsmol l^{-1}. The fluid is diluted further in the DCT, where active transport in the epithelium removes more sodium and chloride from the tubular fluid into the epithelial cells. As the epithelial cell membranes are impermeable to water, the tubular fluid is hypo-osmotic by the time it reaches the cortical collecting duct.

As in the PCT, basal and apical membranes of the epithelial cells of the collecting ducts have aquaporins. Normally, these membrane channel proteins are configured to limit water reabsorption. If the blood osmolarity rises, antidiuretic hormone (ADH) is released from the posterior pituitary. This hormone acts directly on the aquaporins in the collecting duct epithelia so that the membrane channels are fully 'opened' and water moves by osmosis out into the interstitial tissues. This reduces the urine flow. If the blood osmolarity decreases, secretion of ADH stops and the membrane channels close, so water is retained in the collecting ducts. The blood vessels in the medulla, the vasa recta, are arranged as hairpin loops that run close to and parallel to the loops of Henle and collecting ducts. Water reabsorbed from the collecting ducts enters the blood capillaries and leaves the kidneys in venous blood, which maintains the concentration gradients in the medulla. As water is reabsorbed along the entire lengths of the medullary collecting ducts, fluid emerging from the medullary collecting ducts has the same osmolarity as the interstitial fluid around the bend of the loop of Henle at the bottom of the medulla. The longer the loop of Henle relative to the overall depth of the cortex, the higher is the osmolarity of the fluid in the bend. The kidney thereby retains as much water as possible, minimizing loss of water during water shortage.

The relationship between the ability to concentrate urine and the length of the loops of Henle is not straightforward. There is no clear relationship between actual loop lengths and urine concentration in mammals. Average lengths of loops of Henle are not directly proportional to urine concentration when comparing large with small species of mammals. *Notomys* has a loop length of 5.2 mm and produces urine of up to 9000 mOsmol l^{-1} in contrast to the horse with a loop length of 36 mm producing urine of 1900 mOsmol l^{-1}. How can we explain this data in relation to the hypothesis that the length of the loop of Henle *does* affect final urine concentration?

Small mammals have much higher mass-specific metabolic rates than large mammals. Compared with large mammals, the apical membranes in the kidney tubules of small mammal epithelial cells have more infoldings, increasing surface area for absorption.

Increased metabolism will lead to more waste products and a greater demand on the filtration capacity of the kidneys. Therefore the number of nephrons must increase as body size increases, which in turn increases the relative amount of cortex (the area of the kidney where most of the nephron is located), at the expense of the medulla. The relative thickness of the medulla is related to urine-concentrating ability because the medulla contains the loops of Henle. Hence larger animals, even the camel, cannot produce urine as concentrated as that of smaller mammals, because their kidney medulla is relatively small compared with its cortex. Small mammals such as rodents and bats tend to have relatively thicker medullas than larger mammals, which can be correlated with their production of concentrated urine (Figure 3.31).

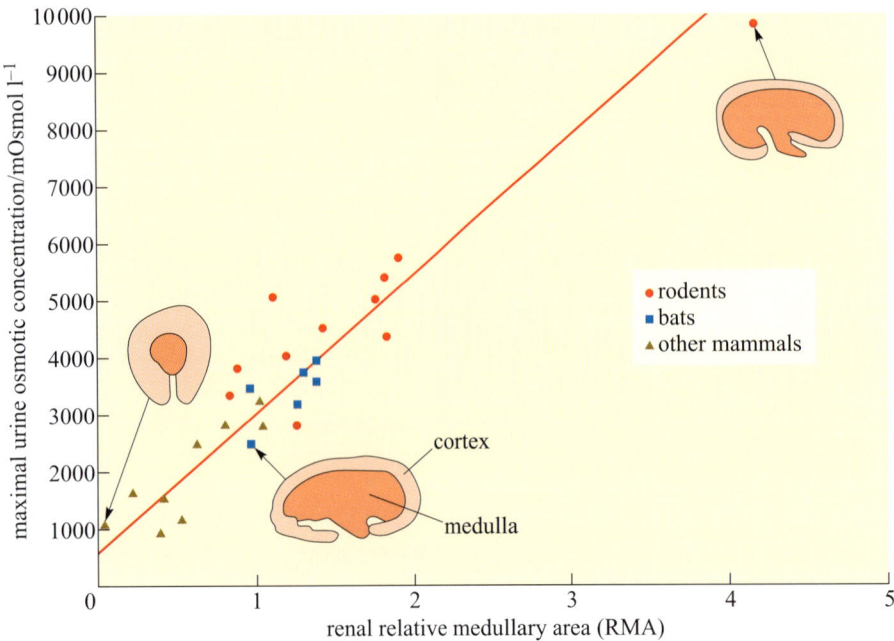

Figure 3.31 The relationship between relatively medullary area in the mammalian kidney (taken at the midline in sagittal section) and the maximal urine concentration that can be produced. Inserts show diagrammatic cross-sections for three kidneys (not to scale).

■ What conclusions could you draw from the data shown in Figure 3.31?

Rodents generally have kidneys with a larger medullary area and produce more concentrated urine than bats and other mammals.

The thicker medulla of small desert rodents could therefore be viewed as a desert adaptation superimposed on a basic body-size-dependent pattern. Most loops of Henle in desert rodents are of the juxtamedullary type, and the epithelial cells have densely packed mitochondria with more cristae per unit volume than a horse's loop of Henle.

■ What is the significance of the greater concentration of mitochondria and more cristae per unit volume of mitochondria in the epithelial cells of loops of Henle in desert rodents compared with those of the horse?

The greater numbers of mitochondria and cristae in epithelial cells of the loops of Henle of desert rodents suggest a higher capacity for ATP synthesis and therefore active transport of Na^+ and Cl^- ions in the kidney of desert rodents.

Conservation of water by the kidney is of crucial importance for the kangaroo rat, which does not drink and can obtain water only from catabolism.

Other desert rodents obtain water from their diet. The degu (*Octodon degus*), found in Northern Chile, lives in semi-arid desert country, known as matorral, which is characterized by evergreen scrub plants. Degus survive on limited amounts of water obtained primarily from their food, which comprises scrub foliage, grass and seeds. There is seasonal variation in the water content of plants; in summer the plant foliage dries out and contains just 3–6% water; in winter, foliage contains 70–80% water. Bozinovic et al. (2003) studied the phenotypic flexibility of water flux rate in *Octodon degus*. Water intake and efflux were measured by use of the doubly-labelled water technique (Section 1.6; Section 2.4.2) in degus kept in a secure enclosure within the matorral. Urine osmolality* was measured in wild-captured degus using microhaematocrit capillary tubes to obtain samples from the urethra.

Table 3.4 Measurements of body mass, water intake and urine osmolality in degus in winter and summer in Chile (data from Bozinovic et al., 2003).

Measurement	Winter (June–August)	Summer (Dec–March)
Mean rainfall/mm	245	12
Body mass/g	119.7	124.8
Water intake/ml day^{-1}	40.4 ± 9.1*	10.3 ± 2.3
Urine osmolality/mOsmol kg^{-1}	1123 ± 472*	3137 ± 472

*Difference between means ± SD for winter and summer statistically significant.

■ Drawing on the data provided in Table 3.4, summarize the physiological strategy for water economy in the degu.

In winter when water content of plants is 70–80%, the rate of water intake is relatively high at 40.4 ml day^{-1}. Urine osmolality is correspondingly low at 1123 mOsmol kg^{-1}. In contrast, in the summer the rate of water intake is relatively low at 10.3 ml day^{-1} and the degus produce a more concentrated urine, with an osmolality at 3137 mOsmol kg^{-1}. The kidney is able to concentrate urine, thereby reducing water loss in the summer when the diet provides very little water.

Bozinovic et al. interpret the ability of the kidney of the degu to concentrate urine to 3137 mOsmol kg^{-1} as an example of phenotypic flexibility in the degu, in response to a lack of water during the summer. Variation in the osmolality of urine is not in itself unusual. After drinking a large volume of water, humans produce a dilute urine; the average osmolality in water-loaded volunteers has been measured at 101 mOsmol kg^{-1}. Following 20 hours of dehydration, urine osmolality in the volunteers increased to 1004 mOsmol kg^{-1}. Such responses result from the physiological regulation of body water content. Recall that the permeability of the epithelium of cortical and medullary collecting ducts is controlled by the hormone ADH (antidiuretic hormone, also known as vasopressin). (Figure 3.39 in Section 3.3.3 shows the feedback control of secretion of ADH, which results in the regulation of body fluid volume.)

3.3.2 Integration of anatomical features and biochemical and physiological strategies in evaporators

Birds and larger desert mammals that use evaporative cooling risk dehydration because of the difficulty of finding sufficient drinking water. For mammals, evaporative heat loss includes panting and sweating.

---

* Whereas osmolarity measures the number of osmotically active particles of a particular substance in a volume of fluid, *osmolality* measures the equivalent number in a mass weight of fluid. For most biological systems the molarity and molality of a solution are nearly exactly equal. For our purposes osmolarity and osmolality can be regarded as equivalent.

In small mammals and birds the temperature of exhaled air is often lower than T_b, resulting in condensation of water on the nasal mucosa. Small desert mammals rely on this mechanism for water conservation, while resting in their cool burrows during the heat of the day. However, for mammals and birds exposed to high T_a, the nasal counter-current heat exchanger minimizes water loss, and so works against the need to increase heat loss by evaporation of water (Figure 3.32).

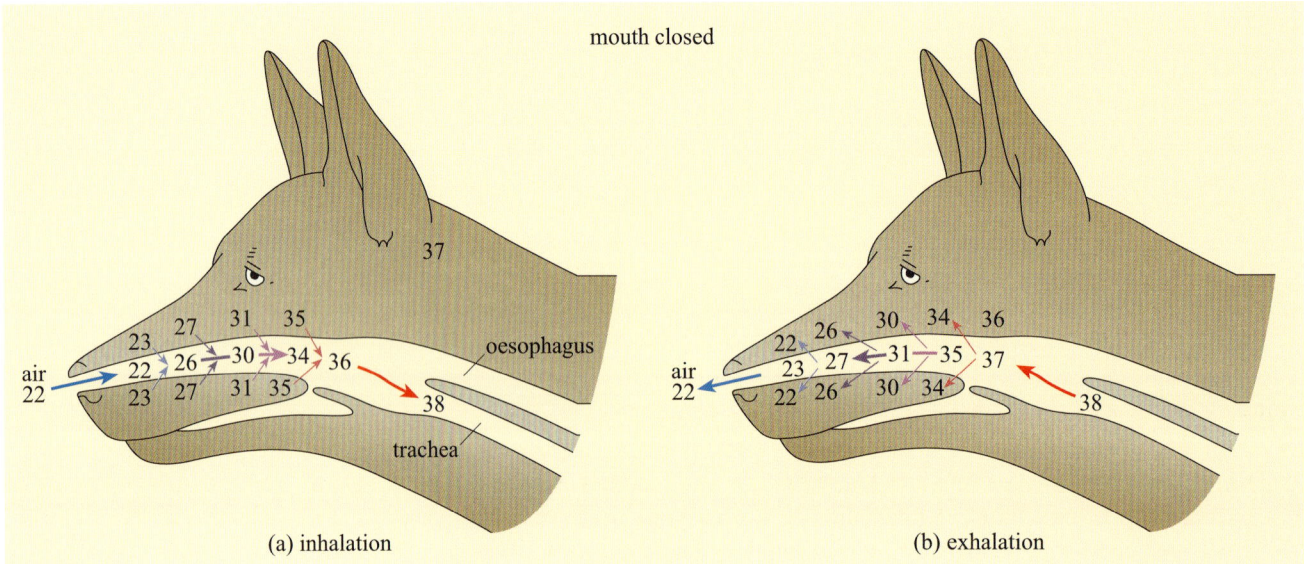

Figure 3.32 The operation of the nasal counter-current heat exchanger. Temperatures (°C) within the dog's nasal passages and at the mucosa during inhalation and exhalation indicate the conservation of heat when the mouth is closed. The small arrows show the direction of heat transfer from the mucosa to the air on inhalation (a) and in the reverse direction at exhalation (b).

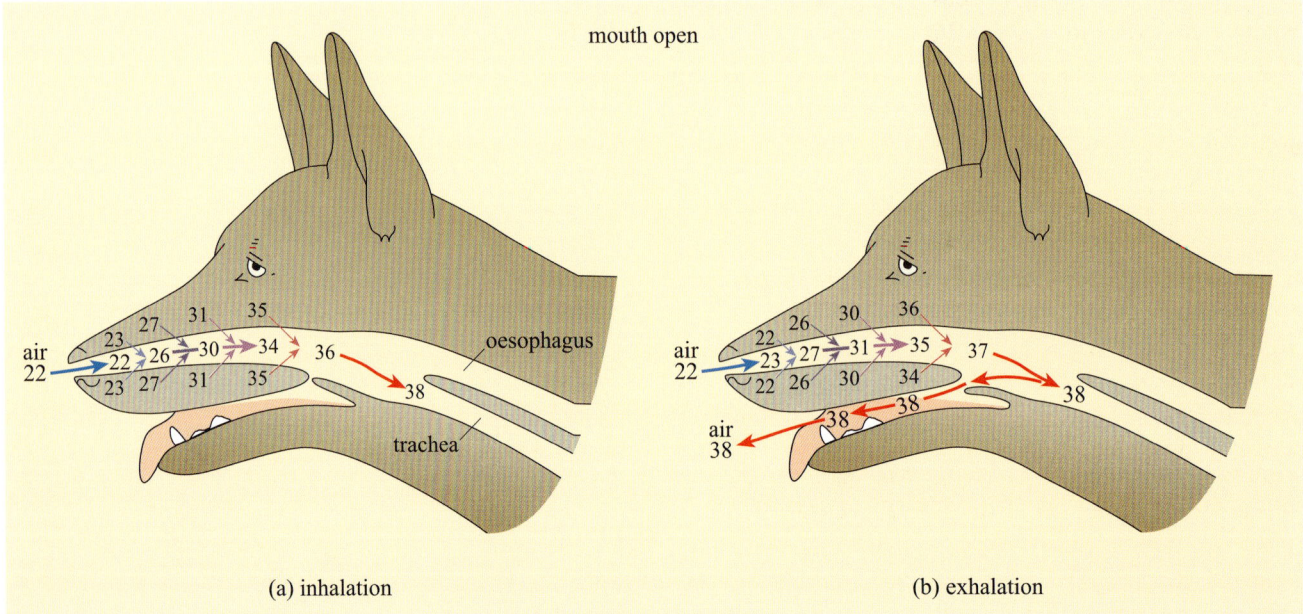

Figure 3.33 Bypass of the dog's nasal counter-current heat exchanger by opening the mouth, an indication of how heat loss is enhanced. Figures are temperatures in °C.

Only when T_a approaches T_b and the temperature of the inspired air reaches that of the body core, is the heat exchanger abolished. Even then, and for dry air, only about 12% of metabolic heat is dissipated at T_b of 38 °C. As humidity increases, the proportion of metabolic heat dissipated declines.

Increasing the rate of ventilation of the nasal mucosa increases the rate of evaporation, but risks over-ventilating the lungs and blowing out too much carbon dioxide.

In dogs, foxes and other species that pant, evaporative cooling is promoted by opening the mouth; a simple mechanical device further increases the effectiveness of cooling by respiratory water loss. A valve at the back of the throat, driven by breathing movements, directs a large proportion of the air that was inhaled through the nose out through the mouth, thereby bypassing the nasal heat exchanger (Figure 3.33).

This mechanism can be used to modulate the rate of evaporative heat loss without affecting the respiratory frequency or volume and is exploited to the full during thermal panting under most heat loads.

During severe heat stress, breathing changes to a slower deeper second-phase panting, in which air passes out through nose and mouth. The dog's tongue is richly vascularized and the rate of blood flow increases with a rise in body temperature. Exposure of the buccal area means that there is no significant heat exchange in the mouth (unlike that shown in Figure 3.32) so the rate of evaporation is maximized during second-stage panting. Second-stage panting normally occurs in the dog during exercise, when both enhanced cooling and increased gaseous exchange in the lungs are required.

Panting is an important cooling mechanism for foxes and dogs that chase prey. The fennec fox (*Fenecus zerda*), a species found in the Sahara desert, is reputed to pant at 690 times per minute after chasing prey. Kit foxes reduce the need for panting by staying in dens during the day and hunting at night, or at dawn and dusk (Section 3.2.3). Rüppell's foxes (*Vulpes rueppelli*; Figure 3.34) live in the Rub' al-Khali of Arabia, the largest existing sand sea, which is an extremely arid desert with no permanent sources of drinking water.

Rüppell's foxes do not drink, but obtain all their pre-formed water from their food, supplemented by metabolic water production. By avoiding the need for panting during the day, Rüppell's foxes might be expected to have a reduced total evaporative water loss (TEWL) in comparison to fox species living in mesic habitats. Resting in a den during the day would reduce TEWL, but Rüppell's foxes would have to travel long distances at night while hunting prey, mainly rodents, birds and arthropods which would increase the need for evaporative cooling. Williams et al. (2002) measured TEWL of individual foxes in a specially designed metabolic chamber. Field water flux (the water flux under natural conditions in the field) was determined in individual foxes using the doubly-labelled water technique. The mean whole body TEWL for six foxes at 35 °C at their basal metabolic rate was measured as 50.1 g water day^{-1}. This value is about 55% lower than expected from comparisons with other mammalian species of similar body mass. The researchers suggest that Rüppell's foxes are particularly efficient at reducing either cutaneous or respiratory water loss. Mean field water flux (FWF) per day in the six foxes was 123 ml day^{-1} with 26.1 ml water day^{-1} provided from catabolism.

Figure 3.34 Rüppell's fox (*Vulpes rueppelli*).

Comparison of mass-adjusted values for water flux in Rüppell's foxes with values obtained for swift foxes living in grass prairie where water is more readily available, showed that water flux is 34% greater in this mesic species. However, care is needed when making such comparisons between just two species, as we shall see in Section 3.4. The water flux of Rüppell's foxes is about 30% less than that predicted by physiologists for a desert mammalian carnivore; the prediction assumed that desert carnivores would have higher rates of water flux than mesic species because of their higher rates of TEWL. It is tempting to suggest that the nocturnal habits of Rüppell's fox with consequent reduction in TEWL account for the low TEWL in this species.

Because birds of all sizes tolerate hot arid conditions, physiologists considered that desert birds, being diurnal animals exposed to extremes of ambient temperature and aridity in deserts, are successful because of their avian physiology, not because of specific adaptations. For example, as the normal range of core body temperature in birds (41–42 °C) is higher than that in mammals, the need for evaporative cooling may not be as great as that in mammals. Because birds are **uricotelic**, that is, they excrete uric acid rather than urea, relatively little water is required for the excretion of nitrogenous waste. Uric acid is excreted as a paste, with a very low water content. It is relevant to note that carnivores, which have a high protein diet, produce relatively large quantities of urea as a waste product. This urea increases the osmolarity in the kidney and helps reduce water loss via the urine.

However, recent work suggests that specific adaptations for life in hot and dry desert climates may have evolved in desert birds. Tieleman and Williams (2000) compiled available values for BMR, field metabolic rate (FMR) and field water flux (FWF), for 21 small bird species living in Old World deserts, and compared these data with the equivalent values for 61 species living in mesic habitats with higher rainfall and denser vegetation cover. The desert species included desert larks, sparrows and finches; the mesic species included owls, finches and sparrows, so there was a wide spread of groups. Two different methods of analysis were used, one for simple comparisons between the two groups, and the other based on phylogenetic contrasts involving the inclusion of phylogenetic relationships in the analysis. Significant differences in BMR and FMR between desert and mesic birds support evolution of reduced BMR and FMR in desert species. Low BMR and FMR might be expected to be an advantage for desert birds because of the associated lower energy demand, and lower release of metabolic heat, and hence lower TEWL, required for dissipating metabolic heat at high T_a. Low FMR values can be linked to the habit in many desert species of resting in the shade or in burrows during the hottest part of the day (recall dune larks from Section 3.2.3). Values for FWF were significantly reduced in desert birds in comparison to mesic birds, but the difference was not related significantly to phylogeny.

Desert birds use panting for cooling, thereby incurring increased evaporative water loss. Female dune larks incubating their eggs pant during midday to regulate their own body temperature and hence that in their eggs. Desert grouse (*Pterocles* spp.), use **gular flutter**, a rapid vibration of the floor of the mouth that provides rapid evaporative heat loss with up to 2 °C cooling in the mouth. *Pterocles* spp. can afford to lose water in this way, as these birds fly long distances every day to drink water from pools. Other birds such as desert larks do not show this behaviour and they may rely entirely on water obtained from their food, so they could not afford to lose so much water by evaporation.

Caution is advisable when using interspecific comparisons to support the view that physiological traits are adaptations. Physiological traits measured in species at different times of year or in different areas may vary, not because of genetic differences, but because of acclimatization. A study of phenotypic flexibility of BMR and TEWL in 12 hoopoe larks (*Alaemon alaudipes*; Figure 3.35), captured from the Arabian desert provides a salutary example.

Two groups of six wild-captured larks were kept respectively at ambient temperatures of 15 °C and 36 °C, fed *ad libitum* and exposed to 12-hour day and 12-hour night regimes. After 3 weeks of acclimation, each bird was placed in a metabolic chamber at 35 °C, a temperature within the thermoneutral range for the hoopoe lark. BMR was measured as the basal rate of oxygen consumption, and TEWL determined from water content of air expired from the chamber. The results are summarized in Table 3.5.

Figure 3.35 Hoopoe lark (*Alaemon alaudipes*).

Initially there was no significant difference ($P > 0.25$) in the mean body mass of the two groups of larks. After 3 weeks of acclimation, mean body mass of the group acclimated at 15 °C was significantly higher than that of the group acclimated at 36 °C ($P < 0.04$).

Table 3.5 Mean values ± SD for BMR and TEWL measured in two groups of hoopoe larks acclimated at 15 °C and 36 °C (data compiled from Williams and Tieleman, 2000).

T_a for acclimation	Body mass pre-acclimation /g	Body mass post-acclimation /g	BMR/kJ day^{-1}	BMR/kJ day^{-1} g^{-1}*	TEWL at 35 °C/ g H$_2$O day^{-1}	TEWL at 25 °C/ g H$_2$O day^{-1}
15 °C	41.3±7.0	44.1±6.5	46.8±6.9		3.55±0.60	3.11±0.4
36 °C	37.2±4.7	36.6±3.6	32.9±6.3		2.23±0.28	2.17±0.7
P for difference between means	>0.25	<0.04	<0.03		<0.003	<0.008

*This column to be completed as part of in-text question below.

■ Compare the mean values for whole body BMR in the two acclimated groups of larks.

Larks acclimated at 15 °C had a greater mean BMR, 46.8 kJ day^{-1}, than the mean BMR, 32.9 kJ day^{-1}, measured for birds acclimated at 36 °C. The increase in BMR is statistically significant ($P < 0.03$).

The BMR of hoopoe larks acclimated to T_a = 15 °C approaches that reported for a temperate species, the woodlark (*Lullula arborea*): 49.4 kJ day^{-1}.

■ You may argue that BMR expressed as kJ day^{-1} g^{-1} is likely to be the same for the two groups of hoopoe larks because body mass for the 36 °C group is greater than that of the cold-acclimated group. Fill in the values for BMR as kJ day^{-1} g^{-1} in the empty column in the table and state whether the BMR expressed per gram is still lower in the warm-acclimated larks.

Mean BMR value, 0.89, for the 36 °C group is still lower when expressed as kJ day^{-1} g^{-1} than the equivalent mean value for the 15 °C group, 1.06 kJ day^{-1} g^{-1}.

■ Compare the mean values for TEWL in the two acclimated groups of larks.

Larks acclimated at 15 °C had a greater mean TEWL at 35 °C, 3.55 g day^{-1}, than the mean TEWL 2.23 g day^{-1}, measured for birds acclimated at 36 °C. The increase in TEWL is statistically significant ($P < 0.003$). Similar results were obtained for TEWL measured at $T_a = 25$ °C.

Those larks acclimated to 15 °C had significantly larger liver, kidney and intestine than larks in the 36 °C group. Birds in the 15 °C group consumed about 420 g food per day, more than three times as much as the 120 g food per day eaten by the birds kept at 36 °C. The overall picture is that the hoopoe lark has high phenotypic flexibility, an advantageous feature for an animal living in a very variable environment. The environment of the Arabian desert has long periods of drought with scarce food resources available but unpredictable periods of rain temporarily increase food supply. The ability to minimize energy expenditure and requirement for water is important for survival of the birds. You may argue that the reduction in TEWL at 36 °C derives from a lower BMR and therefore reduced respiratory evaporative water loss (REWL). Williams and Tieleman (2000) determined that REWL accounts for 31.7% of TEWL at 35 °C with cutaneous evaporative water loss (CEWL), accounting for the remaining 68.3%. If it is assumed that the increase of 42.2% in BMR in 15 °C-acclimated birds results in an equal increase in REWL but no increase in CEWL, then TEWL would have increased by 13.4%. The finding that TEWL increased by 59.2% in the birds in the cold-exposure group indicates that they altered the permeability of their skin to diffusion of water vapour. Williams and Tieleman suggest that desert birds may reduce their CEWL by increasing their skin resistance, either by varying diffusion path length across the skin, or by altering the permeability of skin to water vapour. Diffusion path length can be reduced by vasodilation of subcutaneous capillary beds. Changes in lipid of the skin and increased epidermal thickness may reduce permeability of the skin to water in desert birds.

Studies on small desert birds suggest that lower BMR, FMR and TEWL are typical physiological responses to hot arid environments. Williams and Tieleman suggest that reduced BMR and TEWL may have evolved in desert birds, even though phenotypic adjustments in these physiological variables may be considerable, as demonstrated by their work on hoopoe larks. Whether reduced BMR and TEWL in desert birds results from physiological acclimation, phenotypic plasticity (Section 2.10.2) or an inherited feature is investigated further in Section 3.5. We should also be aware that phenotypic plasticity and the ability to acclimatize physiologically are under genetic control.

3.3.3 Integration of anatomical features and biochemical and physiological strategies in endurers

The endurers, large animals with a relatively low surface area: volume ratio, have problems in losing heat from the body when exposed to high T_a. Certain large lizard species behave like endurers, but they are evaders and evaporators too, a salutary reminder that we should not apply classification criteria too rigidly.

Dipsosaurus dorsalis, the desert iguana (Figure 1.5), lives in the Sonoran desert and is found most commonly in dry sandy areas where creosote bushes grow (Section 3.1).

Dipsosaurus is a plant eater, and it feeds on creosote bush leaves and flowers during the day, often being exposed to high solar radiation for up to 45 minutes.

The species was a puzzle to physiologists because it can attain a T_b of up to 46 °C, without any apparent ill-effects.

- Why would a T_b of 46 °C appear to be incompatible with life in a complex vertebrate species such as *Dipsosaurus*?

The complex globular and fibrous proteins of most vertebrates are denatured at temperatures greater than 40 °C.

Comparison of ATPase activity at different temperatures in a number of different lizard species, showed that for a temperate species, *Gerrhonotus multicarinatus*, optimal temperature for ATPase activity is 30 °C whereas for *Dipsosaurus*, it is about 41 °C (Figure 3.36) and is still functional up to 43 °C or 44 °C.

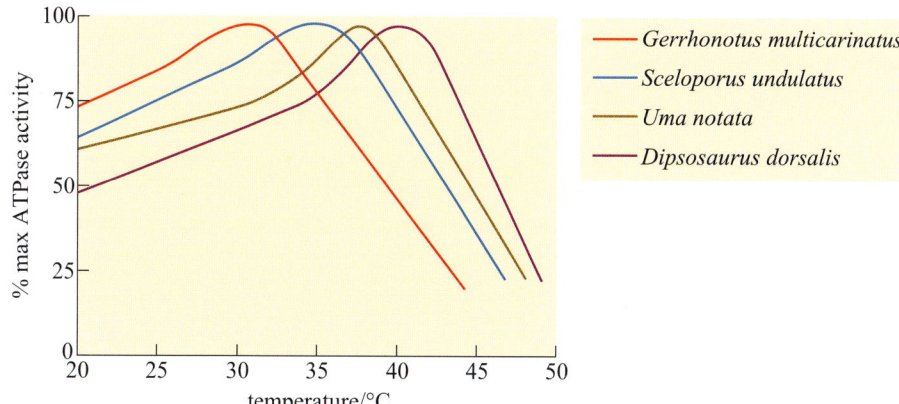

Figure 3.36 ATPase activity plotted against temperature for four species of lizard.

Since *Dipsosaurus* has enzymes that are stable even at high T_b it needs to expend less energy for thermoregulation, e.g. by shuttling, and can forage for longer during the day. By foraging during the day, *Dipsosaurus* avoids nocturnal predators such as foxes. *Dipsosaurus* relies on the moisture content of its diet for water. Reptiles do not have loops of Henle, but most are uricotelic, excreting uric acid and urate salts, not urea, which means that little water is excreted in these end products of protein metabolism.

Reptiles have some degree of physiological capacity to control their rate of change of body temperature. In *Dipsosaurus dorsalis*, radiant heating results in local cutaneous vasodilation, and local cooling results in cutaneous vasoconstriction. Panting and gaping have been observed in heat-stressed lizards such as *Dipsosaurus dorsalis* and *Sauromalus obesus*. When T_b of *Sauromalus* exceeds 40 °C, the mouth gapes, there is a five-fold increase in breathing rate, lung tidal volume decreases by two-thirds, and TEWL increases. Panting has significantly greater cooling effects on the brain than on the rest of the body, probably because of the proximity of the mouth to the carotid arteries in the neck that supply blood to the brain.

The best-known examples of desert endurers include the camel, the oryx and desert sheep; in fact most desert endurers are large mammals. Their relatively low surface area: volume ratio means that they have more difficulty than small animals in losing heat from the body at high T_a. Mammalian and avian endurers are too large to shelter in burrows, and if no shade is available, they may be forced to remain exposed to solar radiation during the day. The hair of large

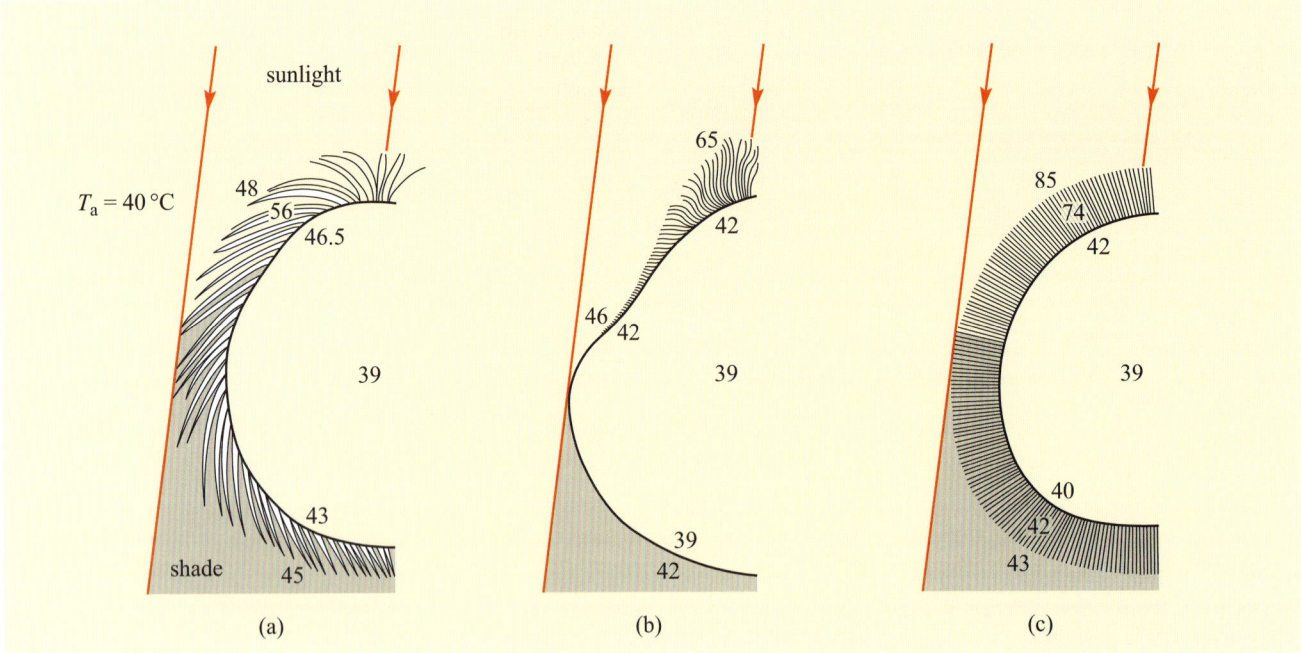

Figure 3.37 The gradients of temperature across the coat of (a) open-woolled Awassi sheep, (b) short-coated camels and (c) Merino sheep with dense fleece. All the animals are exposed to sunlight as shown and T_a of 40 °C.

desert mammals can play an important role in insulation, both from solar heat and nocturnal cold. Figure 3.37 compares the thermal properties of the coats of two breeds of heat-tolerant sheep with that of the camel.

For each example, the raised temperatures of the body core to 39 °C suggest that some heat is being stored during the day. The long loose coat of the Awassi sheep is penetrated by solar energy which heats up the middle layers, making the skin quite hot. Merino sheep have a dense fleecy coat, and lose long-wave radiation from the hot tips of the hairs, maintaining a gradient of up to 43 °C across 4–5 cm of fleece so the skin is protected from overheating. In short-coated camels, a dorsal ridge of long dense hair provides shading and insulation for the skin, while all of the coat, most of it short and smooth, reflects solar energy. Sweating keeps the camel's skin cool.

Sweating, an extreme form of CEWL in mammals, is important for cooling in many species. Sweating is the secretion of water plus some salts from special sweat glands in the skin, which occurs as a response to an increase in T_b. The glands are of two types. Atrichial (without hair) glands are found in primates and also on the pads of cats and dogs. They develop from the epidermis independently of the hairs and open on the free surface of the skin. Atrichial sweat glands are at their densest on human palms and soles, and elsewhere on the body they are at a density of 100–300 cm^{-2} (Figure 3.38a). Epitrichial (around hair) sweat glands develop only in association with hair follicles (Figure 3.38b). They are found in many mammalian species including cattle, sheep, horses and camels.

Epitrichial sweat glands play an important role in thermoregulation in these species. In cattle there are about 1800 cm^{-2}, and in sheep, about 300 cm^{-2}. Sweat is an ultrafiltrate of plasma, containing sodium chloride and other salts, lactic acid and urea. As the sweat glands absorb much of the electrolytes, sweat is hypotonic to plasma; the salt content of sweat falls with acclimatization. Evaporation of sweat, promoted by input of heat energy from the skin, cools the body. One

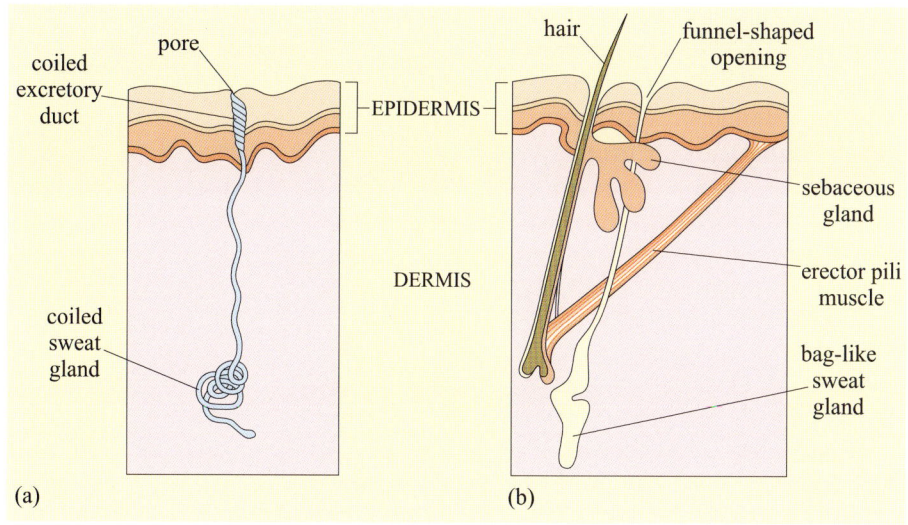

Figure 3.38 The main structural features of sweat glands known to function in temperature regulation. (a) A human atrichial gland. (b) An epitrichial gland of the ox with its accompanying structures.

problem for thermoregulation by sweating in furry mammals is that water evaporating from a fur coat takes a significant proportion of its heat from the air rather than from the skin, and is therefore less effective in cooling the body.

Sweating has a physiological cost in that it involves loss of salts and organic molecules as well as water from the body. Initially fluid lost from the plasma is replaced from various 'non-essential' reserves, such as the digestive glands and the gut, but eventually the osmotic pressure of the blood increases. The change is detected by special neurons, osmoreceptors, in the hypothalamus. In turn the osmoreceptors stimulate release of ADH from neurosecretory neurons in the hypothalamus. Secreted ADH enters the capillary network supplying the posterior pituitary, and from there the hormone is secreted into the bloodstream. ADH binds to receptors in cells lining the collecting ducts of the kidney and promotes resorption of water back into the bloodstream (Figure 3.39). Later on some fluid is withdrawn from the intracellular compartments of the body and there is an inevitable reduction in plasma volume if the lost water is not replaced by drinking.

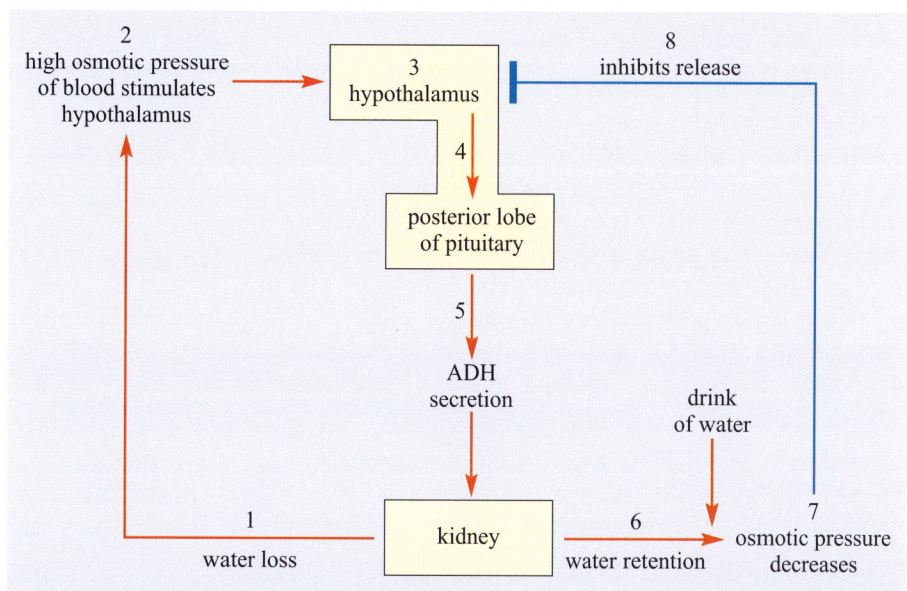

Figure 3.39 Feedback control of the secretion of antidiuretic hormone, ADH. Secretion of the hormone from the posterior pituitary is controlled by the hypothalamus responding to signals received from its receptors, which detect the osmotic pressure of the blood.

It is worthwhile working through the series of events in Figure 3.39. Stage 1 represents a situation where water lost via urine is not replaced by drinking water. The osmolarity of the blood increases, which raises the osmotic pressure of the blood. At stage 2, osmoreceptors in the hypothalamus detect the increased osmotic pressure of the blood. Hypothalamic neurosecretory neurons (stage 3) respond by transferring ADH along axons into the blood circulation of the posterior pituitary (stage 4). ADH secreted into the bloodstream (stage 5) reaches target cells, epithelial cells in the collecting ducts of kidney. The response to ADH is an increased permeability of the basal membranes of epithelial cells in the collecting ducts, which promotes resorption of water. The kidney thereby produces more concentrated urine and water is retained (stage 6). Say the individual now has a drink of water. Absorption of the water into the bloodstream decreases the osmolarity of the blood so the osmotic pressure decreases (stage 7). In stage 8, the hypothalamic receptors respond to the decline in osmotic pressure by inhibiting the release of ADH.

■ Is the control of ADH secretion an example of negative or positive feedback?

It is negative feedback because an increase in the secretion of ADH leads to a change in osmotic pressure, which in turn decreases ADH secretion.

The hormonal control of physiological variables by means of negative feedback loops maintains a constant level of body fluid volume, an important element of homeostasis, the maintenance of a constant internal environment.

Unfortunately in humans, the plasma contributes a disproportionate share of the overall fluid lost. A level of dehydration equivalent to a 4% loss of body weight is accompanied by a reduction in plasma volume of about 10%. Since there is no corresponding fall in blood cells or plasma proteins, the viscosity of the blood rises just at the time when the circulatory system, which bears the main burden of maintaining the body's heat balance, is already under strain as a result of the maximal cutaneous vasodilation that transports heat to the surface and plasma to the sweat glands. The demand on the heart is therefore great as it tries to maintain the blood pressure and peripheral circulation. Hypotension (low blood pressure) and fainting may follow. If correct action is not taken quickly, the result is a rapid rise in body temperature and death through hyperthermia.

The amount of water lost during sweating in humans can be considerable. A man marching in the desert may take on a solar and terrestrial radiation load of 2000 kJ h^{-1} even allowing for the reflectance of his skin and clothing. On the basis of a latent heat of vaporization of 2.4 kJ g^{-1}, about 800 cm^3 of sweat would be needed per hour to dissipate this heat load even on the assumption that all the heat of vaporization is drawn from the body itself. This calculation does not take into account any direct heat loading from air hotter than the body, or the substantial metabolic generation of heat. Therefore, figures of 2 or 3 litres of sweat per hour are feasible. An adult man carrying out moderate work in hot dry conditions may secrete 0.5 litres of sweat per hour and when under high heat stress, up to 3 litres per hour. The efficiency of the human sweating mechanism depends critically on the ability to drink. However, under prolonged heat stress, even if water is available, the limited capacity of humans for physically drinking water leads to a 'voluntary' dehydration of 2–4% of the body mass, producing a chronic condition of decreased plasma volume and a raised blood osmotic pressure, coupled with a fall in urine output of up to 80%.

The camel handles its water balance problems more effectively than humans. Schmidt-Nielsen et al. (1967) showed that in a 290 kg camel, a 17% weight loss due to dehydration was accompanied by only an 8.8% reduction in plasma volume and a 38% fluid loss from the gut. A camel has up to 75 litres of fluid in its rumen (85% of it is water) and another 8 litres in the intestine. The 38% of water loss in the gut is therefore about 30 litres, which minimizes strain on the blood circulation during dehydration. Overall, camel tissues seem to be more resistant to high osmotic pressures than those of many animals. The camel can cope with up to 30% water loss. When provided with water after such a high level of dehydration, the camel can drink rapidly taking in up to 200 litres of water in just a few minutes. Much of the water taken in is stored temporarily in the gut, preventing excessive dilution of the blood which in itself would be harmful.

Schmidt-Nielsen et al.'s research added detail to the common perceptions that camels store free water in the rumen and utilize water derived from metabolism of the lipids released from the adipose tissue that forms the fatty hump. Although camels have a great deal of water in the rumen and intestine, it is proportionately no more than is present in other ruminants. It is true that since the oxidation of 1 g of fat yields 1.07 g water, a 40 kg hump could yield 43 litres of water that can be drawn on during a long journey. Schmidt-Nielsen pointed out that as this mechanism requires oxygen, which can only be obtained by ventilating the lungs, there must be a net loss of water from the respiratory tract when the camel breathes dry air. Whether the fat reserves make a positive or negative contribution to the animal's overall water balance depends on conditions prevailing in the upper respiratory tract.

Research by Schmidt-Nielsen and his colleagues investigated this question. Working on feral camels under natural conditions in central Australia, the researchers measured body temperatures, oxygen consumption and respiratory frequency and volume, as well as the temperature and relative humidity of the inspired and expired air by means of sensors inserted loosely into a nostril. The results showed that during the daytime, severely dehydrated camels exhaled air that was at or near the core temperature and fully saturated with water vapour. At night, however, there was a marked difference, the exhaled air being at or near the ambient temperature with a relative humidity of about 75%.

We can best gain an impression of the camel's water balance by taking an example which uses hypothetical but realistic figures. Consider a camel resting at night in air at 28 °C and 40% r.h. Every 2 minutes it metabolizes 1 litre of oxygen, extracted from 20 litres of inhaled air. This volume of air, under these conditions, contains 216 mg water. This same volume exhaled at core temperature (35 °C) and fully saturated, contains 784 mg water, an overall loss to the camel of 568 mg. If 20 litres of saturated air were exhaled at ambient temperature (28 °C, water content 538 mg), the loss would be reduced to 322 mg, while 20 litres of exhaled air at 28 °C and only 75% r.h. (the humidity measured by Schmidt-Nielsen) would have a water content of 403 mg, further reducing water loss to 187 mg. The maximum saving of water could then be calculated as follows:

$$\text{saving} = \frac{\text{mass } H_2O \text{ in saturated air at } 35\,°C - \text{mass } H_2O \text{ in exhaled air}}{\text{mass } H_2O \text{ in saturated air at } 35\,°C - \text{mass } H_2O \text{ in inhaled air}} = \frac{(784-403)}{(784-216)} = \frac{381}{568} \times 100 = 67\%$$

Although this mechanism only seems to operate at night, it clearly makes an important contribution to water conservation in the dehydrated camel and probably ensures that there is a gain of water from the oxidation of fat.

Camels also conserve water by reducing urine flow, and can do so to a far greater extent than humans. The camel kidney can produce urine twice as concentrated as that of humans (Table 3.3), and can reduce the amount of urea it excretes. According to Schmidt-Nielsen, urea in the blood may be secreted into the rumen, where bacteria incorporate the nitrogen into amino acids and then into protein. This protein is later digested and absorbed, and some of it is deaminated in the liver, releasing urea back into the blood again. This cycle may help to retain nitrogen, and hence the water that would have been needed to excrete it as urea, within the body during periods of water shortage.

Relaxed homeothermy is important for water conservation in the camel, which uses the capacity of its immense bulk (up to 500 kg) to store heat. If water is short in the summer, the camel may maintain a normal core temperature throughout the night and then at around 06.00 h, allow its body temperature to fall to about 34 °C. Throughout the day, its temperature rises due to muscular work, solar radiation or both, but the camel does not begin thermoregulating by sweating until the rectal temperature reaches 40 °C. Thus the camel absorbs sufficient heat to raise its body temperature over a range of 6 °C, compared with only 2 °C in humans. By this means the camel may save about 5 litres of sweat in a day, which is significant, considering that a 500 kg animal resting in the sun loses a total of about 10 litres of water per day by sweating, breathing and excreting. There is no evidence that the camel's lethal body temperature is significantly different from that of humans, but it certainly is more tolerant of changes in body temperature.

Overall the anatomy, physiology and biochemistry of the camel are geared towards conservation of water. Camels survive without drinking for up to 6 weeks during colder periods but in summer, they must drink every 4 days at least. In contrast, Arabian oryx (*Oryx leucoryx*; Figure 3.5) live in the Arabian desert, with no access to free-standing water. A group of wild oryx living in Mahazat as-Sayd, a nature reserve in central Saudi Arabia, have been studied extensively by Williams et al. (2001). Mahazat as-Sayd has no standing water apart from puddles after infrequent rain showers. Plants, mostly grasses and small trees (*Acacia tortillas* and *Maerua crassifolia*), cover only about 21% of the reserve; the rest is sand. Summers are hot, with maximum and minimum temperatures of 41.5 °C and 24.5 °C, respectively, and mild winters, with maximum and minimum temperatures of 23.4 °C and 10.6 °C. The annual rainfall is low, and falls mainly in winter: typical annual rainfall values were 129.6 mm in 1996, 84.3 mm in 1997. Tracking and observations of oryx showed that they feed on three species of grasses (Table 3.6).

Table 3.6 Percentage water content of grasses (means ± SD) in the diet of Arabian oryx (data from Williams et al., 2001).

Species	Summer 1998: % water content of grasses				Spring 1999: % water content of grasses			
	June	July	August	Mean	February	March	April	Mean
Stipagrostis sp.	4.6±2.5	12.7±10.5	4.4±0.4	7.2	38.6±8.1	45.7±4.9	34.5±8.5	39.6
Panicum turgidum	45.4±3.4	44.5±5.3	41.4±5.5	43.8	47.1±4.9	51.9±5.5	49.5±4.4	49.5
Lasiurus scindicus	35.7±5.3	30.4±5.6	27.0±14.5	31.0	40.5±8.2	55.6±9.5	52.5±5.9	49.5

BMR and TEWL at 30 °C were measured in oryx resting in specially constructed metabolic chambers (Table 3.7a). FMRs and field water influx rates (FWIR) in wild Arabian oryx living at Mahazat as-Sayd were measured by use of the doubly-labelled water technique (Table 3.7b). FWIR give a measure of the water intake of the animal. In the summer, six oryx returning in the morning from night foraging to rest in shade were darted with anaesthetic, then weighed and injected with doubly-labelled water, $^2H_2^{18}O$. Blood samples were taken on the injection day and at a mean of 8 days post-injection. The results of assays of ^2H and ^{18}O content of blood and fecal samples were used to calculate rates of water intake and field metabolic rates for each animal (Table 3.7). The mean body mass of 12 oryx studied in 1998–1999 was 85 kg. In summer 1998, when grasses were parched, field metabolic rates of free-ranging oryx were $11\,076 \pm 3070\,kJ\,day^{-1}$ in contrast to $22\,081 \pm 3646\,kJ\,day^{-1}$ in spring 1999, after rains. The difference between metabolic rate in summer and spring was significant ($P < 0.0001$) and Williams et al. suggest that oryx cut their energy expenditure in summer by changing both behaviour and physiology. In summer, oryx forage at night and rest during the day. In spring, following the rains, increased energy expenditure may derive from walking longer distances for foraging, and also the costs of thermoregulation because of reduced T_a.

■ Compare the mean water intake rates in the oryx in spring and summer, and state whether you think the differences you identify are statistically significant.

In spring, mean water intake is much greater, $3438 \pm 1006\,cm^3\,day^{-1}$ compared with $1310 \pm 1019\,cm^3\,day^{-1}$ in the summer. The difference in water intake certainly looks statistically significant (in fact, it is, $P < 0.007$).

■ Looking at the data in Tables 3.6 and 3.7, can you suggest a reason for the difference in water intake rates for oryx in spring and summer?

As no free-standing water is available to the oryx they must have obtained all or most of their water from their diet. Water content of grasses eaten by oryx varied significantly between summer and spring (Table 3.6). Water content of *Stipagrostis* was just 7.2 % in summer 1998, but in spring 1999 rose to 39.6%. Water content of *L. scindicus* rose from 31% in summer 1998 to 49.5% in spring 1999. Water content of *P. turgidum* was similar in spring and summer. As oryx only obtain water from their diet, presumably the lower water content of two of the food species determines a lower water intake in the summer.

Table 3.7 (a) Values of BMR and TEWL for oryx in metabolic chamber at 30 °C and (b) values of FMR and FWIR in free-ranging oryx calculated from the doubly-labelled water technique. Values are means ± SD.

Number and sex of animals	Mean body mass/kg	BMR/kJ day^{-1}	TEWL/g H$_2$O day^{-1}	FMR/kJ day^{-1}	FWIR/cm^3 day^{-1}
(a) 6 females	89.2	9160±732	898±126	—	—
(a) 6 males	79.0	8674±565	829±181	—	—
P/data pair	P>0.06	P<0.05	P<0.05		
(b) 3 males + 3 females (in summer 1998)	81.5	—	—	11076±3070	1310±1019
(b) 5 males + 1 female (in spring 1999)	89.0	—	—	22081±3646	3438±1006
P/data pair	P>0.5			P<0.0001	P<0.007

Herbivores usually have a high water intake rate because of the high water content of plants. Williams et al. were surprised by the low values for water intake in oryx even during periods when the plants contain high percentages of water. Survival of a large herbivore on such low water intake suggests effective behavioural and physiological mechanisms that reduce water loss are in place. The simple strategy of resting in shade during the day and foraging at night (Section 3.2.4) plays an important role in reducing water loss.

Oryx, like camels, allow their body temperature to increase during the heat of the day thereby reducing the need for cooling by evaporative water loss. The temperature gradient in the nasal passages of both oryx and camel is used to protect the brain from overheating during periods of hyperthermia. The anatomy of the blood vessels in the head permits this protective mechanism. In both mammals and birds, blood returning from the nasal regions (and much of the brain) drains into a capacious collecting vessel, the cavernous sinus, from which it flows in the neck veins back to the heart. In many species including dogs, sheep, pigeons, and gazelles, a network of small arterial vessels, the rete mirabile ('wonderful net') passes through the cavernous sinus, branching from the two main arteries supplying the brain, the two carotid arteries. In species where nasal heat exchange is unimportant, e.g. monkey (Figure 3.40a) there is no rete mirabile and the carotid artery passes intact through the cavernous sinus providing little opportunity for exchange of heat. In the oryx, where the rete is present, (Figure 3.40b), its large surface area permits exchange of heat between the warm arterial blood and the cooler blood in the cavernous sinus.

So blood passing out of the arterial rete and entering the circle of Willis, from which the brain receives its blood supply, is at a lower temperature than it was in the neck arteries themselves. As a result, the brain can be maintained at a lower temperature than the trunk and its essential function is maintained during hyperthermia in the rest of the body. The oryx and Thompson's gazelle can maintain a brain–body temperature difference of up to 3 °C during sustained hyperthermia. This facility must be of enormous advantage to desert species when heat storage is used to minimize the physiological consequences of water shortage.

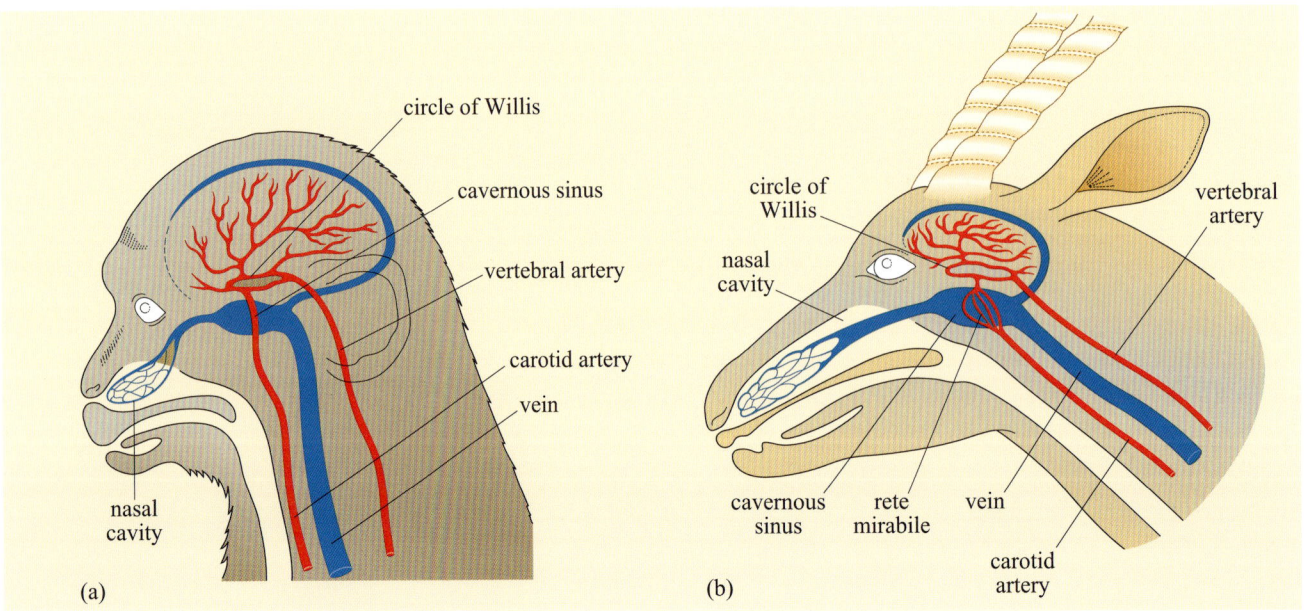

Figure 3.40 Simplified diagrams of the venous drainage from the nasal regions and the arterial supply to the brain and nose in (a) a monkey and (b) an oryx.

Large birds such as the ostrich allow their body temperature to rise when short of water and exposed to high T_a. In fact both desert and non-desert species of bird can tolerate T_b of 32–44 °C, and reduce evaporative cooling when water is in short supply. The ostrich, like other ratites, has a long nose with complex turbinates and the nostrils are relatively far forward on the beak compared with their position in other birds. Expired air is cooled as it passes through the turbinates.

Ostriches can exhale unsaturated air. At $T_a = 36$ °C, exhaled air is at 85% r.h. and by this means about 35% of the water evaporated into the air during inspiration is saved. Thus the ostrich saves about 500 g water per day. However, you may be thinking that while water is conserved, the body does not lose any heat overall. The cooling effect in the turbinates does achieve brain cooling, with a 1.5 °C difference measured between brain and body temperature. When at serious risk of overheating the ostrich supplements behavioural cooling strategies with panting by up to 40 times per minute.

Our study of a few representative examples of desert vertebrates has demonstrated a range of integrated anatomical, biochemical, and physiological adaptations and responses that enable the animals to cope with problems of high T_a and water shortage. Yet there is nothing spectacular about desert animals; no unique and 'amazing' feature has been identified for desert vertebrates. Desert vertebrates rely heavily on behaviour, and we should ensure that we integrate behaviour with the overall package of biochemical and physiological adaptations for each species. Behavioural responses to environmental stresses reduce the need for physiological responses, which may be costly in terms of evaporative water loss or use of energy in an arid environment that has low productivity.

Summary of Section 3.3

Behavioural mechanisms for reducing water loss are integrated with physiology. While *Dipodomys* rests in a cool burrow, the nasal counter-current heat exchanger cools exhaled air, conserving water vapour evaporated from respiratory surfaces. Long loops of Henle operate as counter-current multipliers, producing highly concentrated urine. Desert foxes use panting for evaporative cooling, but high rates of evaporative water loss cannot be sustained; hence the crucial importance of dens for cooling. Research suggests that desert birds have reduced BMR and FMR in comparison to mesic species. Such data need cautious interpretation, as studies on hoopoe larks demonstrated phenotypic flexibility in BMR, FMR and evaporative water loss.

Large endurers have difficulties losing body heat. *Dipsosaurus dorsalis* allows T_b to rise: its enzymes function at 47 °C. Overheated mammals cool by sweating, losing water and salts. Camels reduce the need for sweating by relaxed homeothermy, allowing T_b to rise to 40 °C, but they can function after a 30% water loss from the body. *Oryx leucoryx* obtain water from eating plants. In summer, the water content of grasses is much lower than in winter, highlighting the importance for oryx of foraging at night when T_a is relatively low, and resting in shade during the day.

The brains of oryx and camels are protected from overheating by the operation of a rete mirabile. In the ostrich, blood cooled during passage through complex nasal passages flows to the brain, keeping it 1.5 °C lower than T_b, a significant reduction when the bird allows T_b to increase to 44 °C.

3.4 Integrating across disciplines

3.4.1 Heat-shock proteins

Molecular biology provides further insights into the biochemical and physiological responses of vertebrates to extreme temperatures and aridity in the desert environment. Animals living in hot deserts are at risk of overheating, which in turn results in denaturation of enzymes and other essential proteins. Physiologists were puzzled for a long time about how desert reptiles such as the desert iguana (*Dipsosaurus dorsalis*) function normally at T_b = 44–46 °C. Such high temperatures would be expected to result in complete denaturation of enzymes and other important proteins such as haemoglobin. Yet we saw in Figure 3.36 that ATPase of *Dipsosaurus* continues to function even at 43 °C or 44 °C.

Discovery of heat-shock proteins (Hsps), suggested an explanation. The name 'heat-shock proteins' is applied because levels of Hsps in cells rise rapidly after exposure to abnormally high temperatures, e.g. around 40 °C for mammalian cells. Hsps comprise at least 12 families of related proteins that are found in cells of all animal and plant species so far investigated, so they are molecules that have been highly conserved in evolution, and are not specific to desert species. Nevertheless, Hsps are important for species at risk of exposure to high T_a because of their role as chaperone proteins that 'rescue' proteins whose tertiary structure has been disrupted by overheating. Figure 2.20 in Chapter 2 showed how chaperone proteins bind to denatured regions of a protein and alter the misfolded structure so that the correct three-dimensional structure is regained.

Heat-shock proteins maintain the biologically active conformation of enzymes and other essential proteins in cells during heat stress. For example, exposure of the fruit fly, *Drosophila*, to high temperatures, results in rapid transcription of a heat-shock gene, *Hsp*70. To begin gene transcription, various transcription factors, then RNA polymerase bind to special DNA sequences known as promoters; the latter determine where RNA polymerase binds, e.g. TATA (Figure 3.41), and starts transcription. Each gene has its own promoter. Nucleotide sequences in promoters tend to be highly conserved and are therefore the same in many species.

Figure 3.41 Control of mRNA synthesis during transcription of *Hsp*70. (a) RNA polymerase II pauses after synthesizing about 25 nucleotides of the transcript as CHBF binds to HSE. (b) After a heat shock that produces part-denatured proteins heat-shock transcription factor (HSTF) is converted from an inactive into an active DNA-binding form. This occurs in response to the presence of part-denatured proteins. Monomers of HSTF link to form trimers that enter the nucleus. (c) Binding of activated (trimerized) HSTF to the heat-shock regulatory element (HSE) of the promoter of the *Hsp*70 gene, releases the paused RNA polymerase II, leading to rapid transcription of the *Hsp*70 gene.

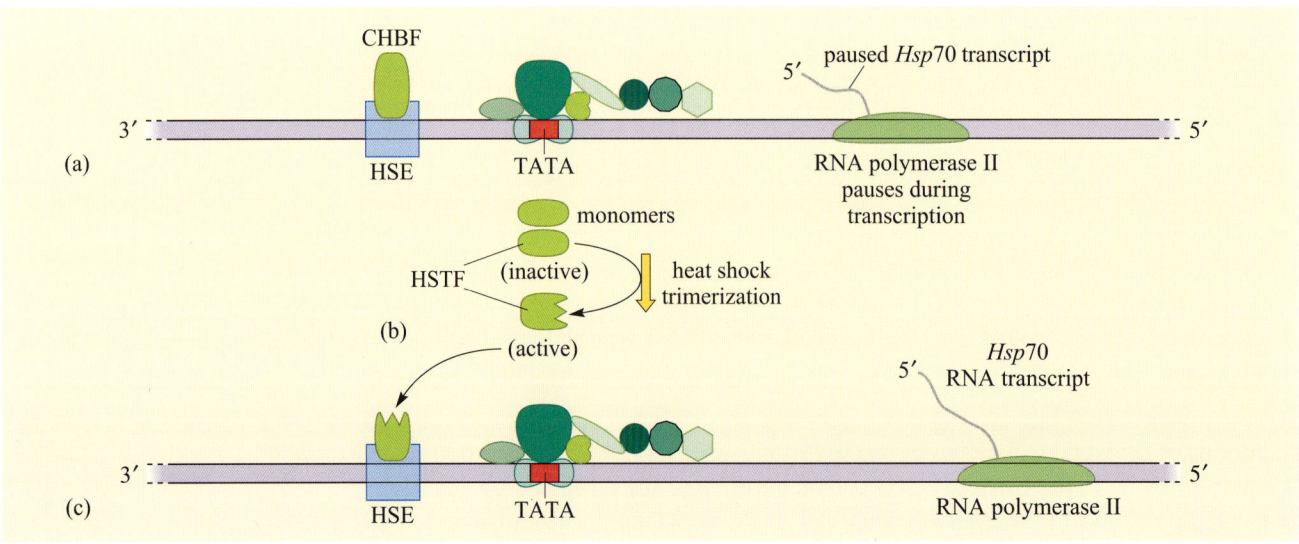

Gene sequences that regulate the promoter may be located adjacent to the promoter, or upstream or downstream of the gene. Regulatory elements control the transcription of genes that are not active all the time. A heat-shock regulatory element (HSE) has been identified in the promoter region of a heat-shock gene. Heat-shock transcription factors (HSTFs) are transcription factors that control *Hsp* gene expression through interaction with HSEs. The complex interlinking functions of the components involved in control of transcription of heat-shock proteins are shown in Figure 3.41.

Given the number of stages involved in *Hsp* gene transcription, researchers were puzzled initially by the rapid response, just a few seconds, to heat shock. Research on *Drosophila* demonstrated that such a rapid response is possible because *Hsp*70 is normally partly transcribed by RNA polymerase II, producing an RNA transcript about 25 nucleotides long. Mechanisms that pause the transcription are not understood well, but appear to involve constitutive HSE binding factors (CHBF) that bind to the promoter and halt transcription of the *Hsp* gene (Figure 3.41).

To understand how transcription of *Hsp* is resumed, we must begin with the role of HSTF. Normally, before heat-induced activation, HSTF exists as monomers in the cytoplasm. Research on heat-shocked mammalian cell cultures suggests that the signal for activation of HSTF is contact with the hydrophobic domains of proteins that have been denatured (Figure 3.42). HSTF monomer extracted from unstressed mammalian cells is bound to Hsp molecules. As the Hsp molecules bind to denatured protein during heat stress, it is likely that as a consequence, HSTF monomer is released from Hsp molecules. The released HSTF monomers trimerize and translocate to the nucleus where they stimulate resumption of transcription of *Hsp* genes by binding to the HSEs in the promoter regions.

Most work on Hsps has been carried out using cell lines and tissue cultures. Few studies have been carried out on vertebrates, and most of those have concentrated on fish. However, Zatsepina et al. (2000) studied the synthesis, properties and activation of Hsp70 in three species of desert lizard:

Figure 3.42 Regulation of transcription of heat-shock protein genes by heat-shock transcription factor.

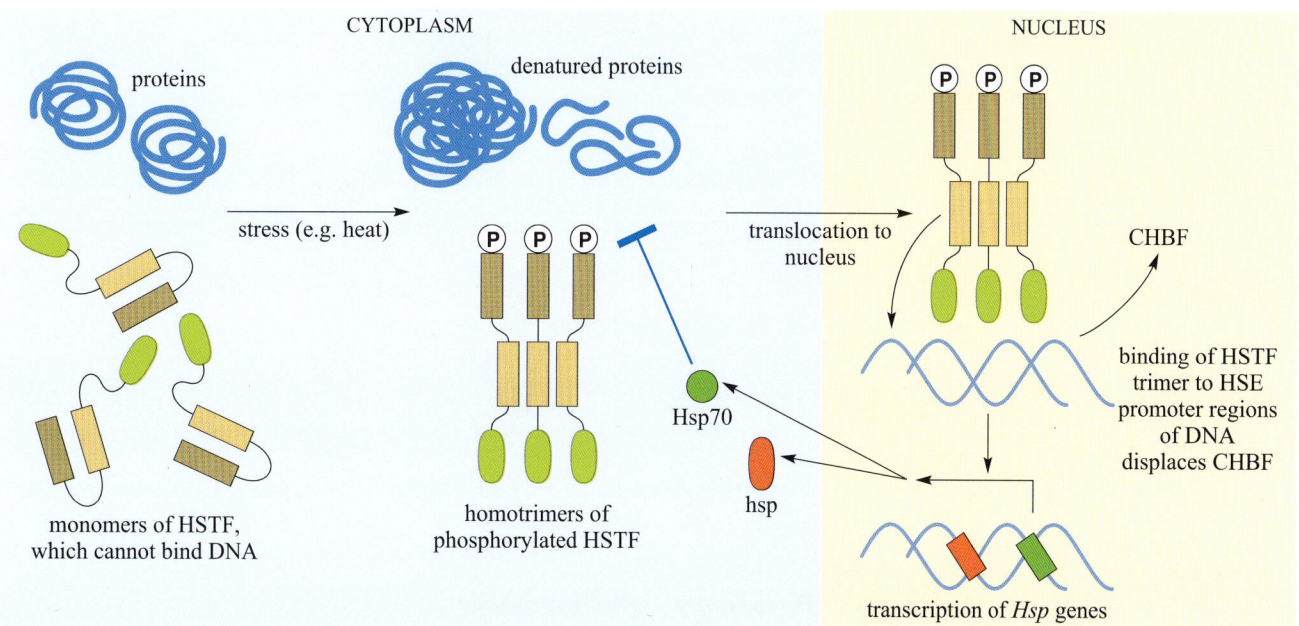

Phrynocephalus interscapularis, a highly thermoresistant diurnal species, and *Gymnodactylus caspius* and *Crossobamon eversmanni*, both nocturnal species. All three species were captured in a sand desert in Turkestan. Studies on a temperate species, *Lacerta vivipara*, provided comparisons with the desert species. All lizards were acclimated for 2 weeks at 25 °C. Heat-shock treatment involved one hour exposure of lizards of each species to a specific T_a at or greater than 39 °C. Following heat shock the animals were killed and cell extracts from the whole body were prepared. Samples of the extracts were mixed with ^{32}P-labelled HSE and incubated at 20 °C for 20 minutes, during which time any HSTF or CHBF present would complex with the ^{32}P-HSE. The free ^{32}P-HSE was separated from the ^{32}P-HSE-HSTF and ^{32}P-HSE-CHBF complexes by gel electrophoresis. The gels were dried and exposed to X-ray film. Figure 3.43 shows the results of the analysis of binding of ^{32}P-HSE to HSTF (complex III) and to CHBF (complexes I and II), in lizards kept at T_a 25 °C and lizards heat shocked at 42, 45 and 49 °C.

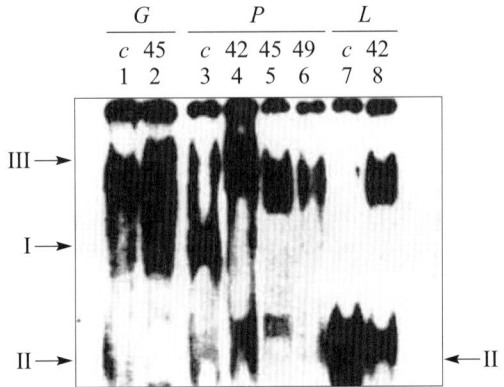

Figure 3.43 Analysis of heat-shock regulatory element (HSE) binding activity in lizard species from different ecological niches. Gel-mobility-shift analysis of whole-cell extracts from control (c) and heat-shocked animals. The extracts analysed by gel mobility shift were prepared from *Gymnodactylus caspius* (G) control animals (lane 1) and from individuals heated to 45 °C for 60 min (lane 2), from *Phrynocephalus interscapularis* (P) control animals (lane 3) and from individuals heated to 42, 45 or 49 °C for 60 min (lanes 4, 5 and 6, respectively), and from *Lacerta vivipara* control animals (lane 7) and from individuals heated to 42 °C for 60 min (lane 8). The locations of the ^{32}P-HSE-CHBF complexes (I and II) and heat-shock-induced ^{32}P-HSE-HSTF complex (III) are indicated by arrows.

At T_a 25 °C for *Lacerta* (c, lane 7) the levels of complex II were high, but they were low for *Gymnodactylus* and *Phrynocephalus*. In contrast, levels of complex I were high for *Gymnodactylus* and *Phrynocephalus* but complex I was absent in *Lacerta*. Complex III, consisting of trimerized and therefore active HSTF bound to ^{32}P-HSE, was present in both *Gymnodactylus* and *Phrynocephalus*, but not in *Lacerta*. So both desert species have activated HSTF even when acclimated to 25 °C. Following heat shock, complex III, activated HSTF, was present in all three species (lanes 2, 4, 5, 6 and 8).

■ What is the significance of the presence of complex III in cell extract?

Complex III consists of HSTF bound to ^{32}P-HSE which means that the HSTF was trimerized and therefore in a form that combines to HSE, and consequently activates transcription of *Hsp* genes.

Zatsepina et al. isolated mRNA from samples of cell extract prepared from control and heat-shocked lizards, and carried out hybridization, probing with *Xenopus laevis Hsp*70, lizard *HSTF1* and *αβ-actin* genes, the latter acting as a standard for comparison (Figure 3.44).

■ Compare the levels of HSTF mRNA and Hsp70 mRNA in *Phrynocephalus* and *Lacerta*, before and after heat shock.

Prior to heat shock, levels of HSTF mRNA were lower in *Phrynocephalus* than in *Lacerta*; in contrast, constitutive levels of Hsp70 mRNA were higher in the

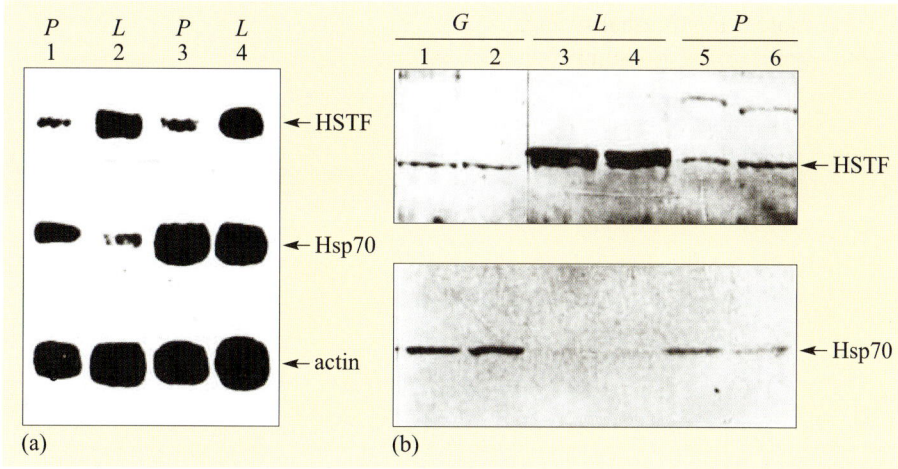

Figure 3.44 Expression of Hsp70 and HSTF in different lizard species. (a) Northern blot analysis of mRNA present in the cells of *P. interscapularis* (P) and *L. vivipara* (L) at 25 °C (lanes 1 and 2) and after heat-shock treatment for 1 h at 42 °C (lanes 3 and 4). HSTF, Hsp70 and actin are indicated by arrows. (b) Western blot analysis of proteins in *G. caspius* (G) (lanes 1 and 2), *L. vivipara* (L) (lanes 3 and 4) and *P. interscapularis* (P) (lanes 5 and 6). Lanes 1, 3 and 5, animals at 25 °C, lanes 2, 4 and 6, individuals heat shocked at 42 °C for 60 min. HSTF and Hsp70 are indicated by arrows.

thermoresistant species *Phrynocephalus*, than in *Lacerta*. Following heat shock both species showed strong induction of Hsp70 mRNA, demonstrating increased transcription of *Hsp70* genes (Figure 3.44a). Neither of the two species showed increased amounts of HSTF mRNA after heat shock, in contrast to the large increases in Hsp70 mRNA.

Zatsepina et al. interpret these data as showing that *HSTF* genes in lizards are not induced by heat shock. Western analysis using antibodies to human Hsp70 showed that levels of cellular Hsp70 are higher in both desert species than in *Lacerta* (Figure 3.44b). Earlier work had measured a three- to fivefold higher Hsp70 content in desert species than in *L. vivipara* kept at T_a 25 °C.

Comparison of *Lacerta vivipara* with the nocturnal desert species *Crossobamon eversmanni* showed that the latter has 2–3 times higher levels of Hsp in its cells. After heat shock, liver cells of *Lacerta* synthesized high levels of Hsps 68 and 85. In the desert species, normal synthesis of all proteins in liver continued after heat shock at 39, 42 and 43 °C, whereas in *Lacerta*, protein synthesis reduced after heat shock at 37 °C and 39 °C and almost ceased at 42 °C.

High levels of Hsps in cells of desert reptiles may stabilize protein structure sufficiently in the absence of thermostable proteins, and enable continuation of normal rates of protein synthesis at T_b up to 45 °C. Zatsepina et al. demonstrated a high level of *Hsp70* transcription (Figure 3.44a) linked to activated HSTF bound to HSE (Figure 3.43, complex III), in the desert species *Phrynocephalus*, even in animals kept at relatively low T_a. Tight regulation of *Hsp* gene expression ensures a response appropriate for the level of heat stress and a subsequent repression of the response when the stress is over. As T_a was increased for *Phrynocephalus* (Figure 3.43), relative amounts of complex III increased as complexes I and II decreased, suggesting removal of suppression of *Hsp* expression. Desert and temperate lizards differed in the quantity and state of HSTF and constitutive HSE-binding activity (CHBA), both under normal and heat-shock conditions.

In the desert species induction of Hsp synthesis at 3–7 °C higher than in temperate forms may link to high constitutive levels of Hsps. Temperate lizards, in contrast, have low constitutive levels of Hsps but high levels of HSTF which probably expedite intense synthesis of Hsps after a brief exposure to heat shock. Severe heat shock is lethal for the temperate lizards but is tolerated by the desert species. *Phrynocephalus* maintains cellular Hsps, and therefore can continue to function normally even when T_b rises to 45 °C. This capacity is of great advantage to a diurnal desert reptile as foraging times can be prolonged. By foraging during the hottest parts of the day diurnal desert reptiles avoid predators.

As you have read in Section 3.2.2, many desert vertebrates avoid overheating by behavioural means, so you may conclude that Hsps are of no more importance in desert species than in temperate species. However, the need for behavioural thermoregulation at any one time may conflict with other needs, e.g. finding food or escaping from predators. A spurt of intense physical activity may raise T_b sufficiently to initiate *Hsp* transcription. Species such as the camel that use relaxed homeothermy as a means of reducing TEWL, experience T_b as high as 41 °C. Desert lizards that forage during the day may routinely experience T_b values high enough to trigger a heat-shock response. Although desert animals are not unique in having Hsps, life in the desert environment where animals are at risk of overheating, is probably facilitated by efficient functioning of Hsps.

Summary of Section 3.4

The integration of physiological and molecular responses links an environmental signal to a physiological response. Heat-shock proteins (Hsps) are chaperone proteins that maintain the structure and function of proteins in cells exposed to high temperatures. The initial response of cells to high temperatures is rapid transcription of heat-shock genes, e.g. *Hsp*70, which is possible because *Hsp*70 is normally partly transcribed by RNA polymerase II. Resumption of transcription requires only three steps: release, then trimerization of HSTF (heat-shock transcription factor), followed by binding to HSEs (heat-shock regulatory elements) in the promoters.

At T_a 25 °C, the desert reptiles *Phrynocephalus interscapularis* and *Crossobamon eversmanni*, have high constitutive levels of Hsps in their cells, in contrast to a temperate species, *Lacerta vivipara*. Cells from *P. interscapularis* and *Gymnodactylus caspius* kept at 25 °C contained high levels of active HSTF bound to HSE in contrast to cells from *L. vivipara*, suggesting that in desert species *HSTF* genes are expressed constitutively. Protein synthesis in liver cells of desert species continued normally following heat shock up to 45 °C, whereas in *L. vivipara* protein synthesis in liver plummeted after heat shock at 37 °C. Cells of desert species are well prepared for heat shock.

3.5 Integrating across species

Populations of related species occupy similar niches in different environments. A big question for environmental physiologists is whether differences in biochemistry and physiology between related species living in different environments derive from physiological acclimatization (sometimes referred to as phenotypic flexibility), phenotypic plasticity or evolutionary adaptation.

Recall from Section 3.3.2 how hoopoe larks, wild-captured from the Arabian desert and kept at T_a 25 °C for just 3 weeks, showed increased body mass, increased food intake and increased BMR in comparison to hoopoe larks kept at 36 °C. Clearly, interspecific comparisons of BMR should be designed with the possibility of phenotypic plasticity and or/flexibility in mind.

Comparisons between vertebrate species of similar body mass within a particular taxonomic group but living in different environments have shown substantial differences in metabolic rate, e.g. metabolic rates of some desert mammals are relatively low in comparison with mammals of similar size within the same taxonomic group. The association between food intake, diet and environmental factors can be expressed in deceptively simple terms. Net primary productivity of the environment is determined to a major extent by environmental factors. Plant growth and therefore availability of food for animals is affected by climate, especially rainfall and T_a. However, metabolic rates of mammal species living on low-energy diets, e.g. herbivores eating large quantities of fibrous plants, are relatively low, in comparison with species eating foods with high-energy content such as fruits and nuts. Mueller and Diamond (2001) proposed that the low net primary productivity (NPP) of deserts and consequent low availability of food can only support mammals with relatively low metabolic rates. In contrast, relatively high metabolic rates may have evolved in species living in environments with abundant food; they 'run and idle fast'. BMR can be regarded as the metabolic rate during 'idling'. Mean field metabolic rates represent the total energy costs of an animal's normal daily activities and rest periods.

To test the hypothesis that there is a direct association between environmental NPP, BMR and FMR, Mueller and Diamond studied five species of deer mice (*Peromyscus* spp.). They are the most common North American mammals and are found in habitats ranging from Alaska to Central America. The cactus mouse (*Peromyscus eremicus*) is found in the North American deserts, including the Sonoran and Mojave deserts. It is nocturnal and emerges from its burrow at night to feed on seeds, and occasionally leaves and insects. Other species of *Peromyscus* live in woodland, the prairies and scrubland (Table 3.8). The five species selected all have similar diets, but live in diverse habitats of varying NPP. All five species are omnivores that feed on seeds, flowers, fruits and also insects and fungi. Breeding colonies of all the species, obtained from the *Peromyscus* Genetic Stock Center, were maintained in the laboratory under the same conditions and provided with water and the same diet *ad libitum*. The mice were kept at 27 °C on a 16 h/8 h light/dark cycle. The mice in these colonies had been in captivity for 10–40 generations. Thus any physiological acclimatization to the environment would have lapsed and selection for adaptation to the natural environment would have been relaxed.

Table 3.8 *Peromyscus* species studied and their natural habitats (Mueller and Diamond, 2001).

Species	Body mass/g	Ancestral site	Habitat type	NPP/g C m^{-2} yr^{-1}
P. eremicus	22.2 ± 2.8*	Nr Tucson Arizona	Sonoran desert	48
P. melanophrys	45.0 ± 6.3	Zacatecas in Mexico	*Yucca*/agave desert	67
P. californicus	43.5 ± 4.5	Santa Monica Mts, CA	Chaparral/coastal sage scrub	340
P. maniculatus	19.0 ± 1.4	Nr Ann Arbor, MI	Deciduous woodland and meadow	600
P. leucopus	19.1 ± 3.5	Nr Linville, NC	Deciduous/coniferous forest	604

* ± SE.

BMR was measured for individual male mice aged 8–15 months during daylight, while the mice were resting and had completed digestion. The dry mass of food consumed, dry mass of faeces produced and body mass of the mice were recorded daily. Values for NPP were calculated for the sites where the founders of the captive *Peromyscus* populations were collected. Figures 3.45a and b show, respectively, mass-adjusted BMR and daily food intake plotted against NPP for the five species of mice.

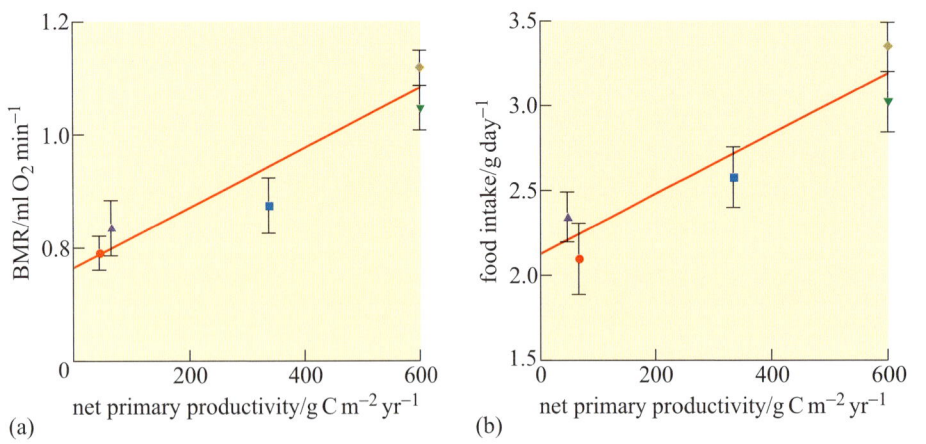

Figure 3.45 (a) BMR, measured as oxygen consumption (body-mass adjusted $\dot{V}_{O_2}$ in units of ml of O_2 min^{-1}) of five species of *Peromyscus* as a function of habitat NPP. Each point related to mean values (± SE) of 10–15 mice per species.

■ Describe the data in Figures 3.45a and b.

There is positive relationship between NPP and BMR. Where the NPP is 50 g C m^{-2} yr^{-1}, animals have a BMR of about 0.8 ml O_2 min^{-1}. As NPP increases to 600 g C m^{-2} yr^{-1}, animals have a BMR of about 1.1. Food intake also positively correlates closely with NPP. With NPP levels of about 50 g C m^{-2} yr^{-1} food intake (g day^{-1}) is around 2.3, rising to around 3.0 when NPP is 600.

Net productivity of each species' habitat of origin, BMR, and total energy intake (the latter equivalent overall to total energy expenditure) differed in the same rank order in the five *Peromyscus* species.

The data support the researchers' hypothesis that the well-provisioned species had evolved to 'run and idle fast'. *Peromyscus eremicus* and *P. melanophrys*, both desert species, have the lowest values for BMR (the 'idling' metabolic rate) and food intake (a measure of FMR). In contrast, two species living in environments with NPP > 600 g C m^{-2} yr^{-1} had the highest values for BMR and daily food intake. *Peromyscus californicus* was intermediate in both BMR and daily food intake. It is important to correlate observations of behaviour of species subjected to physiological measurements and the researchers noted that *P. leucopus* and *P. maniculatus*, species with the highest values for BMR and food intake, were jumpy and ready to escape from their cages, biting people who handled them. In contrast, the two species with the lowest food intake and BMR, *P. eremicus* and *P. melanophrys*, were docile and easy to handle. It is tempting to interpret the behaviour of the desert species as being geared to saving energy. Mueller and Diamond conclude that food availability determines which species of *Peromyscus* do best in a particular habitat.

Summary of Section 3.5

Researchers proposed the hypothesis that the low primary productivity of deserts and low availability of food correlates with low BMR and FMR of desert species. They tested their hypothesis by studying five captive-bred species of deer mice, *Peromyscus*: two originating from deserts with low NPP, one from chaparral and two from woodland with high NPP. The five species have similar diets, and were fed *ad libitum*. The desert species, *P. eremicus* and *P. melanophrys*, had the lowest BMR and lowest food intake. The two species from woodland, *P. maniculatus* and *P. leucopus*, had the highest values for BMR and food intake, while *P. californicus* from chaparral had intermediate BMR and food intake. This investigation was stringent as all the deer mice were descended from stock that had been captive for 10–40 generations. Therefore, the results could not be explained by phenotypic plasticity or flexibility.

3.6 Phylogeny and cladistic analysis

In Section 3.3.2 the point was made that many physiologists consider that desert birds are successful because of their avian physiology, not because of any specific adaptations. While Williams and Tieleman's research on hoopoe larks demonstrated that desert species are capable of flexibility in metabolic rate and evaporative water loss, it suggested that adaptation is important too. The selective advantages of lowered BMR and TEWL for desert birds include reduced energy demand, and lower production of metabolic heat reducing the need for EWL. Both BMR and TEWL appear to be at least partly genetically determined, so it is likely that they are subject to natural selection. We have seen from the hoopoe lark study in Section 3.3.2 that phenotypic flexibility may account for differences between populations of the same species. Phenotypic plasticity (Section 3.3.2) may also provide confounding variables for such studies.

The power of the comparative approach in which a physiological feature in a range of related species is studied can be appreciated from the study of *Peromyscus* species outlined in the previous section. Studies of clades comprising related genera within a family can be more powerful still. Irene Tieleman et al. (2003) tested the hypotheses that both BMR and TEWL are reduced along an aridity gradient within a single family of birds, Alaudidae (larks). Their study also investigated the role of **phylogenetic constraint**, the situation where similar traits seen in closely related species may be there not because they are adaptive, but because they were inherited from a common ancestor. The researchers chose larks for the study, because larks live in a vast geographic area, covering three continents, with species found in a wide range of environments, from hyper-arid desert to mesic grassland. Nevertheless, all larks eat a similar food, seeds and insects, so diet was not a confounding variable in this study.

Aridity is directly related to primary productivity, and is therefore a sound measure of the selection pressures that animals experience with increasing aridity. These selection pressures include decreased food and water availability and increasing air temperatures. The researchers investigated eight species, and also used data in the literature for six other species for the study. BMR and TEWL were measured at night using standard respirometry and hygrometry methods. The aridity index selected, log Q, was appropriate to the habitat of the birds in this study. Log Q is low in hot dry deserts and high in relatively cool wet areas.

Table 3.9 summarizes the results for the 14 lark species. Analysis showed that both body mass and aridity had significant effects on BMR ($P < 0.0001$).

Tieleman et al.'s analysis of the data in Table 3.9 supports their view that BMR of larks increases as the environment becomes more mesic (Figure 3.46a). Also, TEWL of larks decreases as the environment becomes more arid (Figure 3.46b).

Tieleman et al.'s study involved the comparison of two traits across lark taxa, BMR and TEWL. The statistical analyses of the data in Figures 3.46a and b assumed phylogenetic independence, but the researchers were aware that the positive correlation obtained between BMR and TEWL with aridity could be explained by **phylogenetic autocorrelation,** a relationship derived from the phylogeny itself, rather than adaptation.

The researchers constructed a cladogram for 22 species of lark. DNA samples were extracted from blood or tissue samples, and polymerase chain amplification was used to amplify the *cytochrome b* gene (975 base pairs) and the 16S rRNA gene (566 base pairs). DNA was sequenced for each species and statistical analysis carried out to determine the most probable evolutionary tree for the clade. The cladogram (Figure 3.47) indicates a number of clades within the lark family. The lengths of the branches in the cladogram indicate phylogenetic distance between the clades.

Table 3.9 BMR and TEWL for 14 species of lark compared with body mass and annual precipitation in country of origin and environmental aridity index, log Q.

Species	Body mass/g	BMR/kJ day^{-1}	TEWL/ g H$_2$O day^{-1}	Annual precipitation/mm	Log Q
Grey-backed finchlark (*Eremopterix verticalis*)	15.1	—	1.31	33.0	1.76
Stark's lark (*Eremalauda stark*)	15.6	—	1.31	57.2	1.76
Desert lark (*Ammomanes deserti*)	21.5	20.1	1.60	89.6	1.78
Dunn's lark (*Eremalauda dunni*)	20.9	24.7	1.69	89.6	1.78
Hoopoe lark (*Alaemon alaudipes*)	36.9	32.8	2.59	89.6	1.78
Black-crowned finchlark (*Eremopterix nigriceps*)	15.2	16.5	1.34	209.1	2.24
Crested lark (*Galerida cristata*)	31.2	32.2	2.44	209.1	2.24
Calandra lark (*Melanocorypha calandra*)	50.6	49.5	3.03	250.0	2.26
Horned lark (*Eremophila alpestris*)	26.0	28.6	2.08	309.9	2.41
Lesser short-toed lark (*Calandrella rufescens*)	23.6	31.6	—	281.0	2.59
Short-toed lark (*Calandrella brachydactyla*)	24.0	35.6	—	281.0	2.59
Spike-heeled lark (*Chersomanes albofasciata*)	25.7	29.1	3.33	420.4	2.60
Skylark (*Alauda arvensis*)	32.0	62.4	3.17	750.0	3.20
Woodlark (*Lullula arborea*)	25.6	49.4	2.41	750.0	3.20

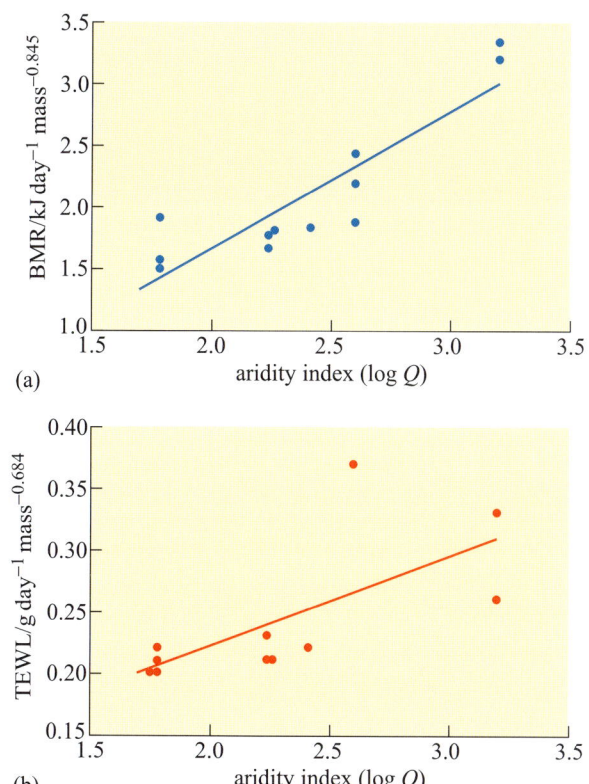

Figure 3.46 Mass-adjusted (a) BMR and (b) TEWL of 12 species of lark as a function of environmental aridity expressed as the aridity index, log Q. Each point represents the mean value for each species. (*Note:* the exponent of mass is different for the data in a and b because although each figure relates to 12 species of lark, data on BMR and TEWL were not available for all species and a different set of 12 birds was used for each graph.)

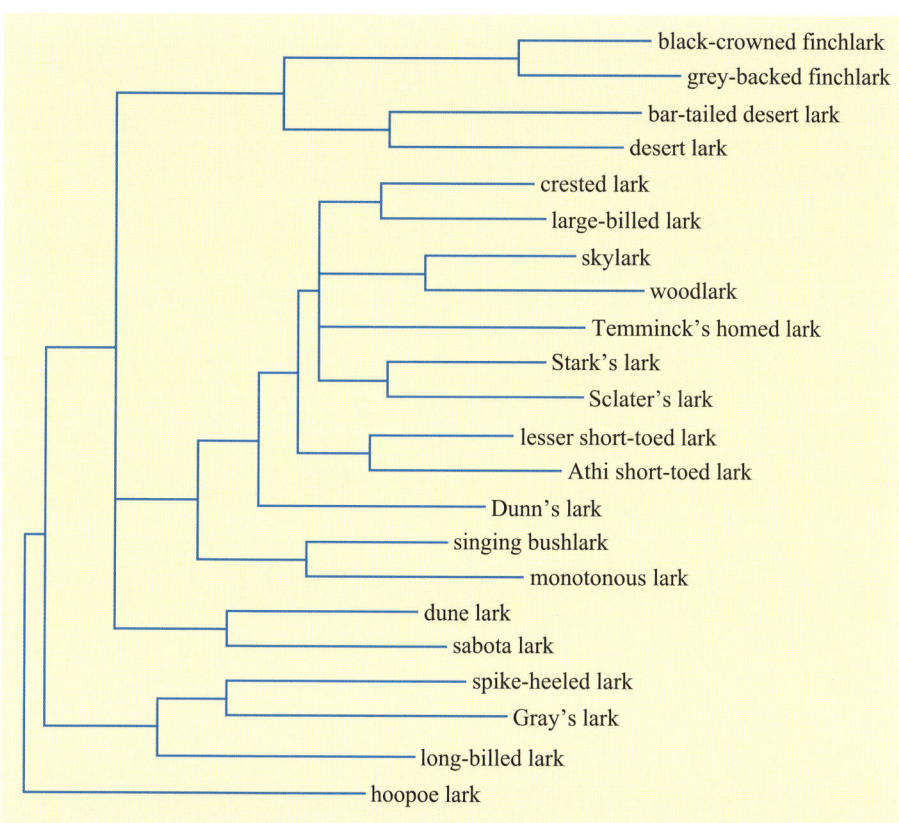

Figure 3.47 Phylogenetic tree of 22 species of lark based on *cytochrome b* and 16S rRNA sequences, and analysed using maximum parsimony criteria.

■ Compare the lengths of the terminal branches of the cladogram with those of the internal (initial) branches.

The terminal branches are long in comparison to the short internal branches.

■ Suggest an outgroup within the cladogram and justify your choice.

The position of the hoopoe lark, suggests that it may be an outgroup. The hoopoe lark is placed at the bottom of the cladogram suggesting that this species might be representative of the common ancestor.

Tieleman et al. interpret the short internal branches, coupled with long terminal branches, as suggesting that lark species underwent rapid adaptive radiation and have lived for a major part of their evolutionary history in diverse habitats. It is possible that variation in TEWL and BMR relates to phylogenetic distances, so that the more closely related genera within a clade have similar TEWL and BMR. For example, if variation in BMR could be explained simply by phylogenetic distance, members of one clade within the lark family would have similar values for BMR, in contrast to members of two other clades, which might have lower and higher values, respectively. In that case, BMR would be correlated with phylogeny itself, i.e. phylogenetic autocorrelation, which may be attributable to phylogenetic constraint or to ecological factors.

There was a significant phylogenetic effect in the data linking mass-corrected BMR with aridity.

Specialized statistical analysis ruled out phylogenetic autocorrelation as the sole explanation of the relationship between BMR and aridity. Since the BMR of larks decreases along a gradient of increasing aridity, Tieleman et al. therefore argued that this relationship is due to adaptation rather than phylogenetic constraint.

Mass-corrected TEWL correlated with aridity with no significant influence of phylogeny.

The decreases in BMR and TEWL with increasing aridity in larks suggest that BMR and TEWL have adaptive significance and that natural selection is the relevant process.

Summary of Section 3.6

A cladistic analysis of 14 lark species (family Alaudidae) investigated the hypotheses that BMR and TEWL are reduced along an aridity gradient. Although larks are found over a huge geographic area, all species have similar diets so diet was not a confounding variable. The researchers collected data on body mass and BMR and calculated the aridity index for the environment of each species. Analysis of the data showed that aridity had a significant effect on both BMR and TEWL. The shape of an evolutionary tree for the clade suggests radiation of larks soon after their origin. No correlation was found between mass-corrected TEWL and phylogeny, suggesting that correlation between TEWL and aridity is not due to phylogenetic constraint. Statistical analysis eliminated phylogenetic constraint as an explanation for the relationship between BMR and aridity although mass-corrected BMR correlated with both aridity and phylogeny.

3.7 Conclusion

In this chapter we have studied animals in the context of their own habitat rather than using the traditional comparative physiology approach of comparing organ systems in different species. Although we have looked at extreme habitats, specifically deserts, it has become clear that, for many species, extreme physiological adaptations are not present and that even endotherms, birds and mammals rely on behavioural strategies, thereby reducing the need for physiological strategies that are costly in terms of water usage and energy. Furthermore, we have learned that desert animals do not always maintain homeostasis, a constant internal environment. There are species of birds and mammals, groups defined as homeotherms, that allow their T_b to rise when T_a is high; this relaxed homeothermy is seen in the camel, the oryx and ostrich. Physiological adaptations that we have seen in desert animals are also present in non-desert species, e.g. panting, sweating and the rete mirabile for brain cooling. Only by integrating biochemical, physiological and behavioural strategies can we understand how an animal survives and exploits its environment. Molecular biology is providing new insights and the little information that is available so far, has already provided another level of analysis to feed into integrative animal physiology. The puzzle of how desert lizards function normally at T_b as high as 46 °C may have been solved, at least partially, by the finding that these species have high cellular levels of heat-shock proteins that function as chaperone proteins.

We have also touched on the relatively new science of evolutionary physiology. In this context identifying the differences between acclimation, phenotypic plasticity and adaptation is of crucial importance as we saw from the studies on desert larks and *Peromyscus*. Once physiological traits shaped by natural selection have been identified, research on the evolution of such traits across species or higher taxa can be linked to evolutionary trees. One important issue is how does the physiology of an ancestral species affect what is possible in its descendants? Similar traits seen in closely related species might be there not because they are adaptive, but because they were inherited from a common ancestor – the phenomenon of phylogenetic inertia. Williams and Tieleman eliminated phylogenetic inertia as being the cause of the lower TEWL in desert lark species. Although the researchers identified a significant effect of phylogeny in the mass-corrected values for BMR there was also a significant effect of aridity on BMR that was independent of phylogeny. Statistical analysis of the data supported the researchers' view that natural selection is likely to be the process that explains the correlation between decreasing levels of BMR and TEWL with increasing aridity. Mueller and Diamond's study of *Peromyscus* species, obtained originally from diverse habitats but captive for 10–40 generations, showed that desert species had significantly lower BMRs than the species from more temperate habitats. This study provides evidence that lower BMR in the desert species is a genetic trait, not the result of acclimatization or phenotypic plasticity.

Learning Outcomes for Chapter 3

When you have completed this chapter you should be able to:

3.1 Define and use, or recognize definitions and applications of, each of the **bold** terms.

3.2 Provide examples that show there is a continuum of desert climates and environments that links to diversity of flora and fauna.

3.3 Explain, with examples, the thermoregulatory strategies of evaders, evaporators and endurers, and interpret relevant data.

3.4 Describe the importance of integration of behaviour, anatomy, physiology and biochemistry in the study of animals that live in deserts.

3.5 Explain physiological mechanisms of water conservation and cooling in named evaders, evaporators and endurers, and interpret relevant data.

3.6 Recognize potential ambiguity and uncertainty in attributing observed physiological or biochemical features and responses to high T_a and aridity to genotypic adaptation, phenotypic plasticity or acclimatization.

3.7 Explain how the role of heat-shock proteins (Hsps) in cellular responses to temperature extremes links to the molecular mechanism for control of transcription of *Hsp* genes and interpret blots that track *Hsp* transcription.

3.8 Explain the use of integration across related species in designing and interpreting experiments to investigate whether features such as BMR and reduced TEWL in desert species are adaptive, or are derived from phylogenetic constraints or phenotypic flexibility.

Questions for Chapter 3

Question 3.1 (LOs 3.1 and 3.3)

Figure 3.48 illustrates the activity of the antelope ground squirrel *Ammospermophilus leucurus* during a typical day in the Nevada desert.

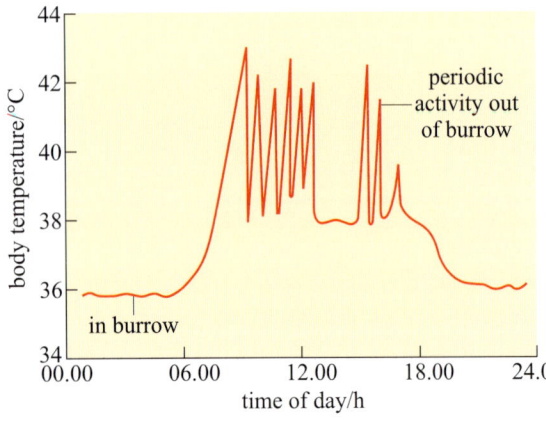

Figure 3.48 Temperature and activity in the antelope ground squirrel, a small desert mammal. The peaks represent periods of activity on the surface, outside the burrow.

(a) Describe the pattern of activity suggested by the data.

(b) Explain how the behaviour of the ground squirrel relates to thermoregulation.

(c) Is *Ammospermophilus* an evader, an evaporator or an endurer?

Question 3.2 (LOs 3.1, 3.3, 3.4 and 3.6)

Classify the following statements as *true* or *false* and write a brief explanation of your answer.

(a) Many small desert mammals are able to endure hot and physiologically stressful ambient temperatures by storing heat for a number of hours.

(b) The coat of camels has an insulating effect in hot environments and camels shorn of their coat are likely to have a higher evaporative water loss.

(c) Measurements of BMR in the hoopoe lark have demonstrated that this desert species always has a lower BMR than mesic lark species, and that low BMR in the hoopoe lark is adaptive.

(d) Panting is a thermoregulatory response shown exclusively by large mammals.

(e) When faced with severe heat stress, mammals resort to autonomic responses, whereas reptiles and birds respond behaviourally.

Question 3.3 (LOs 3.1–3.3)

The following strategies (a)–(f) are adopted by various desert-dwelling species.

(a) The avoidance of heat by burrowing or moving into shade;

(b) the production of concentrated urine and/or uric acid;

(c) a pelage or feathers providing substantial insulation;

(d) a high-threshold T_b for onset of sweating;

(e) heat storage;

(f) radiative or conductive heat loss from extensive surfaces of bare skin.

Which of these strategies do the following animals employ?

1. jack rabbit
2. oryx
3. hoopoe lark
4. desert reptiles
5. kangaroo rat
6. camel

Question 3.4 (LOs 3.1, 3.6 and 3.8)

(a) Researchers measured significantly lower whole body BMR in wild-caught individuals of a desert-dwelling rodent, compared to BMR measured in a closely related species living in temperate woodland. Why would a lower BMR be an advantage to a desert-dwelling species?

(b) It is tempting to assume that lower measured values of BMR in desert species represent an evolutionary adaptation. What other processes could result in lower BMR for a desert mammal?

Question 3.5 (LOs 3.1 and 3.4)

Researchers investigated the hypothesis that the skin of a desert lizard shows certain properties that confer thermal resistance, thereby slowing down the rate of heating of the body when the lizard is in full sun. An experiment was designed in which the percentage of solar heat gain in intact live lizards was compared with heat gain in isolated lizard skin preparations. The researchers found that the percentage solar heat gain in the isolated skin preparations was significantly greater than that in the live animals. They concluded that the skin in live animals has physiological and physical properties that confer resistance to solar heat gain.

Write a critique of this experiment and suggest an experimental design that would improve the validity of the results.

Question 3.6 (LOs 3.1, 3.6 and 3.8)

Williams et al. measured BMR and FMR in oryx living in the Mahazat as-Sayd nature reserve in the Arabian desert. The results are summarized in Section 3.3.3. The researchers then compiled data from the literature for minimum resting metabolic rate for 15 species grouped in the Artiodactyla, ranging from the mouse deer with body mass just 1.61 kg and dik-dik at 43.97 kg to moose weighing 325 kg and camel 407 kg. Values for whole body BMR of each species were plotted against body mass in kg (Figure 3.49).

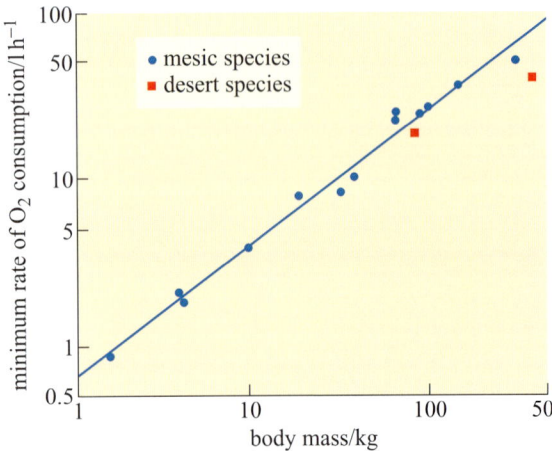

Figure 3.49 Minimum rate of oxygen consumption of Artiodactyla. Red squares represent the oryx (84.1 kg) and the camel (407 kg).

Describe the data in Figure 3.49. Do the data for oryx and camel suggest that the desert species have a lower BMR than the other 13 species? Note that the graph is a logarithmic plot, in which the values on the axes increase by factors of 10. You should interpret this graph as you would for any plot.

CHAPTER 4 HIBERNATION AND TORPOR

Prepared for the Course Team by Laurie Haynes

4.1 Introduction

In Chapter 1 you learnt about the range of strategies for thermoregulation adopted by animals. The huge range of analytical techniques and approaches that are used in the study of thermoregulation was also described. In particular, adaptive mechanisms are increasingly being uncovered at all levels of living organization – molecules, cells, tissues and whole systems. This chapter examines hibernation, a special form of adaptation that animals can make to the ecological demands of remaining in a chosen habitat in winter. Hibernation is a state which enables energy-efficient survival when ambient temperatures are so low that foraging or simply maintaining normal core body temperature and basal metabolic rate are either energetically too costly or impossible.

Polar endotherms can maintain a high T_b even when living actively at sub-zero temperatures. Such animals have very good thermal insulation and may have a plentiful food supply to sustain the increased thermogenesis needed to maintain a large difference between T_a and T_b. For many animals, however, the food supply in a cold environment becomes scarce or inaccessible beneath snow or ice.

■ What are the options for surviving a very cold winter?

1. *To remain active*. This strategy is possible for an animal with appropriate insulation, considerable energy reserves and the ability to compete successfully for a continuing food source. The arctic fox and the emperor penguin are examples of such animals.

2. *To migrate to another habitat* for the duration of an inhospitably cold season. This strategy is possible if the animal has sufficient mobility to leave the extreme latitudes as the available food dwindles. Many birds and bats adopt this strategy.

3. *To endure the periods of low temperature extremes in the chosen habitat at low metabolic cost* – by reducing T_b, locomotion and other life functions at all levels. The use of this approach, to varying extents, is seen in a huge and diverse group of birds and mammals. It has also been adopted by poikilothermic vertebrates and invertebrates, though often for different reasons.

Option 3 is adaptive hypothermia, as seen in torpor and hibernation. Small homeotherms living at latitudes (or altitudes) at which they experience long periods of cold weather and lack of solar radiation have few options. Consider the problems a small animal is likely to encounter compared with a large one. It has a relatively high surface area to volume ratio and therefore a high potential for heat loss (Section 3.2.1): even at moderate T_a, it normally has a high metabolic rate and cannot carry enough really effective insulation, whether of fur, feathers or blubber. Some species can survive in burrows if they can emerge regularly and find enough plant material to eat beneath the snow, but even these animals are likely to use energy-conserving strategies for much of the time. Thus, small rodents and insectivores, for example, have little choice but to exploit option (iii) above, uncoupling their homeothermic mechanisms or re-setting the critical body temperature (T_c): in other words **adaptive hypothermia**.

Thermal adaptation in animals overcomes the ecological and bioenergetic constraints of living in extreme climates. For warm-blooded vertebrates, evolution has generated almost every imaginable approach to this problem, and the adaptations adopted for each species reflect different approaches to the evolutionary cost-benefit analysis of variations on conventional endothermy. Whilst most hibernating endotherms lower the temperature of all or parts of their bodies by between 5 and 25 °C, many ectotherms and some mammals are characterized by their ability to depress their T_b to below the freezing point of water. In either situation, physiological adaptations must include:

1. thermoregulatory systems with control mechanisms different from, or with the ability to override, those which operate in the seasons when the animal is euthermic;
2. biochemical and cellular control mechanisms capable of protecting tissues against damage, and compensating for energetic and metabolic disadvantages which are manifested at low body temperatures.

The terminology used to describe the different forms in which adaptive hypothermia is observed in animals is complicated. However, it is best viewed as a way of identifying a more or less persistent entry into a state of sustained physiological depression and metabolic dormancy. A source of clear definitions which is reasonably contemporary at the date of writing is provided by Körtner and Geiser (2000).

Torpor is best defined as entry of the whole animal into a state of hypothermia which is accompanied by behavioural inactivity, regulated by a combination of external and internal signals. **Hibernation** is defined as a sustained and profound state of torpor, entry to and exit from which is governed by internal signals together with exclusively seasonal external cues.

We can place the methods of reducing body temperature for the purpose of energy saving into the hierarchy below:

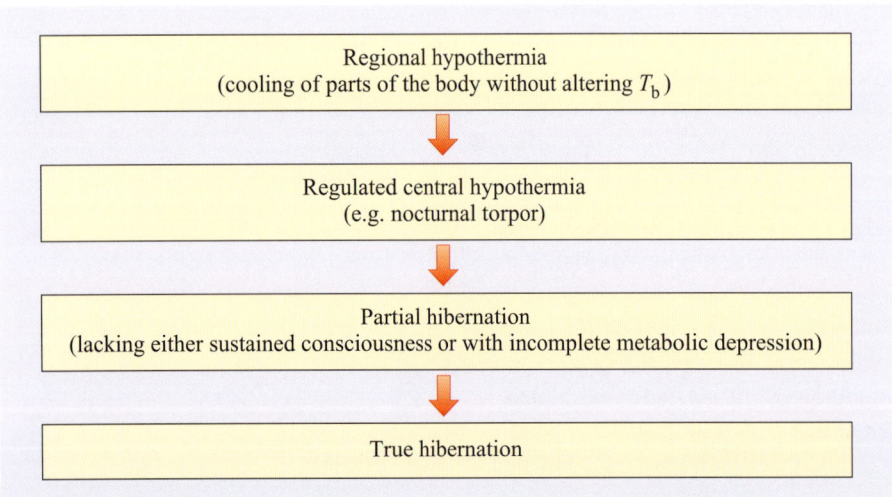

True hibernators undergo three definitive and coordinated physiological changes:

- **Thermal dormancy** – the ability of an animal to operate its biological functions at very low core body temperatures.
- **Behavioural suppression** – the cessation of activity of many muscles, which depends upon the ability of the brain to override sensory inputs and endogenous rhythms such as breathing.
- **Metabolic inhibition** – the ability of an animal to undergo episodic **bradymetabolic** changes: the depression of energy-related and anabolic reactions.

Adaptation to climatic extremes affects organisms at several levels. In plants, which have little coordination at the level of systems, physiological adaptations to extreme heat, cold and dehydration can occur just as they do in animals that have such control. Adaptation is manifested not only at the level of tissue and organ systems but also at the level of genes, proteins, protein complexes and cells. Many of the fundamental responses to hypothermic extremes mirror those seen in aestivation – a state of torpor seen in some ectotherms adapting to arid rather than cold conditions (Section 3.2.2). In both hibernation and aestivation, further adaptations can be seen at the level of cells and tissues: the existence of **protective measures** that enable rapid recovery from very cold temperatures, lack of oxygen and low energy supplies.

■ What protective measures might be required to keep cells alive during periods of torpor?

Such protective measures might include: prevention from freezing of the cytosol and organelles; maintenance of life functions in the absence of oxygen or energy-yielding substrates; delay or neutralization of processes which normally eliminate dying cells that may harm tissues if they remain.

Almost all hibernating animals prepare during the summer and autumn seasons by eating large amounts of food that they convert into fat, providing additional energy stores. Cellular metabolic processes are linked to a central regulatory mechanism, an 'internal clock', that provides the reference point for entering hibernation and when to resume normal behaviour.

Apart from occasional periods of arousal to forage and excrete, hibernating animals are inactive for several months on end. During this period, the lack of food and water means that physiological processes, blood and cellular biochemistry undergo major changes. Animals may be inactive for shorter periods toward the end of hibernation periods, as they emerge to access signals (e.g. light levels, T_a) which trigger the switch to normal activity. Environmental signals are integrated in the brain, pineal gland and other centres implicated in controlling seasonal changes in physiology and behaviour.

The annual cycle that governs entry into and exit from hibernation is also under the control of internal physiological systems. For example, as the bodies of hibernating male squirrels return to normal in the spring, a sustained, increased secretion of sex hormones prevents return to hibernation.

Termination of hibernation is highly sensitive to temperature change. In fact, current global upward trends in ambient temperature are having a measurable effect in shortening the hibernation seasons of a number of species, such as the yellow-bellied marmot living in the Rocky mountains of the northwestern USA. Snowfalls have also increased with the changing climate, so that the ground is still covered with snow when the marmots arouse from hibernation (earlier than used to be the case), making food hard to find.

Summary of Section 4.1

Hibernation is a physiological and behavioural adaptation whose function is to maximize energy efficiency in animals remaining in the same area the whole year round. It is an alternative to the provision of sufficient insulation to remain warm, forage continuously and sustain a constant high metabolic rate.

There are three aspects of coordinated regulation in hibernating or torpid animals: thermal, behavioural and metabolic. They operate independently, at least to some degree, and at the level of the whole organism down to that of individual molecules. There are also adaptations that protect the organism against cell and tissue damage.

Hibernation and torpor are regulated at the level of the whole animal by biological rhythm generators that are adjusted by environmental stimuli which initiate rapid reversal signals at the conclusion of the period of dormancy.

4.2 The nature and extent of hibernation and torpor in endotherms

4.2.1 Degrees of torpor

Adaptive hypothermia occurs in at least six distantly related mammalian orders (Table 4.1) and in several orders of birds. There is a spectrum running from those species which can tolerate a drop in T_b by 2 °C for a few hours, to the seasonal deep hibernators which maintain a T_b as low as 4 °C for weeks on end.

Table 4.1 Groups of mammals which contain species that routinely become torpid.

Group	Sub-group (and example)	Comments
Prototheria	spiny anteater	seasonal
Metatheria	Didelphidae (American opossums)	occasional
	Dasyuridae (insectivorous mice)	occasional
	Phalangeridae (possums)	seasonal
Eutheria	Rodentia* (see Table 4.2)	seasonal (and daily)
	Primates (dwarf lemurs)	seasonal
	Chiroptera* (temperate bats)	seasonal (and daily)
	Insectivora* (tenrec, African shrew, golden mole, hedgehog)	seasonal
	Carnivora (black bear, brown bear, badger)	seasonal lethargy – not deep hibernation

*Includes native British species.

■ What characteristics would you expect to find in seasonal hibernators?

Seasonal (deep) hibernators are likely to be small and to live in an environment where there may be a large difference between T_a and T_b, and their food is likely to be absent or inaccessible for long periods. Most such animals are herbivorous or insectivorous. *Size* is a critical factor in the depth of torpor: for example, black and brown bears inhabit the same territory as several deep hibernators, but provided they have shelter, they can manage on stored energy reserves, mainly of fat, for extended periods by lowering their T_b by only 2–6 °C. *The availability of food* is another factor: a number of small birds, for example the American goldfinch (*Carduelis tristis*; Figure 4.1)[*], can survive in winter temperatures down to −60 °C, remaining active and maintaining T_b with a huge (in excess of 500% over summer levels) increase in thermogenesis. However, other species, such as the redpoll (*Carduelis flammea*), which inhabits the northern United States show, in addition, a nocturnal hypothermic torpor.

Figure 4.1 American goldfinch (*Carduelis tristis*).

If an animal is very small, quite short periods without food may present a problem, and nocturnal hypothermia and torpor can be important for energy conservation. For example, several species of tropical humming-bird undergo nocturnal torpor, even though the difference between T_a and T_b is not huge.

4.2.2 Species showing torpor or deep hibernation

Among the birds, torpor occurs in a number of species in the orders Apodiformes (humming-birds and swifts), Caprimulgiformes (nightjars, nighthawks, goatsuckers and poor wills) and Coliiformes (mousebirds). In all of the humming-birds (family Trochilidae) studied to date, torpor, if it occurs, takes place on a daily (or more usually nightly) basis. They are able to re-warm themselves independently of T_a and show an increased thermogenesis if T_a falls below 18 °C during the time when the bird is not searching for food. The tiny rufous humming-bird (*Selasphorus rufus*; Figure 4.2), weighing only 3.0–5.5 g, has a summer range in North America that extends to Alaska, but it overwinters in Mexico. While undertaking this huge migration it undergoes overnight torpor, especially when breaking its journey for a few days, feeding on nectar to rebuild its energy reserves.

Figure 4.2 Rufous humming-bird (*Selasphorus rufus*).

There is no evidence for long periods of unbroken torpor in the humming-birds. The poor will (*Phalaenoptilus nuttalli*; Figure 4.3), which lives in the southern USA, is perhaps the only bird studied so far that shows bouts of torpor at all comparable to those of seasonally hibernating mammals. Poor wills kept in the laboratory at a T_a of 1 °C without food go into torpor, with a T_b of 6 °C, from which they arouse spontaneously about once every 4 days, showing an exceptionally large rise in their metabolic rate in the process. The pattern of torpor and change in T_b that they undergo in the wild is not known.

Figure 4.3 Poor will (*Phalaenoptilus nuttalli*).

Many birds, including doves and pigeons, enter shallow torpor (with the T_b falling to about 32 °C) when deprived of food. If food is available, the pigeon responds to low T_a by a large (up to 55%) increase in BMR. Indeed, a

---

[*] In this chapter we make reference to a wide range of species and have given both a common and the scientific name. There is no need to learn these scientific names. They do though allow you to check precisely which species is involved in a study. The same species may have a different common name in different locations. Moreover, one common name may be used for different species in different locations.

reduction in food supply seems to be a major factor in the induction of torpor in almost all the bird species studied. The white-throated swift (*Aeronautes saxatilis*) and the mousebird (*Colius* sp.) can tolerate a T_b of −20 °C and spontaneously re-warm at low ambient temperatures.

Among mammals, many groups contain species that undergo different degrees of adaptive hypothermia. Of the placental mammals, the largest number of hibernating species is found among rodents (see Table 4.2) and bats. All temperate-zone bats, including the 15 UK species, undergo daily torpor during certain seasons with some species remaining torpid for extended periods, as also does the European hedgehog (*Erinaceus europaeus*). Another famous example, which was a food valued by the Romans because of its habit of storing fat prior to hibernation, and was the sleepy guest of the Mad Hatter, is the hazel dormouse (*Muscardinus avellanarius*). Most research on hibernating mammals has focused on bats, hedgehogs, hamsters and in particular, the sciurid rodents (squirrels).

Table 4.2 The six (out of 30) families of rodents in which adaptive hypothermia is known to occur.

Family	Examples	Comments
Zapodidae	meadow jumping mouse, Scandinavian birch mouse	deep hibernation
Heteromyidae	pocket mice, kangaroo mice	erratic, seasonal hibernation
Gliridae	dormice (including the native British species)	deep hibernation
Muridae	African fat mouse	daily torpor
Cricetidae	hamsters white-footed mice	deep hibernation daily torpor
Sciuridae	chipmunks, marmots (woodchuck), ground squirrels (at least a dozen species)	deep hibernation

4.2.3 Hibernators as eutherms

Hibernating endotherms are not the easiest animals to study. Thus, until the late 1960s many biologists believed that mammalian hibernation was a process in which thermoregulation was simply 'switched off', following the receipt of a set of 'cues'. These cues included a declining T_a, a shortening daylength, the extent of body fat and a lack of food etc. With this model, the hibernator essentially becomes an ectotherm whose T_b follows the T_a quite closely and who is at great risk if the T_a in the hibernaculum (hibernatory winter retreat) falls below freezing. Indeed, prior to the 1960s, many workers assumed that a poor thermoregulatory ability was a prerequisite for hibernation.

■ What evidence casts doubt on these views?

You may be aware that many animals overwinter in hibernacula that are not very deep or well defended against cold. If they were indeed so dependent on T_a, there would be a very high mortality rate. In addition, the ability to hibernate occurs in species very closely related to non-hibernators; it is unlikely that thermoregulatory ability would vary so widely between closely related genera

Figure 4.4 Golden-mantled ground squirrel (*Spermophilus lateralis*).

of rodents such as the non-hibernating Djungarian or Siberian hamsters (*Phodopus* sp.)* and the hibernating Turkish hamster (*Mesocricetus brandti*) and black-bellied or European hamsters (*Cricetus cricetus*).

It has been recognized for many years that hibernators can arouse at intervals throughout the hibernating season without an apparent rise in T_a and that arousal could occur when animals were handled or disturbed in the cold. Both of these observations imply a remarkably high degree of control by the animal (Table 4.3).

Table 4.3 Body temperature and heat production of a hibernating marmot at different environmental temperatures (derived from Benedict and Lee, 1938).

T_a/°C	T_b rectal/°C	Relative rate of heat production per day/arbitrary units
0.9	4.1	100
3.0	4.5	54
2.2	4.6	54
2.6	4.7	36

■ In what way does the data in Table 4.3 suggest that T_b is controlled in the hibernating marmot?

The T_b in each experiment remained relatively constant, varying only between 4.1 °C and 4.7 °C, but the heat production varies, being very much higher when T_a was held at the lowest temperature.

Nevertheless, it was not until the effects of manipulating hypothalamic temperatures (see Section 4.6) of otherwise cold hibernating rodents were investigated, that researchers began to consider that hibernating mammals might be exercising control similar to non-hibernating (normothermic) ones – a view that is now generally held.

Summary of Section 4.2

Adaptive hypothermia occurs widely in both mammals and birds, but the ability is scattered throughout different families: even within single families, some species show torpor and some do not, suggesting that the ability may have evolved independently many times. Whereas a number of small birds show a daily, shallow torpor, so far only the poor will has been described as showing extended bouts of torpor comparable to those seen in mammals. Species of birds and mammals that hibernate (Figures 4.4 to 4.8) seem to have a highly advanced euthermic ability and in most cases can control T_b closely down to 3–4 °C, contrary to earlier views which assumed that hibernation was a manifestation of poor thermoregulatory ability.

Figure 4.5 Meadow jumping mouse (*Zapus hudsonius preblei*).

Figure 4.6 Black-tailed prairie dog (*Cynomys ludovicianus*).

Figure 4.7 Woodchuck (*Marmota monax*).

Figure 4.8 Little brown myotis bat (*Myotis lucifugus*).

---

* There is some confusion over the common names of the species of *Phodopus*. Until about 1980 *Phodopus sungorus* was considered to have two subspecies – *Phodopus sungorus sungorus* and *Phodopus sungorus campbelli*. *Phodopus sungorus sungorus* was known as the Djungarian hamster. There was no commonly used name for *Phodopus sungorus campbelli*. Then studies showed that these two subspecies were in fact different species and they were renamed as *Phodopus sungorus* and *Phodopus campbelli*. For some reason, *Phodopus sungorus* was then referred to as the Siberian hamster and *Phodopus campbelli* as the Djungarian hamster. Scientists who study these animals may well use either the older or newer common name. The moral here – always use the correct scientific name for a species as well as what might be its common name. Common names are common only to the culture in which they are used. In this text we refer to *Phodopus sungorus* as the Siberian hamster.

4.3 Characteristics of hibernation behaviour

The animal kingdom reveals a bewildering variety of regulated hypothermic behaviours, which are characterized by sustained hibernation at one extreme and regular short bouts of shallow torpor at the other. The many patterns observed and the variety of animal groups that exhibit these behaviours have not made it any easier to work out why different animals adopt their own strategies. Elephant shrews (*Elephantulus myurus*), which live in the relatively moderate climate of southern Africa for example, reduce their T_b to one of the lowest levels seen in mammals in which frequent torpor bouts are observed. Torpor occurs with complete recovery about five times a day over the winter months, with the T_b falling to as low as 7.5 °C at a T_a of 2.5 °C. It seems that elephant shrews, whose body temperature fluctuates closely with environmental temperature cycles, are budgeting their energy by using – as heterotherms do – passive heating to assist their return to normal T_b levels in the spring. Below we consider the physiology of 'typical' hibernators, but there are new extremes of behaviour still to be explained, and no doubt yet to be discovered.

4.3.1 Signals for entry

Despite the fact that hibernation is reflected in a number of profound and operationally distinct physiological changes, changes in T_b continue to be the recognized signs of its onset, interruption and termination, because of the relative ease of monitoring T_b. Onset is triggered both by endogenous and exogenous cues.

■ What would you identify as exogenous cues?

The three most important environmental stimuli initiating torpor are food supply, daylength and T_a, though the order of their importance differs between species and between seasonal hibernation and daily torpor.

Both food supply and the amount of body fat are relevant. In the short term, a diminution of available food towards the beginning of the hibernation period may itself trigger hibernation, and in a laboratory cold room at constant temperatures, torpor can be induced in several species by the removal of food. Examples here include some humming-birds and the poor will, where entry into torpor rapidly follows the removal of food. In the longer term, food supply determines the animal's ability to build fat reserves, and in some species, including some hamsters and ground squirrels, the presence of large fat reserves may be necessary for the animal's entry into hibernation.

However, others may not readily enter torpor in the absence of a store of food. Siberian hamsters (*Phodopus sungorus*) may fail to enter torpor even when injected with doses of insulin that result in a large and long-lasting fall in blood glucose. In this species in the wild, as well as others such as chipmunks, there is evidence that daylength is of greater importance as a cue. In general (and perhaps always), decreasing daylength is part of the stimulus, but the mechanism by which it acts may involve several separate routes. The first, which is not yet well understood, is that decreasing daylength causes an increase in the secretion of the peptide hormone melatonin from the pineal gland in the brain. Melatonin appears to act via a number of routes to predispose the animal to torpor. The second route is via the hypothalamus and the gonads; in most hibernators, entry into hibernation does not take place if there are high levels of androgens in the blood, and in the Turkish hamster withdrawal of the testes into the body cavity is a prerequisite for hibernation in males. Likewise, an injection of androgen into a torpid male hamster provokes arousal. It may also be that melatonin from the pineal gland acts to reduce gonadal secretions, as well as acting directly on the brain. Once again, the importance of ambient temperature as a cue to entry into hibernation may vary between species, though no species enters a bout of torpor unless the T_a is below its thermoneutral level.

Animals such as hamsters and chipmunks are sometimes called **facultative hibernators** because they hibernate in response to environmental conditions. This contrasts with the so-called **obligative hibernators**, such as the ground squirrels and marmots, whose sequence of fattening, hibernating and arousing seems to be strongly driven by an endogenous annual cycle under physiological control. Figure 4.9 is drawn from data on the golden-mantled ground squirrel (*Spermophilus lateralis*; Figure 4.4) (Strumwasser, 1960), kept in the laboratory for 2 years under constant conditions of light (12 hours light and 12 hours dark) and at a constant temperature (22 °C).

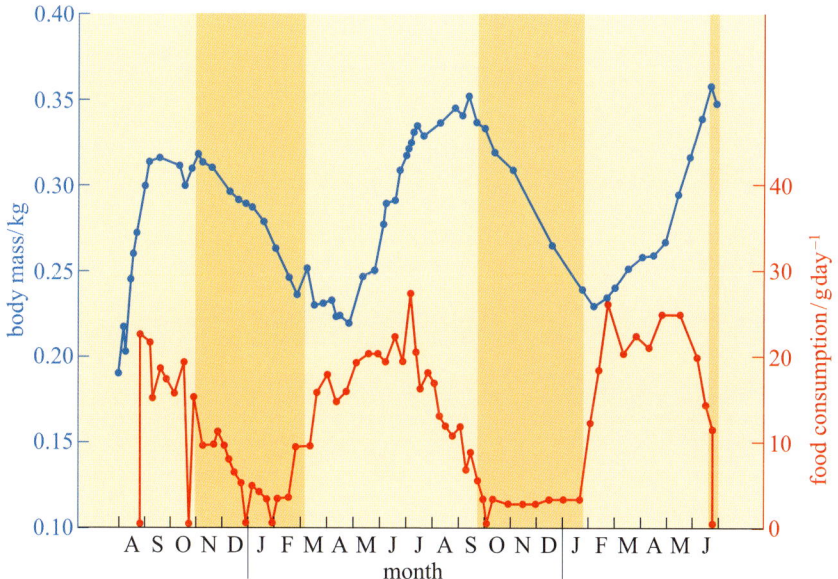

Figure 4.9 The circannual rhythms of body mass (blue), food consumption (red) and hibernation in the golden-mantled ground squirrel (*Spermophilus lateralis*). The shaded orange bars indicate periods of hibernation.

■ What several conclusions on the triggers for hibernation can you draw from this figure?

As the sequence of food consumption, weight gain and hibernation continue as normal under constant environmental conditions, environmental triggers are not essential for hibernation in this animal. However, it does appear that the cycle may be shortening: in the first year the animal entered hibernation in late October, in the second in late September, and at the end of the experiment it was just entering hibernation at the end of June. Thus, this figure suggests that although the timing of the cycle may be primarily due to an endogenous circannual rhythm, in natural circumstances the timing may be re-set annually by environmental factors.

Thus, a sharp distinction between environmental and endogenous cues to entry cannot be drawn: indeed, the importance of the state of gonadal activity to hibernation in the hamster already reveals the interrelationship of these cues.

Many hibernators show a marked cycle in the production of thyroid hormones, with a decrease in their secretion at times when a non-hibernator would be increasing secretion to increase thermogenesis. Further hormonal and neural controls over the hibernation cycle are described in Section 4.6.

4.3.2 Physiological changes during entry

Under normal euthermic circumstances, animals kept in an ambient temperature of 0 °C would be expected to show a marked increase in metabolic rate and adaptive thermogenesis (see Chapter 3). However, the response in hibernators is the opposite. Figure 4.10 shows data from a woodchuck (*Marmota monax*; Figure 4.7) about to enter torpor. Following a period of 2 hours or so when T_b is held more or less constant, but oxygen consumption and heart rate are highly irregular, the woodchuck lowers its oxygen consumption (a measure of its metabolic rate). Within 8 hours of the start of entry into torpor, metabolism appears to be at a minimal base-line, with T_b subsiding smoothly to about 12 °C within 14 hours (Lyman and O'Brien, 1960).

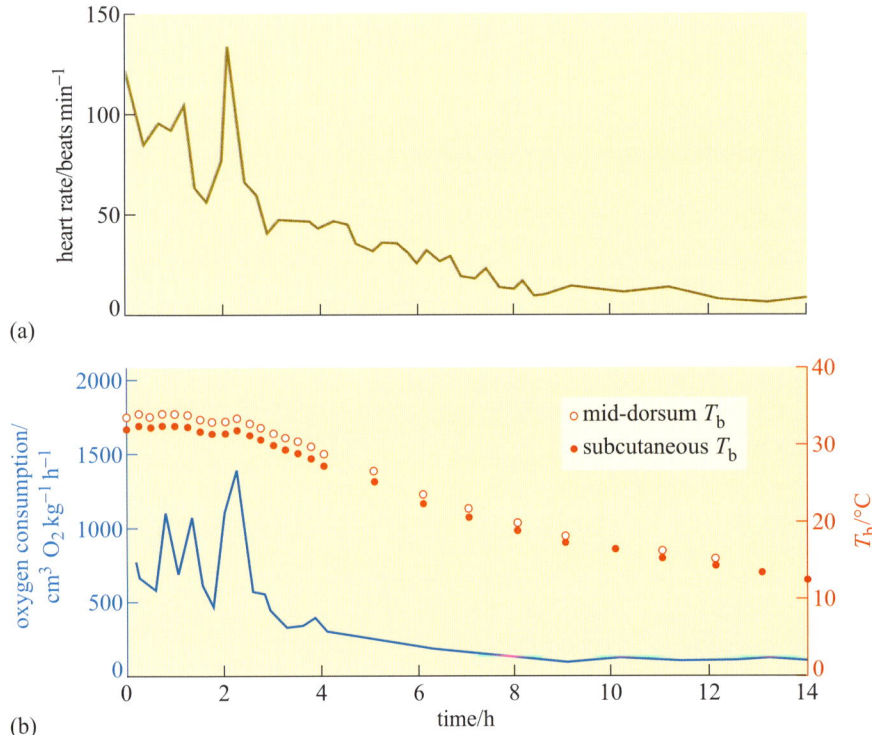

Figure 4.10 The heart rate, oxygen consumption and T_b of a woodchuck entering hibernation.

Entry into hibernation can take a great deal longer than the few hours it takes in the woodchuck. In the 1960s at Harvard University, Felix Strumwasser recorded the brain temperature (T_{brain}) of the Californian ground squirrel (*Citellus beecheyi*) entering hibernation (Figure 4.11). T_{brain} dropped during each dark period but rose again before the period of light. On the first, third and fifth 'nights', the drop was to less than 34 °C, but on the remaining 'nights' the drops were successively greater. Strumwasser (1960) argued that these *test drops* indicated metabolic and neuronal preparation for deep hibernation. Such test drops are quite common in mammals entering hibernation, but are by no means universal.

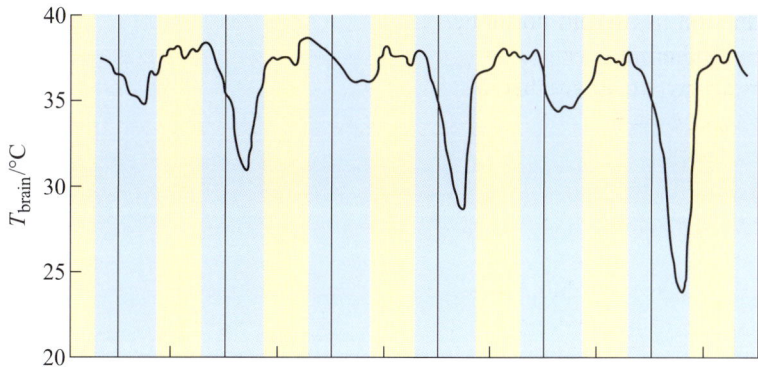

Figure 4.11 'Test drops' of brain temperature in the Californian ground squirrel. Dark periods ('nights') are shown in blue and light periods ('days') in cream. The vertical lines indicate midnight of each day.

The blood pressure of ground squirrels entering hibernation has been measured by inserting a catheter into the aorta. Once the catheter is in place and the animal has recovered from the operation, blood pressure can be measured with no stress to the animal. At first, the mean blood pressure remains within the range of the active animal, but as hibernation deepens the blood pressure decreases. A mild peripheral vasoconstriction takes place and it persists throughout the period of hibernation. This vascular response may be important in maintaining adequate blood pressure to the brain, given the huge reduction in heart rate.

However, during entry into hibernation, there are large fluctuations in vasomotor tone, and periods of superficial vasodilation occur, which may have the effect of accelerating heat loss. Indeed, it is quite likely that vasomotor control largely determines the changes in T_b seen in test drops. These fluctuations alternate with short periods of shivering, suggesting that the rate at which the T_b is allowed to drop is being carefully controlled.

4.3.3 Maintenance

Entering hibernation is not a passive process in response to falling T_a. Nor is deep hibernation a passive process or indeed a uniform state. Figure 4.12 shows the pattern of hibernation (as measured by the heart rate) of an arctic marmot (*Marmota caligata*) kept in the laboratory at a T_a of 10 °C for 18 days in February. Despite being inactive, every one or two days the heart rate rises abruptly, remains high for a number of days, and then falls again. These records are from an animal under laboratory conditions, but similar changes have been recorded from animals in the wild.

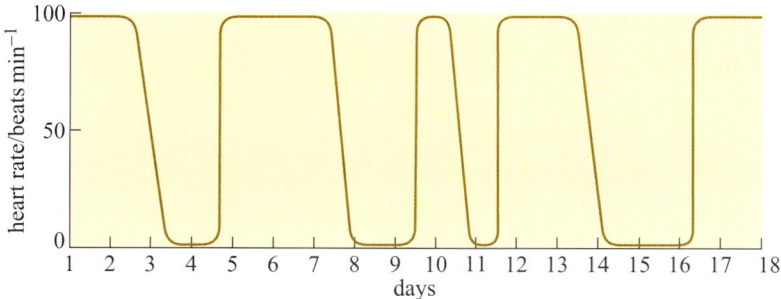

Figure 4.12 The heart rate of an arctic marmot (*Marmota caligata*) undergoing bouts of hibernation in the laboratory during early February.

Most hibernators spend the winter in hibernacula or dens some feet below the ground surface. The importance of the behaviour involved in the selection and 'engineering' of these winter quarters is demonstrated in Figure 4.13, which shows the variation in external temperature in Alaska from September to May. Superimposed on these data are temperatures of the warmest and coolest burrows of arctic ground squirrels (*Spermophilus undulatus*). You can see the burrow temperatures vary by less than 10 °C throughout these winter months, though for most of that time they are below zero (Mayer, 1960).

An exception to the rule is the dormouse (*Muscardinus avellanarius*), which hibernates above ground, usually amongst the leaf litter (which provides some protection) on woodland floors. The most obvious changes in deep hibernation (apart from the lowered T_b) are concerned with metabolism.

- Heart rates of over ten or under three beats per minute are rare. The major cause of these extremely low rates is the lengthening of the time between individual beats.
- Cardiac output is also reduced, to about 1.5% of normal (a mere 1 cm^3 blood min^{-1} in the ground squirrel).
- Respiration is greatly reduced. It may take place at quite evenly spaced intervals, or long periods of *apnoea* (cessation of breathing) may occur followed by several deep inspirations. (The record for holding a breath in torpor is 150 minutes in a hedgehog, though the average for this species is 60 minutes.)

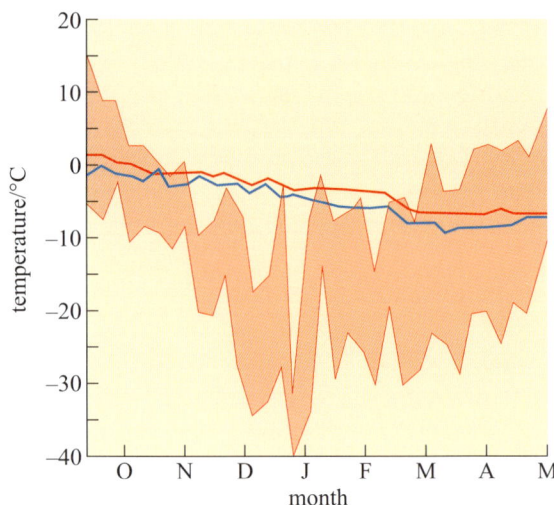

Figure 4.13 Temperatures of the burrows of arctic ground squirrels in Alaska. Readings from the warmest (red line) and coldest (blue line) burrow in the sample are shown. The shaded area shows fluctuations in maximum and minimum air temperature outside the squirrels' burrows.

The lower T_b of hibernating animals and the changes in the respiratory and cardiovascular performance lead to marked acidosis: the pH of arterial blood falls by 0.24–0.48 and P_{CO_2} is increased by a factor of 2.5–4.0. The oxygen supply to the tissues is dependent upon cardiac output, haemoglobin concentration and on the shape of the oxygen dissociation curves. Figure 4.14 illustrates the latter for euthermic (T_b 38 °C) and hibernating (T_b 6 °C) ground squirrels.

■ What do you conclude from Figure 4.14?

The curve for the hibernating squirrel has shifted markedly to the left. Half-saturation (P_{50}) is therefore achieved at a much lower P_{O_2}, indicating a higher oxygen affinity.

Other evidence indicates that there is also an increase in haemoglobin in the blood as an animal enters hibernation. The hibernator's tissues therefore have a high tolerance of hypoxia. These changes explain how a hibernating animal such as a hedgehog can survive the long period of apnoea that is characteristic of intermittent breathing; it draws on the increased oxygen stored in the blood.

In addition to changes in respiration and in the cardiovascular system, there are marked changes in endocrine function. Endocrine gland atrophy is characteristically found prior to the onset of hibernation; particularly atrophy of the pituitary, gonads, thyroid and adrenal glands.

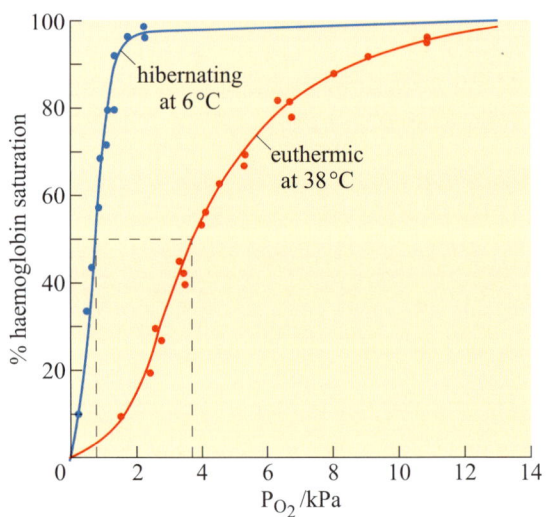

Figure 4.14 Haemoglobin dissociation curve of blood from euthermic and hibernating ground squirrels, determined at the corresponding T_b of 38 °C and 6 °C. Dashed lines show the P_{O_2} at half saturation (P_{50}) (from Musacchia and Volkert, 1971).

The physiological state of an animal in deep hibernation is, however, dynamic (the physiological controls are still working) and not that of a passive animal *made* hypothermic. For example, if there is a decline in the resistance of the peripheral blood vessels and a drop in blood pressure, there follows a compensatory increase in heart rate and cardiac output. The most graphic illustration of the fact that a hibernator retains physiological control mechanisms is its response to the falling ambient temperature. If T_a drops below a particular level (which depends upon the species in question), there is *always* a compensatory increase in heart and respiratory rate, *and* a rise in metabolic rate, and therefore a tendency to raise or at least preserve T_b. If the carbon dioxide concentration of the inhaled air is increased, then hibernating mammals react by increasing their breathing rate. In the hedgehog, the CO_2 threshold is 0.7–1.7%, at which point the periods of apnoea become shorter. Continuous breathing replaces periodic breathing at 5–9% CO_2.

The evidence suggests therefore, that the hibernator is sensitive to changes in its environment and that appropriate physiological responses can still be made. If the change or response is major, then the individual rapidly begins to arouse. It is this ability of hibernators to elevate T_b from 5–10 °C to the euthermic level, even at T_a values below zero, that puts them into a class of their own.

4.3.4 Arousal

We can identify three types of arousal during the hibernation period, on temporal rather than physiological grounds. The first is alarm arousal, in response to a major exogenous stimulus such as a sudden large drop in environmental temperature. The second is a periodic arousal when, in the absence of external cues, the animal spontaneously begins to re-warm. The third is the final arousal in the spring when the animal does not re-enter hibernation but emerges to a sustained euthermia. Physiologically, all three are similar.

Alarm arousal

A potentially life-threatening event, such as a fall in T_a to below zero, elicits a transient metabolic response in a hibernator. If the lowered temperature is maintained, the animal responds not just with transient increases in metabolism, but with a sustained rise in T_b and complete arousal.

Mechanical stimuli as well as temperature changes can evoke arousal. In animals fitted with electrodes just under the skin to monitor muscle action potentials, an externally applied stimulus results in a long-lasting burst of action potentials. The response in the fat dormouse (*Glis glis*) is very striking. This species hibernates with its bushy tail curled over its back. If the erect hairs are gently displaced, a burst of muscle action potentials occurs with a concurrent rise in respiratory and heart rates. Vibration, pressure, locally applied heat or cold, and the infusion of a variety of substances all produce the muscle response. In fact, the responsiveness of receptors in hibernators appears to increase with decreasing temperature, in marked contrast to the situation found when non-hibernators are made hypothermic.

The adaptive value of such a response is obvious. An animal torpid in a burrow seems quite defenceless. By retaining a high degree of surveillance, the animal can still perceive disturbances in air-flow or collapse of the burrow and make the appropriate response.

Periodic arousal

All mammalian hibernators arouse periodically. The frequency of the arousal and the length of the euthermic periods between bouts of hibernation vary widely with species, among individuals, and with the time of year (e.g. in deep hibernators, the larger species seem to have longer periods of wakefulness than the smaller ones). The arctic marmot (*Marmosa caligata*), whose heart rate recording is shown in Figure 4.12, aroused from hibernation every 2–3 days and remained euthermic for 3–4 days at a time. Figure 4.15a shows the average number of days between successive arousals of golden-mantled ground squirrels at various times in one hibernation season and Figure 4.15b shows the frequency of arousals in a Richardson's ground squirrel (*Spermophilus richardsonii*).

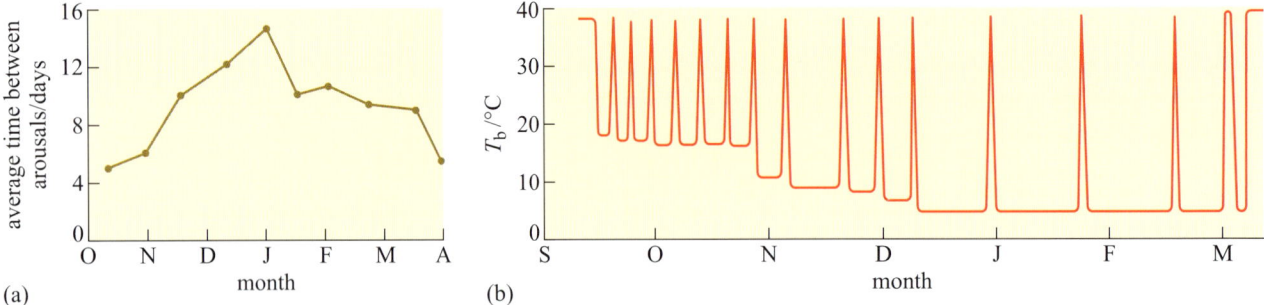

■ What do you deduce from Figures 4.15a and b?

The frequency of arousal changes during hibernation. Early on, the arousals occur every 5 or 6 days. In December and January these arousals occur every 10–15 days in the golden-mantled ground squirrel, and nearer 30 days in Richardson's ground squirrel: as the time for emergence approaches, the arousals become more frequent again.

Evidence points to the likelihood that the internal mechanisms which set circadian rhythms do not operate during deep hibernation. But in studies on hibernation in European ground squirrels kept under conditions as close to the natural habitat as possible, continuous recording of temperature using implanted thermal dataloggers showed that entry and arousal from torpor was synchronized to the time of day in some animals. All animals entered torpor in the afternoon and some even aroused at the same time. Circadian temperature fluctuations occurred at higher temperatures during entry and arousal phases.

What could be the function of periodic arousals? Some hibernators, such as hamsters, store food for the winter in their nests or burrows, and eat during the periods of arousal. Bats arouse to drink on mild nights. For those species that only metabolize fat from their stores during hibernation, the reason for arousal is not so obvious, particularly as the energy expended in a single arousal lasting a few hours can equal that used in 10 days of hibernation. In the arousing golden hamster (*Mesocricetus auratus*), changes in oxygen consumption and temperature in various parts of the body are rapid and extensive (Figure 4.16).

At the start of arousal with an ambient temperature of 5 °C, oxygen consumption is 60–80 cm^3 kg^{-1} h^{-1}. Within 3 hours this rate rises 100-fold, to a level comparable to that of violent exercise. At the start of arousal, cheek pouch temperature is virtually the same as that of the rectum but you can see that, as arousal progresses, the cheek pouch temperature rises more rapidly and there is a difference of more than 20 °C by 160 minutes. A short time later, oxygen consumption reaches its peak and declines rapidly, whereas cheek pouch and rectal temperatures attain the euthermic level. In some species, for example many ground squirrels, the process of arousal may even be much faster than it is in the hamster (Figure 4.17).

Figure 4.15 (a) Average number of days per month between arousals in golden-mantled ground squirrels in the hibernation season, and (b) arousals (as measured by body temperature) throughout the season in a Richardson's ground squirrel. Note the lower T_b in the torpid animal towards the end of the season.

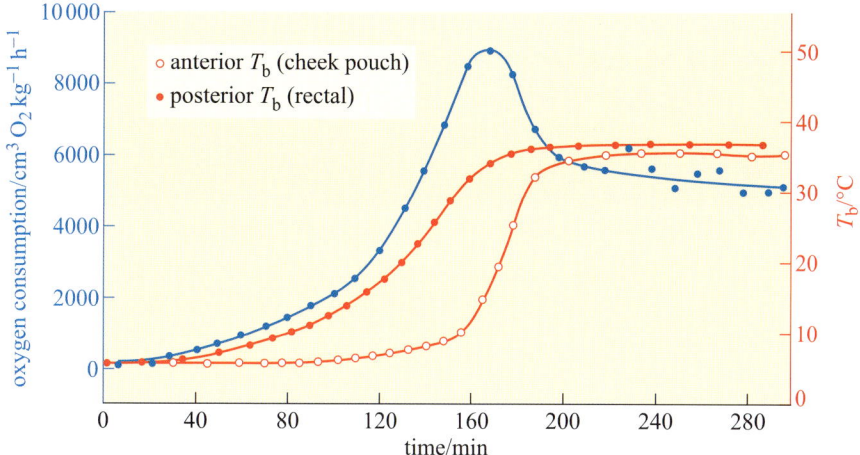

Figure 4.16 Oxygen consumption (blue) (as a measure of metabolic rate), anterior T_b and posterior T_b in the golden hamster (*Mesocricetus auratus*) during arousal (Lyman, 1948).

A difference between anterior and posterior T_b during arousal is seen in virtually all hibernators, including bats. Since the end of the 19th century, it has been known that there are profound circulatory adjustments during hibernation. By using a radio-opaque dye and X-ray equipment, it is possible to show that blood flow to the posterior region of the golden hamster is restricted during hibernation, but increases in the forelimbs, heart, diaphragm, thorax and deposits of Brown Adipose Tissue (BAT) during the initial stages of arousal. Figure 4.18 shows various measures, blood pressure, heart rate, and rectal and heart temperatures, during arousal of a 13-lined ground squirrel (*Spermophilus tridecemlineatus*) (Lyman and O'Brien, 1960).

Peripheral vasoconstriction, and thus the resistance to blood flow, appears to lessen at the start of arousal, indicating vasodilation, but then rises rapidly while the heart rate is also increasing: as a consequence the rapidly beating heart is working against a high blood pressure. In these circumstances, although the heart may be an inefficient pump it is a good source of heat as the animal warms up. As rectal temperature increases rapidly, the blood pressure starts to decline, associated with a decrease in peripheral

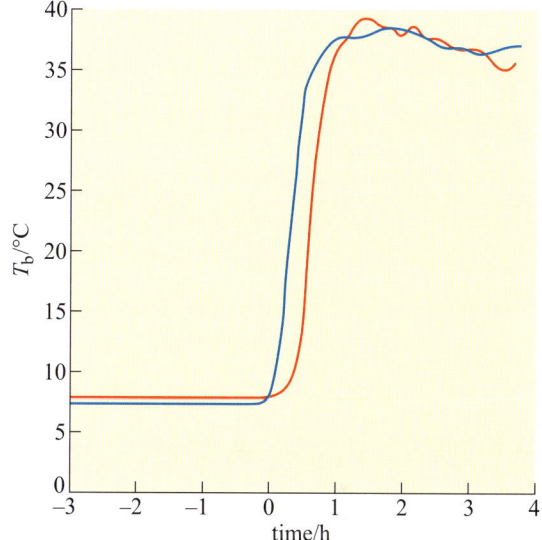

Figure 4.17 Core body temperatures of two ground squirrels before and after spontaneous arousal (Mayer, 1960).

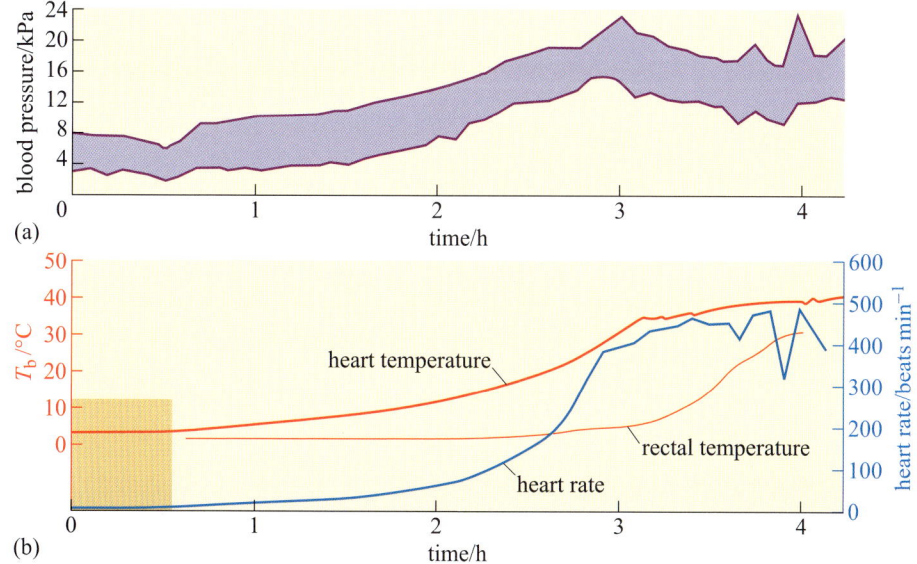

Figure 4.18 Blood pressure, heart rate, and rectal and heart temperatures in a 13-lined ground squirrel arousing from deep torpor after being disturbed, at a T_a of 3 °C. (The band indicating blood pressure shows the range between the systolic (upper value) blood pressure and diastolic (lower value) blood pressure.)

resistance, probably due to a sudden vasodilation in the posterior regions of the body is thought to be responsible. Evidently, vasomotor changes are important in arousal as well as in entry to torpor.

Hibernators arouse without recourse to external heat, so what is the source of the heat required? The violent shivering that accompanies some stages of arousal suggests that contraction of skeletal muscle is important but even animals in which skeletal muscle activity has been inhibited by curare (which blocks transmission at neuromuscular junctions) can re-warm.

■ From what you have read previously in Book 1 (Section 1.3), what alternatives to shivering might act as a source of heat?

BMR is maintained mainly by a number of tissues with high metabolic activity. One of these, BAT, is unique in its ability to adjust its output of heat from being very low to being the body's principal source (see Box 4.1).

BOX 4.1 BACKGROUND TO BROWN ADIPOSE TISSUE (BAT) – THE ROLE OF UNCOUPLING PROTEIN

Experiments using rabbits show that a marked lowering of T_a initially lowers the temperature in many parts of its body, including areas that contain BAT. However, the temperature of the area with BAT rapidly returns to normal (371°C) because of thermogenesis within the BAT. In contrast, muscle temperature continues to fall because it has no capacity for temperature-related thermogenesis. The temperature of organs surrounded by, or near to BAT, show a smaller decrease in temperature. These results explain how core temperature, and hence the functioning of essential organs, is protected.

Mammals contain white adipose tissue (WAT) and, usually, BAT. The proportion of the total mass that is BAT varies from species to species, some having none but some having up to about 5% of body mass. The extent to which BAT produces heat in response to cold stress depends on the amount of BAT. Broadly speaking, mammals with small neonates, small mammals that are cold-acclimatized, and mammals arousing from hibernation, all contain a significant amount of BAT and are substantially dependent on it for thermogenesis. BAT is *very* much more thermogenic than any other tissue, by about a factor of ten (per unit mass), and is located in discrete depots with a good vascular supply. These depots contain brown adipocytes, which have a characteristic size and structure and are well endowed with numerous, large mitochondria with many cristae and a high concentration of cytochromes. It is the degree of vascularization and the concentration of cytochromes that give BAT its brown colour. Brown adipocytes also contain more glycogen than white adipocytes. White adipocytes are mainly unilocular (one large vacuole containing fat), whereas brown adipocytes that are thermogenically active are multilocular (a large number of small vacuoles containing fat). Brown adipocytes uniquely contain uncoupling protein (UCP). BAT is well innervated by neurons of the sympathetic branch of the autonomic nervous system. Neurons innervating brown adipocytes contain the neurotransmitter, noradrenalin; those innervating blood vessels contain another neurotransmitter, neuropeptide Y.

WAT is primarily a store of fat that is mobilized to provide lipid fuel for tissues remote from itself. In contrast, the function of BAT during cold stress is that of thermogenesis. When a BAT depot is cold-stressed, there is rapid cell division that increases the mass of BAT. The diameter of blood capillaries increases, and there are also changes in the histological make-up. Cold stress also leads to an increase in BAT sympathetic nervous activity and in the tissue concentration of UCP.

Non-BAT mitochondria (whether in WAT or any other tissue) are capable of transforming about 40% of the chemical energy of the respired substrates to the chemical energy of ATP; the other 60% appears as heat. However, because the rate of respiration is coupled to ATP production, the rate of respiration, and thus oxygen consumption and heat production, is comparatively low. ATP is produced by a mechanism that depends on the inner mitochondrial membrane being impermeable to hydrogen ions *except* via channels that are part of the ATP synthase complex. In BAT mitochondria however, fuel oxidation is uncoupled from ATP production, thus ensuring that all the energy released from oxidation is released as heat. As a consequence of uncoupling, which is brought about by the uncoupling protein UCP, the rate of heat production is greatly increased (Figure 4.19).

UCP is situated in the inner mitochondrial membrane and acts as a proton translocator providing a route by which hydrogen ions (built up in the intermembrane space as a consequence of electron transport) are able to re-enter the matrix *without* passing through the ATP synthase complex.

Figure 4.19 A schematic summary of proton translocation actions of UCP.

Blood flow to BAT reaches a maximum level in the arousal process, and recent work has shown that, while the uncoupling protein (UCP) in the mitochondria is in a 'masked' or inactive form during hibernation, the amount of active UCP is rapidly increased during arousal (see Section 4.4).

The interscapular region contains the largest mass of BAT near to the body surface. In hibernating mammals, this region is significantly warmer than other parts of the body (except possibly the heart) in the early stages of arousal. In big brown bats, the area of BAT is always warmer than the heart, and arousal is very fast (8 °C to 37 °C in less than 30 minutes). Figure 4.20 shows the thermographic tracing of body heat (infrared radiation) in an arousing bat. The area of BAT is the warmest part which suggests that it is the thermogenic source (Hayward and Lyman, 1967).

We should not infer that BAT is the major or only source of heat in all arousing hibernators. In those rodents which lack BAT, and probably also in the hedgehog, most of the heat in a normal arousal is generated by the shivering of skeletal muscle. Although birds do not seem to possess BAT, they are capable of

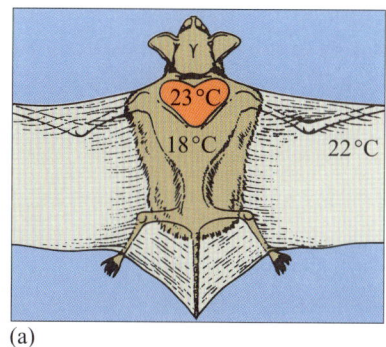
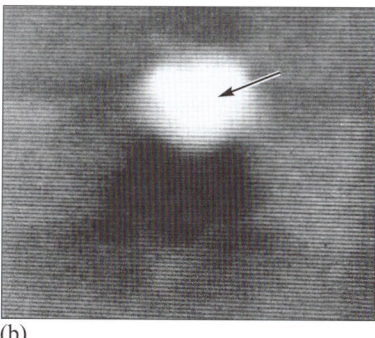

Figure 4.20 (a) A bat in position for thermography (dorsal side uppermost) showing the location of the interscapular region (red), and the major temperatures prevailing at the commencement of thermographic scannings. (b) Thermogram of the dorsal surface of a bat during its arousal from torpor. (The higher the temperature and intensity of infrared radiation from the skin, the brighter the image.)

non-shivering thermogenesis (NST), which is cold-induced heat production that is not due to muscle shivering. This kind of thermogenesis takes place in the muscles of cold-adapted ducklings and emperor penguin chicks, and there is evidence that it depends on free fatty acids liberated from WAT. Molecular evidence reviewed in Section 4.4 is pointing to the possibility that WAT, as well as BAT, can serve as a source of heat. However, it is not yet known what the relative importance of shivering and non-shivering thermogenesis is during arousal from daily (or more prolonged) torpor. Most birds can increase their BMR by a factor of four or five under extreme cold stress.

Final arousal

Emergence can be viewed as the final step in the series of periodic arousals. Instead of re-entering hibernation, the animal maintains the euthermic condition. The cue for maintaining this final arousal is probably not temperature, as some species emerge when T_a is well below zero. It is also difficult to see how arousal could be affected by daylength, since the hibernating animal is usually underground in a cavity or a burrow. Perhaps fat or food stores reach a minimum level or the timing of the final arousal is pre-programmed into the animal's activity cycle.

4.3.5 Length of torpor bouts in hibernation

It is obvious that there is a very high energetic cost to arousal, and an even higher one to the periods of euthermic wakefulness prior to re-entering torpor. If an animal could simply enter torpor once, and arouse 2, 4 or 6 months later, depending on the environment, it would represent a huge energy saving. Thus, it has been assumed that either prolonged torpor is physiologically impossible, or there is some strong selective value to the species in regular arousal. In the case of some small species of mice, which cannot store very much energy as fat and therefore build up a cache of seeds in their hibernacula, periodic arousal to feed is explicable, as is arousal to forage in those species that do not make food stores. For most species, however, there is no such obvious rationale. Larger animals tend to have lower metabolic rates than smaller animals, but tend to have longer euthermic intervals.

Summary of Section 4.3

The physiological details of deep or seasonal hibernation vary widely between species. However, the general pattern is similar, involving controlled entry to torpor, with or without 'test drops', and periodic arousals. The intervals between these arousals depend on size, T_b and other factors. The frequency of the arousals falls off during the deepest part of the hibernation. Entry to hibernation may be triggered by temperature, daylength and shortage of food, especially in facultative hibernators (e.g. hamsters, chipmunks), or by endogenous circannual rhythms, as in some obligative hibernators (e.g. marmots, ground squirrels). In spite of the very low T_b, physiological control is maintained, as is a low level of metabolic activity.

The three types of arousal – alarm, periodic and final – are physiologically similar. Alarm arousal is initiated by external stimulation. Periodic arousal is initiated by endogenous signals. Strategies for increasing T_b during arousal include both muscle activity and NST. BAT is certainly important as a source of heat for arousal in many species of mammals, though apparently not in birds. Final arousal occurs in spring, though it is not known what prevents the animal from re-entering hibernation. Arousal appears to be physiologically imperative at some stage during torpor.

4.4 Physiological adaptations – molecules and cells

Even after many years of research, the phenomenon of hibernation continues to be a mystery to scientists. Despite coming nearer to an understanding of how and why it happens, some fundamental questions remain unanswered. Is there a genetic basis underlying the evolutionary predisposition of animals to hibernate, given its occurrence in many groups of vertebrates and invertebrates? Is the problem of metabolic adaptation in cells separate from thermal regulation which occurs throughout the organism?

4.4.1 Scientific approaches

Faced with an exploration of the unknown, rather than simply testing an existing hypothesis, research can adopt two different approaches in attempting to associate molecular events with physiological functions.

The *analytical approach* seeks to identify all the significant changes that accompany a specific physiological adaptation and then seeks to explain these changes. For example, differences in the patterns of expression of a number of genes that manifest themselves during hibernation can be identified against a background of no change. The processing of very large numbers of genes in this way became possible in the late 1990s with the advent of DNA chip technology. This powerful approach can provide information about the state of expression of thousands of genes in each tissue (Box 4.2, overleaf).

Such an analytical approach has led to the identification of a few genes in a range of species whose expression is either increased or decreased during hibernation (Table 4.4). The table shows that some of the genes are associated with the metabolic, respiratory or control functions that you might predict would be linked to the maintenance of, or recovery from, hibernation. The links revealed in this kind of analysis sometimes appear indirect or tenuous. The benefit of this 'needle-in-the-haystack' approach, however, is that it reveals changes in expression patterns in a handful of proteins amongst a very large number which *do not* undergo such changes – even those that might be *predicted to do so*. Genetic analysis has revealed significant changes in the expression of genes whose function is not known or not directly related to hibernation (for example, the HT20 gene family in Table 4.4).

Table 4.4 Genes whose expression level changes in hibernation, identified in microarray and subtractive expression experiments.

Gene	Tissue	Change in expression	Possible function
apoferritin	liver	increase	iron storage protein – increases availability of iron for cytochromes and haemoglobin
c-fos	brain	redistribution[+]	rapid-response gene – coordinates reaction to physiological change
genes for fatty acid binding proteins	BAT	increase	preparation for rapid thermogenesis during arousal
genes for glyceraldehyde phosphate dehydrogenase	liver, muscle	decrease	reduces glycolysis
HT20 family	muscle, WAT	decrease	function unknown – all related structurally to protease inhibitor α1 anti-trypsin
genes for pyruvate dehydrogenase	heart, skeletal muscle	increase	depresses metabolism by preventing pyruvate kinase conversion into acetyl CoA

[+]*c-fos* expression undergoes a decrease in the hypothalamus during hibernation and a redistribution within specific neuronal nuclei on arousal.

BOX 4.2 DNA 'MICROARRAYS' FOR THE STUDY OF GENE EXPRESSION

To elucidate the entire mRNA content of a cell, or to study all the genes that are being actively transcribed at any one time, every protein-coding gene in the genome would have to be analysed. The enormity of this task can be appreciated only by considering the number of genes that are activated in a single biological process. For example, during the transition from aerobic to anaerobic respiration in cells of the yeast *S. cerevisiae*, changes in the expression of 1740 genes have been recorded. This large change in gene expression in *S. cerevisiae* has been analysed using the technique known as **DNA microarray** (Figure 4.21).

The design of DNA 'chips' allows many hybridization experiments to be performed in parallel. A DNA chip is a very thin layer of silicon, $2\,cm^2$ or less in area, carrying a large number of DNA probes, a microarray, each with a different sequence and each at a defined position on the chip. The probes can be short oligonucleotide sequences, and can be spotted onto the silicon using high-speed robotics to form a microarray, which is then incubated with the labelled target to allow hybridization to take place. To determine which oligonucleotides have hybridized to the target, the surface of the chip is scanned and the positions at which the signal emitted by the label is detectable are recorded.

Studies on cancerous tissue using this technique discovered genes whose expression patterns differed significantly when normal colon epithelial cells were compared with colon cancer cells. In brief the method was as follows (Figure 4.22). Messenger RNA preparations were made from cancerous and normal cells. Each preparation was then labelled with a radioisotope attached to a fluorescent marker and allowed to hybridize to a microarray containing probes for several thousand human genes. The hybridization signal associated with each gene was then used to assess the particular mRNA in each preparation. In pancreatic cancer cells, about half of these genes also showed abnormal expression levels. The implication of this study is that some genes are abnormally expressed in more than one type of cancer while others are expressed more specifically. DNA microchip technology has been one way to enable such studies of gene expression.

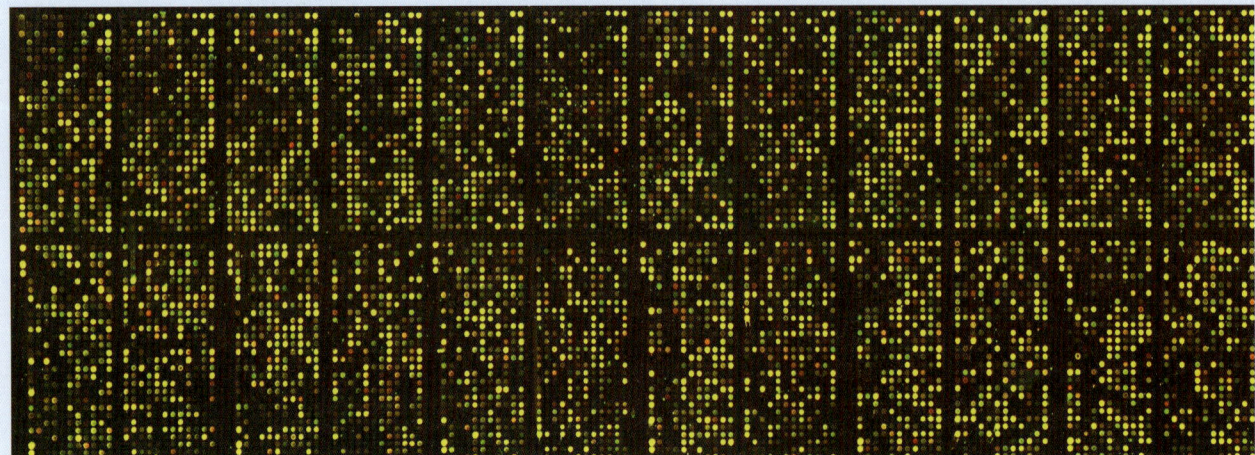

Figure 4.21 Example of a DNA microarray.

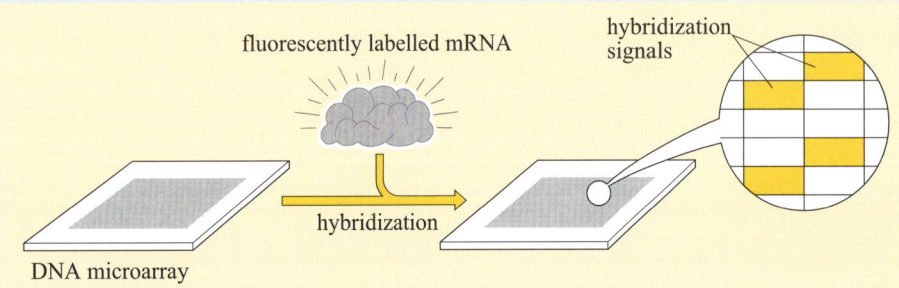

Figure 4.22 Analysis of gene expression using DNA microarrays.

■ How can we attach relevance to an observed change in gene expression, and hence predicted changes in the biosynthesis of the protein which the gene encodes?

First, the nucleotide sequence of each gene is compared with those in the genomic database for the species. High levels of sequence **homology** (similarity) enable us to predict functional analogies with these genes on the basis of predicted protein structure, spatial (tissue location) and temporal (daily or seasonal) expression patterns. If such a search fails to establish any homology with any known gene, it is necessary to establish a function without precedent for the new protein. The most valuable evidence frequently comes from observations on loss of physiological function when one or both copies of the gene are deleted from the genome. Hypotheses on the function of the protein may then be tested by determining the proteins and subcellular structures with which it interacts in order to fulfil its function.

The analytical approach has indicated that different tissue-specific genes are activated or repressed in the brain and peripheral organs of hibernating animals. The genes encode pre-identified proteins which are linked to the role of each tissue in hibernation (i.e. reactive control in the brain, metabolic adaptation in liver muscle and adipose tissue) plus some proteins whose function is unclear.

The *targeted approach* seeks to investigate a possible role for molecules already suspected of participating in physiological regulation. An example is the enzyme arylalkylamine-*N*-acetyltransferase (AA-NAT). This enzyme catalyses the rate-limiting step in the production of a hormone, melatonin, from the pineal gland. Melatonin has the capacity to re-set circadian rhythms in a variety of physiological processes. Entry into hibernation requires neural and endocrine control systems that govern the normal day/night (circadian) cycles of metabolism and T_b, to be overridden.

■ What would you expect analysis of gene expression to show?

Expression of AA-NAT, leading to elevated biosynthesis of melatonin, should increase just prior to onset and during hibernation. Circadian rhythms would then be interrupted whilst the animal is torpid.

This prediction was supported by studies using 13-lined ground squirrels. Messenger RNA for AA-NAT protein biosynthesis increased in the brains of hibernating animals as expected – an accepted indication that the enzyme's activity had increased (Figure 4.23) (Yu et al., 2002).

In a second example, researchers proposed a working hypothesis that the uptake of fatty acids by hibernating tissues, which is normally under tight control, should be deregulated in preparation for the task of storing large reserves of lipid. They expected to see changes in the activity of a key enzyme, acetyl CoA carboxylase. This enzyme catalyses the formation of malonyl CoA, a potent and major inhibitor of mitochondrial fatty acid uptake. In studies on Richardson's ground squirrel, the hypothesis was proved right. Acetyl CoA carboxylase in heart muscle was much reduced prior to and during hibernation, so allowing more fatty acid uptake by mitochondria in cardiac muscle.

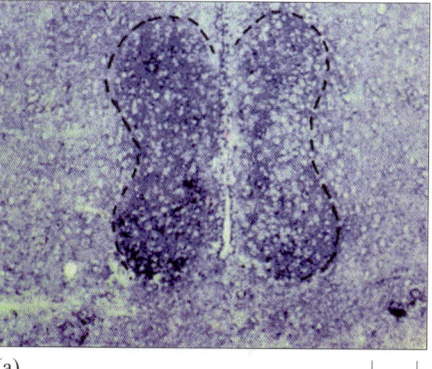
(a)　　　100μm

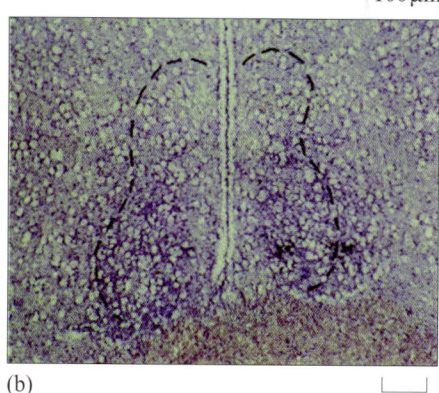
(b)　　　100μm

Figure 4.23 The effect of hibernation on AA-NAT mRNA expression in the brain of the ground squirrel demonstrated by *in situ* nucleic acid hybridization histochemistry on a tissue section. Cells positive for AA-NAT mRNA were detected by a blue-purple colour in (a) hibernating and (b) non-hibernating animals.

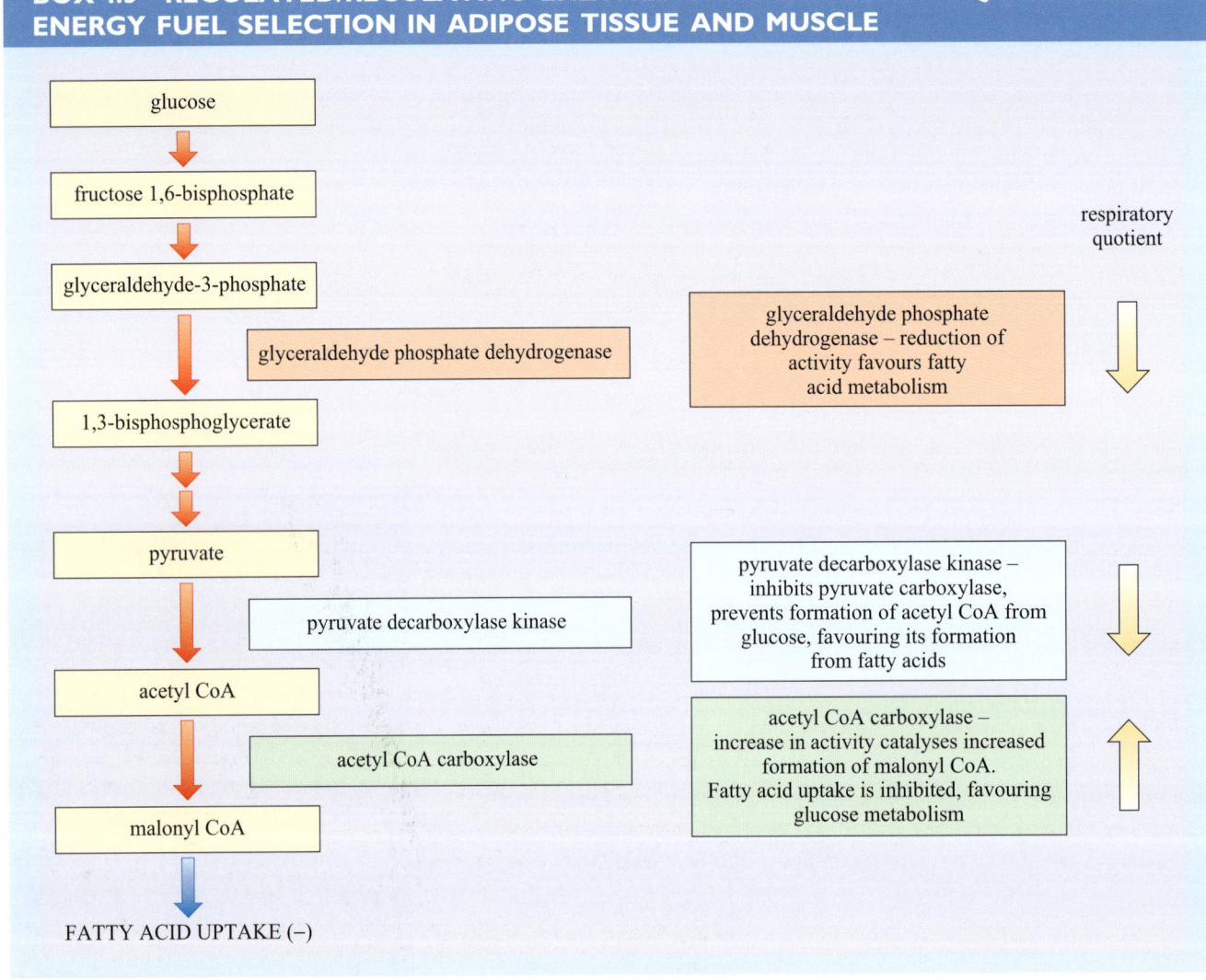

BOX 4.3 REGULATED/REGULATING ENZYMES THAT INFLUENCE RQ VALUE AND ENERGY FUEL SELECTION IN ADIPOSE TISSUE AND MUSCLE

Gene expression research has provided evidence for the enzymes that play an important role in determining the choice of respiratory fuel in adipose tissue and hence the respiratory quotient (RQ; see Box 4.3). As we will see in the next section, changes in energy sources are a characteristic of hibernators.

4.4.2 Arresting protein synthesis

The regulation of T_b in hibernators has traditionally been viewed as the fundamental physiological process in hibernation. But recently, questions have been raised about whether thermal changes initiate or simply accompany metabolic depression. Is the metabolic inactivity of animal tissues during bouts of torpor or in hibernation, the cause or the result of hypothermia? A common-sense view is that temperature directly influences metabolism by regulating enzyme activity. Evidence of separate, temperature-independent regulation of metabolic processes in hibernators would lead us to reconsider the classical view of hibernation.

In greater horseshoe bats (*Rhinolophus ferrumequinum*), in which large changes both in T_b and in body mass occur during hibernation, the duration of torpor is inversely related both to T_a and to the animal's level of body hydration in hibernating ground squirrels. Protein synthesis in brain tissue is actively arrested for several weeks at a time and, at the onset of hibernation, mRNA is translated into protein at a much slower rate in brain tissue extracts even when measured at 37 °C (Frerichs, 1998).

This change in metabolism may lead to the onset of cellular quiescence during hibernation. Protein synthesis *in vivo*, as measured by the incorporation of radioactive leucine into the brain tissue of ground squirrels, is almost undetectable during entry into hibernation, even though T_b is still high. Figure 4.24a is an autoradiogram which shows tissue sections that have been exposed for several days to X-ray film. The colours represent different levels of incorporation that are detected as radioactive emissions. The green areas show relatively high levels of incorporation in an area of the brain called the hippocampus in the active but not in the hibernating animal. The *in vitro* studies in Figure 4.24b confirm that there is relatively little leucine incorporation into the brain tissue of the hibernating animal. The differences in control mechanisms over protein synthesis in hibernating and euthermic brains are subtle. There is no difference in the structure or quantity of mRNA or the functioning of the ribosomes on which it is translated to protein. However the 'transit time', or duration taken for polyribosomes to process each mRNA molecule, is three times longer in hibernating animals. The initiation factor, eIF2, which is involved in the aggregation of ribosomes and the initiation of translation, may be inhibited in hibernating cells. Although this finding does not point to suppression of specific areas of metabolism, we can predict that energy generation, tissue homeostasis and growth are all likely to be suppressed by a general fall in protein synthesis.

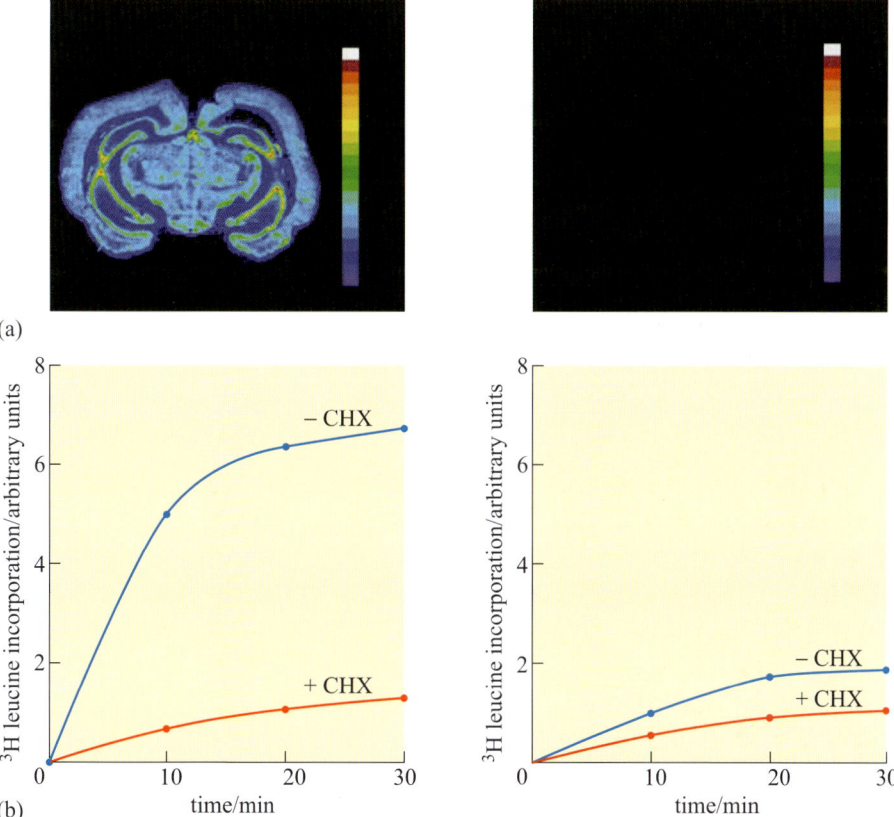

Figure 4.24 (a) Colour-coded autoradiograms, showing rates of cerebral protein synthesis as detected by the incorporation of a radioactive derivative of leucine in an active (*left*) and a hibernating (*right*) ground squirrel in coronal (vertical and left-to-right) sections of the brain at the level of the hypothalamus. The bar on the right of each figure shows increasing levels of leucine incorporation from purple (zero) to red (high). During hibernation, leucine incorporation was not detected. (b) Cytoplasm from hibernating cells translates mRNA to protein at a lower rate than euthermic cells, even at 37 °C. Following introduction of radio-labelled leucine at time zero, the reaction was allowed to proceed for 30 minutes. Incorporation of leucine into the cell extract was several times higher in cells from active (*left*) compared with those from hibernating (*right*) brains. Incorporation was blocked by cycloheximide (+CHX), a specific inhibitor of protein synthesis.

However, the change in protein synthesis in laboratory experiments is just as evident in cells from hibernating animals at 37 °C; it does not appear to be a consequence of thermoregulation. This finding does not mean that the change from protein synthesis is wholly independent of thermoregulation. In golden-mantled ground squirrels, reorganization of polysomes and an increase in mRNA elongation increases abruptly at 18 °C during arousal, as the need for protein synthesis becomes critical.

4.4.3 Cellular changes

Hibernation can result in the deposition of fat in adipose tissue. In tissues of finite size which are important sources of energy and sites for fuel metabolism, changes in cell structure (redistribution of organelles involved in energy metabolism and protein synthesis) are the most likely adaptation to a state of torpor. Liver hepatocytes of the hibernating dormouse (*Muscardinus avellanarius*), are visibly different from those of arousing and euthermic dormice when viewed in thin section with a microscope (Figure 4.25a(i) and (ii)) (Maletesta et al., 2002).

■ Can you see evidence of these differences in Figure 4.25a?

There is a substantial reduction in the cross-sectional area of both the whole cells and the cytoplasm in the hibernating dormice. The number of granular glycogen deposits is reduced which suggests that carbohydrate metabolism is reduced. The Golgi apparatus shrinks dramatically, which indicates that reductions in protein and lipid synthesis as well as carbohydrate metabolism have occurred.

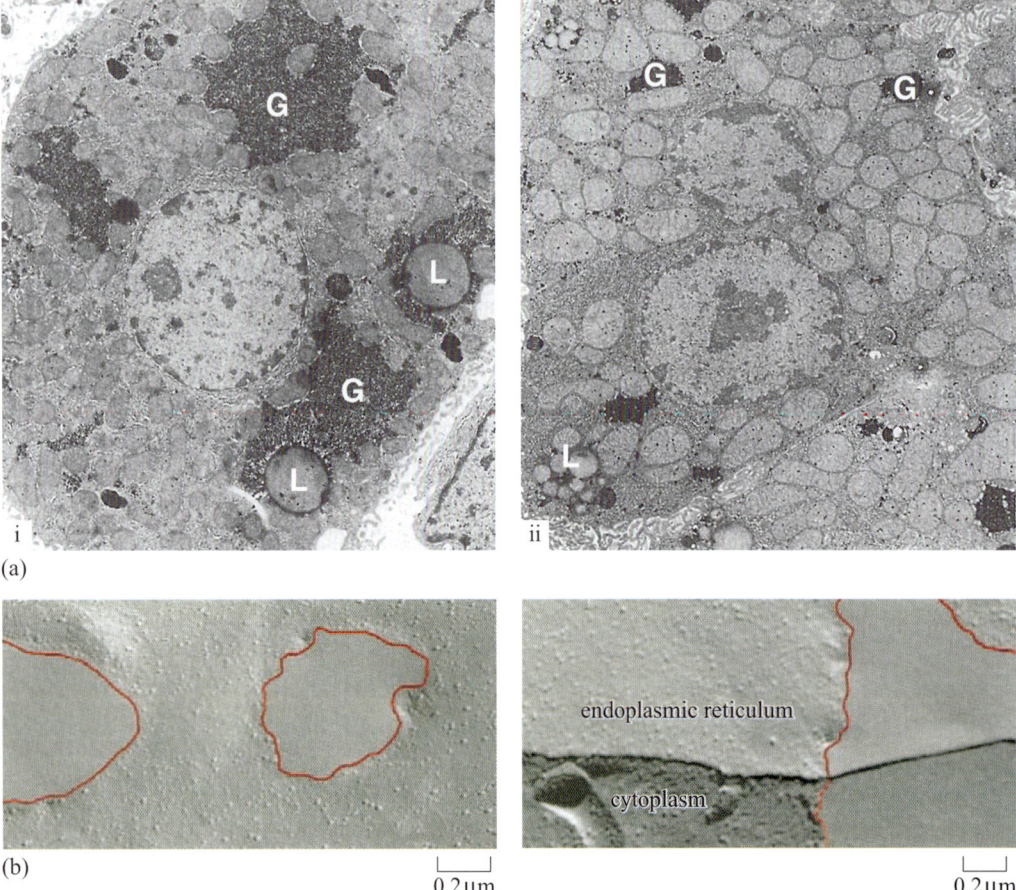

Figure 4.25 Alterations in fine structure of hibernating dormouse cells. (a) Changes in hepatocyte glycogen (G) and lipid storage deposits (L) in (i) euthermic and (ii) hibernating dormice. (b) Redistribution of membrane proteins into 'patches', as seen in freeze-fracture images, may enable membranes to function at lower temperatures and withstand rapid temperature changes.

Glycogen does not reappear in the cell for several hours after arousal. However, the whole cells and the cytoplasm increase significantly in size and start to resemble those of euthermic animals. The reverse changes are true for lipid storage in the cell, with the proportional cross-sectional area occupied by lipid droplets being significantly increased in early hibernation, reduced in deep hibernation and almost disappearing during arousal. Changes in the structure of organelle membranes in neurons from the brain of hibernating hypothermic ground squirrels are also visible when viewed with an electron microscope. Membrane lipids and proteins coalesce to leave patches free of protein (Figure 4.25b) (Azzam et al., 2000). These detailed observations provide clear indications that fundamental restructuring of lipid bilayers can occur as an adaptation to torpor. The reasons for the changes are not fully understood, but presumably they give some protection from cold-damage and permit rapid recovery of cells from temperatures close to zero. The survival of cells is also the subject of molecular adaptations as we will see below.

4.4.4 Cell survival mechanisms

Physical damage is not the only danger that faces cells recovering from low temperatures in the absence of oxygen (due to a 90% drop in blood flow to the brain) and energy supplies. A universal sign of recovery from such conditions is the production of reactive oxygen species (ROS) (Box 4.4). The electron transfer chain that participates in the formation of water from oxygen in mitochondrial respiration can also be used in the production of the free radical superoxide, sometimes called 'singlet oxygen' because the molecule contains an extra unpaired electron. These short-lived molecules can initiate spontaneous chain reactions through electron transfer leading to the generation of highly

---

BOX 4.4 GENERATION OF REACTIVE OXYGEN SPECIES

Normal respiratory pathway:

$$O_2 \xrightarrow[2e^-]{4H^+} 2H_2O$$

Generation of superoxide:

$$O_2 \xrightarrow[e^-]{2H^+} O_2^{\cdot -} \xrightarrow{\text{superoxide dismutase}} H_2O_2$$

Generation of hydroxide free radicals:

$$H_2O_2 + O_2^{\cdot -} \xrightarrow{5e^-} HO\cdot + H_2O$$

Generation of lipid peroxides from hydroxide free radicals:

$$HO\cdot + \text{polyunsaturated FA} \xrightarrow{5e^-} \text{lipid peroxides}$$

Generation of reactive species that damage proteins and nucleic acids:

$$\text{lipid peroxides} \xrightarrow{5e^-} \text{malonaldehyde} + \text{4-hydroxynonenal}$$

reactive hydroxyl free radicals. They then react with free fatty acids forming the compounds malonaldehyde and 4-hydroxynonenal, which cause damage to proteins and nucleic acids, eventually leading to cell death. There are two adaptive mechanisms which hibernating vertebrates have adopted to counter the potentially toxic consequences of a surge in oxygen supply on arousal.

First, the concentration of a number of ROS-neutralizing compounds such as vitamin C (ascorbic acid) and glutathione, and the enzyme superoxide dismutase, increases on arousal from hibernation. Neutralizing compounds are 'scavengers' of lone electrons present in superoxide and hydroxyl free radicals, whilst an increase in the activity of superoxide dismutase results in the conversion of all the superoxide to hydrogen peroxide and hence reduced formation of the hydroxyl free radical. In the ground squirrel, circulating levels of ascorbic acid increase by up to five times. Continuous measurements of blood ascorbate in arctic ground squirrels during arousal show that the levels of anti-oxidant start to decrease at the peak level of oxygen consumption, indicating that ascorbate is being distributed to respiring tissues to counter oxidative damage.

Secondly, proteins that prevent the sequence of events leading from ROS damage to cell death are activated. In the 3-lined ground squirrel (*Lariscus insignis*), cells lining the intestine are particularly vulnerable to ROS damage as they adapt to the absence of dietary nutrients. Lipid peroxides are one of the end-products of superoxide activity and their levels increase during entry into, and during, the early phase of a torpor bout. Their formation is accompanied by a substantial increase in biosynthesis of a protein known as NFκB. The protein is a gene regulator that is only stimulated in affected cells by redox reactions which accompany the generation of superoxide. No stimulation is seen in WAT where no lipid peroxidation is measured. NFκB activates genes that lead to cell death by preventing metabolic pathways initiated in the mitochondria. A hibernation induction trigger (HIT) circulates in the plasma of hibernating mammals (discussed in Section 4.6.5) and is believed to protect cells from death by activating similar protective genes. This area of hibernation research, as you might expect, is of considerable interest to medical researchers seeking the means to protect victims of cerebral ischemia and stroke, characterized by loss of blood flow very similar to that experienced by hibernators, from long-term tissue damage.

Summary of Section 4.4

Molecular approaches to the study of hibernation have combined largely non-hypothetical *analytical* and *targeted* methods. The analytical approach shows changes in the expression patterns of novel genes, often with unforeseen or unknown functions in hibernation, whilst indicating that the scale of the full range of adaptive genetic changes is small. The targeted approach has confirmed an important role for gene products normally involved in maintaining biological rhythms and energy-generating metabolic reactions.

The basis for the slowing of metabolic processes is the arrest of protein synthesis in hibernating cells through changes that inhibit the initiation of messenger RNA translation and polysome assembly. Although the mechanism operates in hibernating cells at any temperature once initiated, it is triggered during entry to torpor at a critical T_b. Changes that occur in the structure of hepatocytes, their fuel deposits and mitochondria during the transition from carbohydrate to lipid metabolism, are indicative of enduring adaptations at microscopic level.

Hibernators have a system of protection against cellular injury or death resulting from the actions of reactive oxygen species (ROS) produced during the respiratory burst that accompanies arousal. Protective mechanisms include an increase in the level of ROS-neutralizing compounds in the blood and regulators that inhibit biochemical pathways leading to the death of individual cells.

4.5 Physiological adaptations – respiration and energy provision

The change in BMR observed in all hibernators has traditionally been viewed as a passive response that is a consequence of hypothermia. However, many studies have provided evidence for temperature-independent regulation of BMR. In the alpine marmot (*Marmota marmota*), a BMR that is less than 5% of summer levels is maintained despite the frequent fluctuations in body temperature between 8 and 18 °C. The mechanism of body temperature regulation in marmots, during long periods of hibernation, has become clearer following investigations of T_b and BMR throughout this phase. Entry into hibernation is facilitated by a precipitate drop in BMR that precedes slower temperature changes, then throughout the winter, bursts of thermogenesis occur quite independently of T_a (Figure 4.26) (Ortmann and Heldmaier, 2000).

It is widely acknowledged that mammals switch to the use of lipid from WAT during hibernation. A period of 'fattening-up' precedes the onset of hibernation, under the control of hormones which stimulate lipid storage. In mice induced to enter a near-torpid state, levels of leptin are low, reducing lipolysis and promoting lipid storage. However, BMR is lowered in the little brown bat (*Myotis lucifugus*; Figure 4.8), despite an increase in plasma leptin during the pre-hibernation period which suggests that weight gain is controlled by other hormones together with a resistance to leptin-induced satiety in this species.

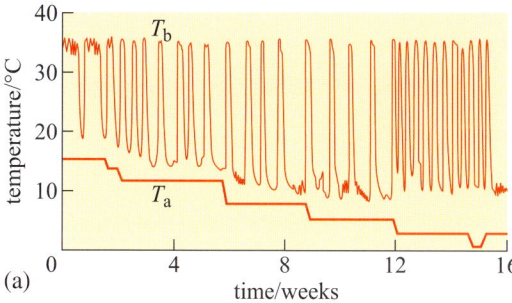

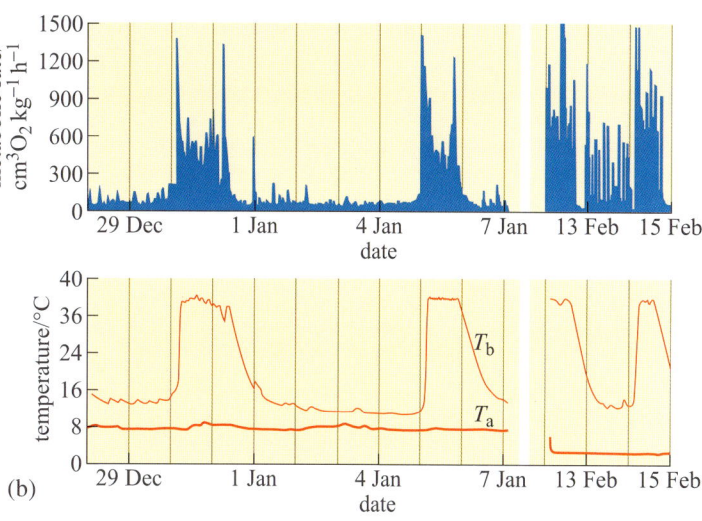

Figure 4.26 (a) Time course of body temperature (T_b) over an entire hibernation season in the alpine marmot (*Marmota marmota*). Ambient temperature (T_a) decreased stepwise from 15 °C in autumn to 0 °C in spring. (b) Time course of weight-specific metabolic rate (*top*) and T_b and T_a (*bottom*) over a time period of 10 days and then 3 days. T_a was decreased from around 7 °C at the end of December through the beginning of January to around 2.5 °C in mid-February. Note that bursts of higher metabolic rate occur during deep hibernation which do not synchronize with changes in T_b.

4.5.1 Energy sources in torpor and hibernation

For animals that show daily torpor, such as Siberian hamsters (*Phodopus sungorus*) and Djungarian hamsters (*Phodopus campbelli*) (Figure 4.27), blood glucose remains the respiratory fuel for several hours following its onset. Thereafter there is a gradual reduction in respiratory quotient (RQ) indicating a change to lipid metabolism as the metabolic rate is reduced.

There are some exceptions to the rule that lipids are the energy source of hibernation. Arctic ground squirrels (*Spermophilus parryii*) overwinter in hibernacula at temperatures that are substantially below freezing, whilst maintaining a T_b at or just below 0 °C. In laboratory studies, an increase in RQ from 0.71 towards 0.85–0.90, together with a fall in the amount of stored glycogen in liver and muscle, has been measured on entry to hibernation. Together, these changes point to a switch to glucose as the principal energy source.

Figure 4.27 Djungarian hamsters, *Phodopus campbelli*.

■ Why might carbohydrates be required as fuels in some hibernators?

First, following arousal episodes, glucose-utilizing tissues such as brain and blood cells would have a specific energy requirement that cannot be met by stored lipids. Secondly, there is a need to replenish lipids within BAT for the substantial needs of NST at the next arousal episode.

The latter need is not met from stored glycogen, but from gluconeogenesis, the biosynthesis of glucose from amino acids. The provision of amino acids requires the breakdown of muscle protein that contributes to the weight loss seen in many hibernators. The pattern of hibernation is frequently related to the combination of lipids with protein as the prime metabolic fuels. Black-tailed prairie dogs (*Cynomys ludovicianus*; Figure 4.6), whose diet is rich in polyunsaturated fatty acids (PUFA) and which use protein as a source of energy in the winter only during periods of reproductive activity, enter shallow torpor only infrequently and do not hibernate continuously between autumn and spring. However, the duration of torpor and size of the reduction of T_b during hibernation can be increased by increasing the dietary intake of PUFA during the period before entry.

Arctic ground squirrels are capable of cooling their T_b to −3 °C, while simultaneously keeping the parts of the body involved in regulation and maintaining energy metabolism – the brain and intercapsular BAT – above zero. This ability shows that metabolic rate is regulated independently of body temperature, a conclusion which is further borne out by the data in Figure 4.28 (Buck and Barnes, 2000).

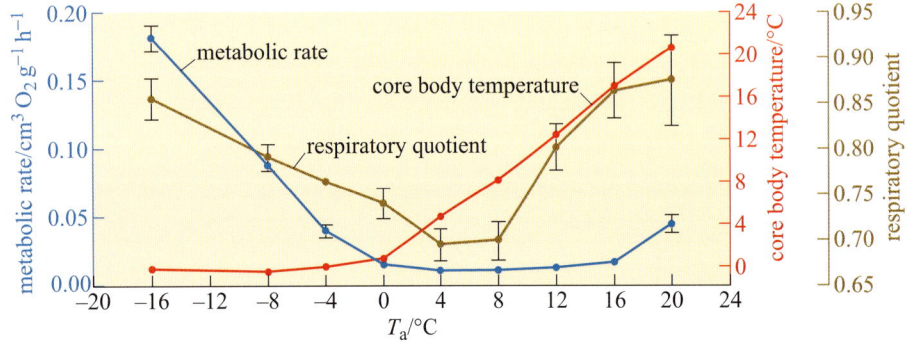

Figure 4.28 Effect of T_a on metabolic rate measured as rate of oxygen consumption, core body temperature, and respiratory quotient of arctic ground squirrels during steady-state torpor. (Values are means ± SE.)

Between T_a values of −16 °C and 0, T_b remains relatively constant but metabolic rate rises over 15-fold as the difference between T_a and T_b increases. At T_a 0 to 20 °C, T_b increases with T_a; however, metabolic rate does not change significantly from a T_b of 0 °C up to a critical temperature of 12 °C, indicating the existence of a temperature-independent inhibition of metabolic rate.

■ How do you interpret the change in respiratory quotient over the range of T_a used in the investigation shown in Figure 4.28?

For most mammals, hibernation requires a shift away from the oxidation of carbohydrates and towards the oxidation of fatty acids released from stored triacylglycerols as the primary source of energy during torpor (See Box 4.1).

In 13-lined ground squirrels, the activity of PDK4 is elevated between three and eight-fold by increased translation of specific mRNA in three tissues important in energy metabolism – heart, skeletal muscle and WAT – during entry to torpor (see Sections 1.1 and 4.4.2). It is important to bear in mind that translation of this gene, as well as that of a small number of other gene products (see Section 4.4), is selectively increased against a background of greatly reduced translational activity. This observation also points to the possibility of tissue-specific differences in the levels of biosynthesis of metabolically critical proteins.

4.5.2 Mitochondrial adaptations

During the winter months, whilst hibernating vertebrates maintain a very low metabolic rate, major reorganization of mitochondrial metabolism occurs. The phenomenon has been studied in some detail in frogs which, although not hibernators in the true sense, can endure very low water temperatures under the conditions of profound hypoxia that exist when they lie dormant for long periods below the surface. In contrast to normoxic conditions, the muscle mitochondria of dormant frogs depress their metabolic rate by up to 75%. Since muscles comprise a large part of the body mass, depression of their mitochondria decreases the overall oxidative metabolism of the frog profoundly. Although uptake of oxygen into mitochondria is decreased both in normoxic hypothermia and in the anoxic conditions of dormant frogs, only in the latter are long-term adaptations observed. Such adaptations include an increased affinity of mitochondria for dissolved oxygen, a reduction in the activity of mitochondrial enzymes, a reduction in the activity of the electron transfer chain and a reduction in the proton leak across the inner mitochondrial membrane.

In normoxic mammalian muscle mitochondria, it has been estimated that over 30% of the standard metabolic rate comprises the movement of protons into the mitochondrial matrix which is uncoupled from ATP synthesis (see Section 4.3.4). The electrochemical gradient (and hence the potential energy available from oxidative phosphorylation) across the inner mitochondrial membrane is maintained by the 'proton-motive force' (PMF), measurable as the potential difference across the membrane. At a time when energy substrate is very scarce, the proton leak is counterproductive for energy conservation.

To reduce energy wastage, the inner membrane ATP-synthase in frog muscle begins to catalyse the reverse reaction, releasing energy by ATP hydrolysis, to maintain the proton gradient across the membrane. The small amount of energy required to reduce the cycling of protons to a level equalling their rate of supply from the electron transfer chain is more than compensated for by the increased efficiency of the anoxic mitochondrion. This principle is illustrated in Figure 4.29. Under normoxic conditions, the passive flow of protons from the intermembrane space of the mitochodrion into the matrix is increased as the potential difference across the membrane rises. This gradient is nearly absent in the anoxic mitochondrion (Boutilier and St-Pierre, 2002).

It is not surprising that this phenomenon is also seen in mammals that hibernate on land, such as the arctic ground squirrel. The supply of lipid to the mitochondria of the heart and BAT is substantially increased during hibernation, as shown by the increased production of fatty-acid binding proteins that deliver fatty acids to the enzyme complex catalysing β-oxidation.

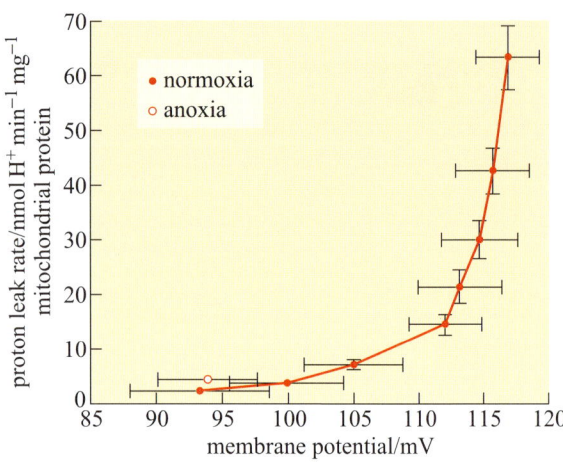

Figure 4.29 Kinetics of frog skeletal muscle mitochondrial proton leak during normoxia and in anoxia.

4.5.3 Inspiratory drive

The supply of oxygen to tissues such as the heart, liver and WAT is, under euthermic conditions, invariably linked to and dependent upon local blood flow and pulmonary function. However, as we have already seen, under conditions in which blood flow is reduced to a trickle, the control of energy supply switches to local adaptations in the capillaries and tissue cells, including the oxygen affinity of erythrocyte haemoglobin, the supply and metabolism of respiratory fuels and the rate of protein synthesis. In contrast to the highly regulated phenomena of hypothermia and bradymetabolism, regulation of lung function is passive and follows directly from the primary regulation of T_b.

Artificially induced hypothermia to mimic hibernation (Figure 4.30) is elegantly illustrated in the golden-mantled ground squirrel (*Spermophilus lateralis*) (Zimmer and Milson, 2002). On entry to a torpor bout, the normal regular pattern of breathing becomes episodic with periods of around 25 breaths being followed by up to 30 minutes of apnoea (Figure 4.30a). Although breathing frequency becomes lower and periods of apnoea lengthen, the basic pattern persists into steady state-hibernation from a T_b of 20.2 °C down to 5.3 °C. Thereafter, as T_a decreases and T_b falls below 6 °C during hibernation, the episodic pattern converts to one of slow, evenly spaced single breaths. Cooling is induced in euthermic squirrels using a combination of two anaesthetics which increase thermal conductance and suppress NST. The animals are cooled whilst anaesthetic levels are reduced and then withdrawn as T_a reaches 10.2 °C.

Severe hypothermia reproduced the patterns of breathing seen on entry to hibernation with a fall in oxygen consumption and carbon dioxide excretion to around 0.2% of their euthermic levels (Figure 4.30b). These changes were reflected in a similar reduction in heart rate, suggesting that hypothermia triggers changes in both the cardiovascular and inspiratory control centres of the brain. In hibernation, however, respiratory responses to reduced circulating levels of oxygen (hypoxia) and increased carbon dioxide (hypercapnia) are still evident, whilst in induced hypothermia, only the latter is present. These data indicate that the ability to sense tissue oxygen levels in hibernation is regulated independently of T_b.

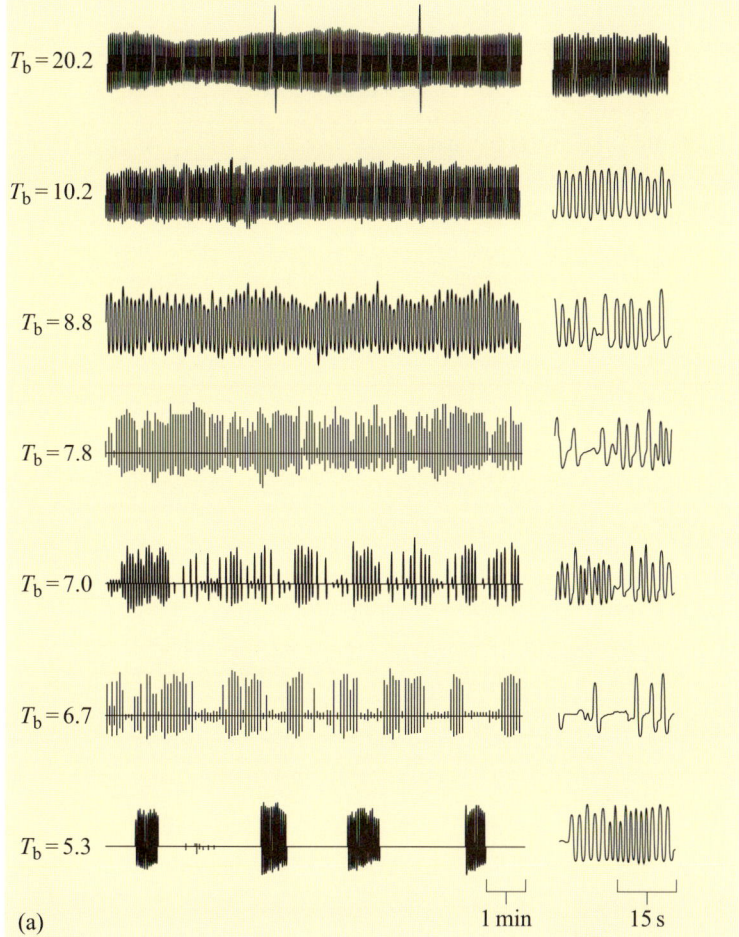
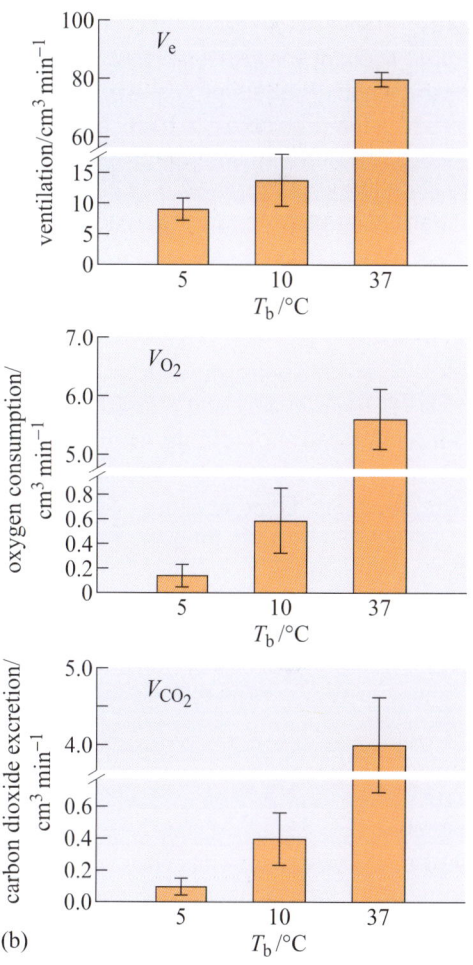

Figure 4.30 (a) Respiratory airflow traces illustrating the breathing patterns of an individual golden mantled ground squirrel during progressive cooling in hypothermia. Anaesthesia was used during initial cooling to 20.2 °C and removed at 10.2 °C. Note the change in breathing pattern to an irregular pattern when anaesthetic was removed, followed by waxing and waning (8.8 and 7.0 °C) and finally, episodic breathing, at 6.7 and 5.3 °C. The panel at the right hand side of each trace is an expanded 30 s view from the second minute of the trace. (b) Histograms showing total ventilation, oxygen consumption and carbon dioxide excretion. Data are expressed as means ± SE.

4.5.4 Energy budgeting – the benefits of hibernation and torpor

Studies performed on ground squirrels in the wild and in the laboratory have allowed estimates to be made of energy expenditure in hibernating and euthermic animals over similar periods (Wang, 1987). The average time spent by Richardson's ground squirrel in a periodic arousal in the wild is about 10 hours and the frequency of arousal decreases during November–March, when animals are spending more than 90% of their time in torpor. Monthly total oxygen consumption in January is about 35% of that in August, and rises again in February and March. Between July and March, entry into hibernation accounts for 27% of total energy expenditure, torpor for 33% and arousal for 40%. However, if the time between periods of torpor is taken into account, the figures are 12% for entry, 17% for torpor and 19% for actual arousal with 51% for the euthermic periods between arousals.

Entry into hibernation seems to cost little. Entry and the hibernation period together account for less than one-third of the total energy expenditure averaged over the whole 9-month period. Arousal in itself is not that expensive (less than 20%). It is the time between bouts of hibernation, when the animal is euthermic, that consumes more than 50% of the energy expended, though the amount varies between months. In December and January for example, entry and torpor account for more than 35%, and time in euthermia for about 40% of the energy expended.

We have estimates of how hibernators apportion expenditure throughout the hibernating season. But how do the figures compare with figures from similar individuals that are euthermic over the same period? Wang's estimates are given in Table 4.5. From August through to February, the savings due to hibernation are over 80%. Even in July and March, months in which arousal frequency is high and time in torpor is short, the savings are significant. On average over 9 months, hibernation endows a small mammal, such as this ground squirrel, with an 88% saving of energy. Over an entire year, the hibernating habit saves an animal 60% of the energy used by an individual of similar size that continuously maintains the euthermic condition.

Table 4.5 The reduction of metabolic cost by hibernating, in Richardson's ground squirrel.

Month	Total energy expenditure with torpor/cm^3 O_2 $month^{-1}$	Total energy expenditure if animal remained euthermic/cm^3 O_2 per month	% energy saved by exhibiting torpor
July	382	162	38.0*
August	592	323	81.6
September	452	313	85.4
October	372	372	89.9
November	312	432	92.7
December	262	542	95.1
January	221	555	96.0
February	332	501	93.3
March	762	394	51.2*
		mean for 9 months	87.8

* During parts of July and March the animal is not in hibernation, hence these values are low. When calculating the % energy saved by exhibiting torpor, it was assumed that the animals were hibernating for the equivalent of only half of these 2 months, i.e. a total of 8 months in the year.

A study of a close relation, the golden-mantled ground squirrel, showed that, during the hibernating season, energy expenditure in this particular species was also only 20% of the expected value if the animal remained euthermic. Furthermore, the energy consumption throughout the 7 months of the hibernating season was only 15% of the total annual consumption.

Both these species fit our earlier definition of obligative, deep hibernators; the picture may be a little different in facultative hibernators responding more directly to environmental cues. Using one such animal, the golden hamster (*Mesocricetus auratus*), the energy budgets of a whole population in the laboratory were estimated. The hibernating behaviour of individuals varied widely. The average for the group was that only 18% of the 'hibernating season' was spent in torpor, though this value might not be representative of a wild population. However, when in torpor their energy consumption was only 9% of normal euthermic consumption at that temperature. Overall, the group saved 23% of the energy they would have used without torpor, still quite a significant figure.

■ What are the disadvantages to an animal in entering torpor, or full seasonal hibernation?

The animal becomes very vulnerable to predators. A deep hibernator may take quite a long time to respond even to an alarm arousal. Also, unless the 'cold alarm arousal' is efficient, the T_b may go below the point from which it can recover.

The energy cost of hibernation must also take into account the special requirements of females in gestating and caring for the young. Their T_b fluctuations, torpor bout duration and BMR often differ from those of males of the same species. In pregnant female black bears (*Ursus americanus*) in the North American Rocky Mountains, the cost of winter reproduction, including gestation and lactation, was found to be 1432 kJ day^{-1} to produce two young. Fat provides 92% of the total energy for lactation and gestation during early winter. Consequently, these animals have 89% larger fat depots than have non-reproductive females entering hibernation. The rate of fat loss was 37% greater, and protein loss was about 2.4 times higher for reproductive females than for non-reproductive females. Protein utilization is a last resort in any hibernating animal due to the danger of nitrogen toxicity and the disadvantages of muscle weakness on arousal. However, a source of transplacental amino acids is essential even when the mother is not feeding. Black bears minimize this cost by having a shorter period of post-implantation gestation together with metabolic pathways that enable them to hydrolyse urea and recycle their nitrogen more effectively than other mammals (Chapter 5 covers this in more detail).

4.5.5 The importance of size and habitat

The use of hibernation to gain energetic advantage must be weighed against a number of considerations, particularly animal size and behaviour, biogeographic distribution and habitat. Small animals, which can carry less fat and have a higher surface area to volume ratio and BMR, are more likely to lose energy as heat and in maintaining life functions if they do not use hypothermic strategies in winter. Few hibernating mammals have a total body mass greater than 5 kg. Indeed, in large animals there may be a cost advantage in *not* hibernating as less energy is likely to be needed to see them through the winter in a prolonged state of fasting close to a thermoneutral T_a, than in hibernation with re-warming. For example, an adult bear's re-warming energy cost is 10^5 greater than that of a mouse and the energy required for re-warming is equivalent to the energy needed to remain euthermic for nearly 3 days (compared to 3 hours in the mouse).

The importance of size is balanced by the importance of the habitat, and by the ability of animals to evolve their thermoregulatory strategy to match ecological energy demands, enables them to diversify just as much as their foraging and locomotion techniques. Pigeons and doves (Columbidae) for example, belong to a large (300 species) family of birds that shows remarkable ecological diversity. Hypothermic strategies are used by species whose adults range in mass from 35 to 800 g. An implanted temperature-sensitive transmitter was used to measure T_b, and a flow-through respirometer for V_{O_2} and V_{CO_2}. Cloven-feathered doves (*Drepanoptila holosericea*) are fruit-eating birds from the island of New Caledonia, and the effects of food on the energy metabolism of the only captive pair in the world were compared with the effects on grain-eating African Namaqua doves (*Oena capensis*) which are adapted to arid desert habitats. Namaqua doves show shallow torpor in response to food deprivation, with T_b falling to 30–37 °C: there was a fluctuation of T_b with T_a, but a metabolic defence at T_a below 25 °C. This regulation is reflected by the increase of metabolic rate at low T_a (Figure 4.31a). This exercise saves them around 10% of their daily energy requirements. In contrast, cloven-feathered doves (Figure 4.31b) show a more pronounced onset of deep torpor (with T_b falling to between 24 and 30 °C), induced by darkness at T_a lower than 15 °C, and leading to a substantial reduction in metabolic rate of up to 62%. However, this torpor only lasts up to 3 hours as the process of cooling is also several hours in duration. Total daily energy consumption is reduced by a comparable amount (10–15%), including the cost of re-warming. The very different form of thermoregulation in the cloven-feathered dove therefore, has very little energy budget advantage and may instead be an adaptation to its island habitat together with its fruit diet, that may make it difficult to find food at certain times of the year.

Further evidence that energy budgeting varies with habitat comes from work on mouse-eared bats (*Myotis myotis*) (Koteja, 2001). Using a non-invasive method called Total Body Electric Conductivity (TOBEC), lean body mass and fat content were measured in hibernating animals either in buildings or caves in Poland (Figure 4.32). Although food availability was similar in each habitat, fat content was reduced from 19 to 6% of total body mass between December and April in both males and females. Calculations showed that the bats need about 4.9 g of fat (191 kJ) to sustain a 165-day hibernation. However, the rate of fat usage varied considerably between different sites and at different phases of hibernation. Although the average amount of fat remaining in April would be sufficient to support at least six more weeks of hibernation, the level of reserves was close to zero in some individuals.

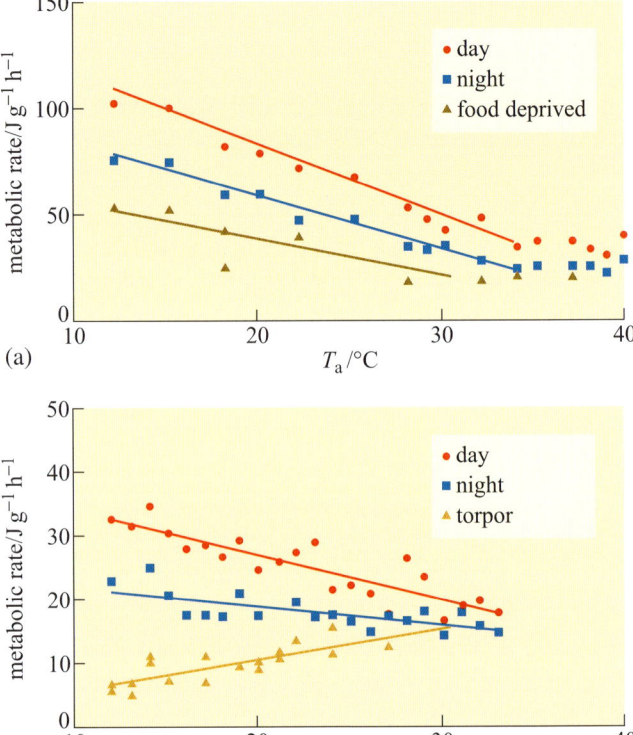

Figure 4.31 Metabolic rates versus T_a in (a) Namaqua doves (*Oena capensis*) and (b) cloven-feathered doves (*Drepanoptila holosericea*) (Schleucher, 2001).

A model has recently been drawn up to predict the relative energetic advantages of hibernation in little brown bats (*Myotis lucifugus*; Figure 4.8) living in different latitudes. T_a at each latitude influences total winter energy requirements. Hibernation is only likely to confer bioenergetic advantages within a fairly narrow range of winter durations for animals living at favourable hibernaculum temperatures. As it turns out, *M. lucifugus* lives no further north than predicted by the model and so hibernation energetics could be the most important factor governing its geographic range, of this species at least. It follows that the consequence of global warming might be a northward expansion of the species within only two or three generations.

Summary of Section 4.5

BMR is regulated independently of T_b at least in hibernating mammals. Entry into hibernation is characterized by a gradual fall in RQ, which indicates a switch from carbohydrate to lipid metabolism for energy provision (through the phosphorylation of pyruvate dehydrogenase, the inhibitor of mitochondrial fatty

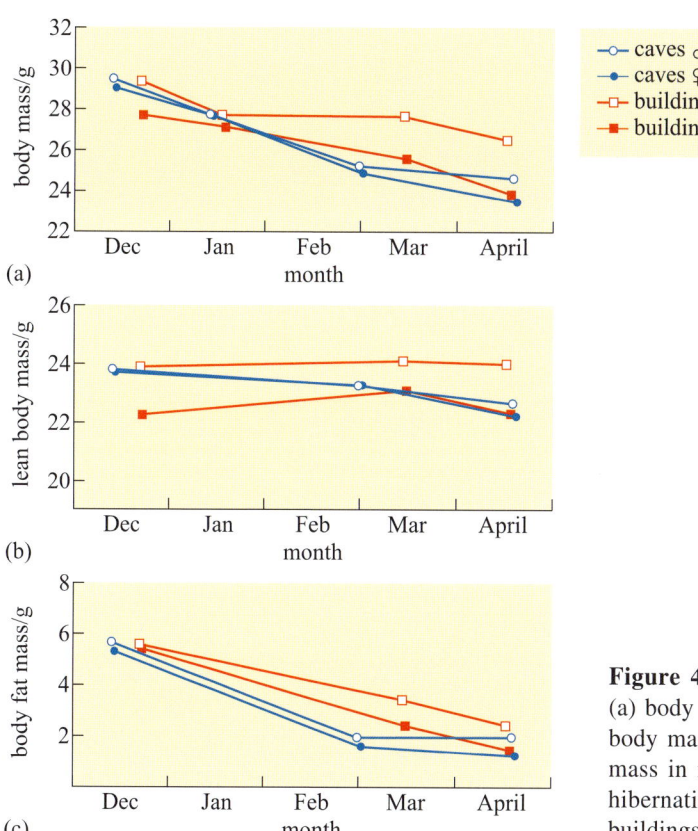

Figure 4.32 Changes in (a) body mass, (b) lean body mass and (c) body fat mass in mouse-eared bats hibernating in caves and buildings in Poland.

acid uptake). There is evidence that some other vertebrates, such as hibernating frogs, may continue to use carbohydrate catabolism or activate gluconeogenic pathways when arousal episodes completely deplete stored triacylglycerols.

The most important adaptive modifications of the mitochondrion are reduced oxidative metabolism, the reversal of the proton leak across the inner membrane as an energy-saving mechanism during hibernation, and an increase in production of uncoupling proteins (UCP) in preparation for metabolic thermogenesis during arousal episodes.

Lung ventilation is linked to oxygen supply in euthermic animals, but in hibernation where tissue blood flow is almost absent it becomes less important. The characteristic inspiratory patterns of hibernation, as well as the fall in heart rate, have been shown to follow passively from the depression of T_b.

The energy budget of mammals for euthermia, torpor and arousal can be estimated both in the laboratory and natural habitats. For small mammals that are obligate hibernators, energy saving is in excess of 80% for the winter months and about 60% averaged over the year. Lactation and gestation can stretch the energy budget of female mammals considerably, and they may display further adaptations to minimize the cost. Hibernation should not, however, be seen as the universal solution for energy saving in winter and considerations of size, behaviour pattern and habitat all have an effect upon the useful duration of torpor bouts or whether torpor is an appropriate strategy at all. Hibernation physiology may be a major factor in determining the biogeographical and ecological diversity of mammals and perhaps other vertebrates.

4.6 Control systems

Measurements of thermoregulation, respiration and metabolic depression in the edible dormouse (*Myoxus glis*) during the early stages of torpor, hibernation and aestivation, indicate remarkable similarities in the profile of physiological changes for all three adaptive phenomena, suggesting that they are controlled by essentially the same mechanism. The capacity for adaptive hypothermia in animals is clearly determined genetically and is manifested in cells from many different tissues. Nevertheless, we have known for a long time that control centres in the brain exist which coordinate behavioural (feeding, drinking, reproductive) as well as physiological (circulatory, respiratory) functions, and establish cyclical rhythms for these functions which relate them to daily, monthly and seasonal changes in an animal's habitat.

■ What three functions must the brain fulfil to achieve this coordination?

1. It must act as a central receptor, integrating signals such as ambient light levels, daylength and temperature with internal indicators of physiological state.
2. It must contain an internal 'clock' that can be set to operate with reference to environmental changes.
3. It must contain command centres, which adjust body temperature, metabolism, respiration, circulation and behaviour to adapt to prevailing conditions in the habitat.

4.6.1 The hypothalamus as central regulator

Research in the past 30–40 years has established that the hypothalamus, which lies below the thalamus and above the optic nerve chiasma and the pituitary gland in the brain, fulfils all of the functions listed above, at least in part. The main function of the hypothalamus is *homeostasis*. Factors such as blood pressure, body temperature, fluid and electrolyte balance, and body weight are held to constant values called the set-points. Although set-points can vary over time, from day to day they are remarkably fixed.

The hypothalamus is an internal regulator of biorhythms. The supraoptic nucleus (SON) is found just above the optic tract in both brain hemispheres (Figure 4.33). It is the site of the 'clock' in mammals that sets an internal daily (circadian) and annual rhythm by which body functions such as metabolism, core body temperature, reproductive physiology and locomotive behaviour are governed. The endogenous rhythms of SON nerve cells are entrained to environmental cues such as daylength with reference to sensory signals from peripheral sense organs, particularly the eyes and light-sensitive receptors within the brain itself. The onset of torpor and hibernation seem to occur in golden hamsters (*Mesocricetus auratus*) during the active phase of the circadian cycle, but hamsters of the same species but with genotypes causing longer or shorter natural circadian rhythms did not experience different bout-lengths of torpor. Thus the hypothalamic oscillator relinquishes control over the timing of hibernation bouts once they have started. It seems likely that during hibernation, the ventromedial hypothalamus, close to the midline of the brain, becomes more important in thermoregulation.

Figure 4.34a shows an experiment with a marmot (*Marmota flaviventris*) in its euthermic phase, in which whole-body metabolic rate is measured using a

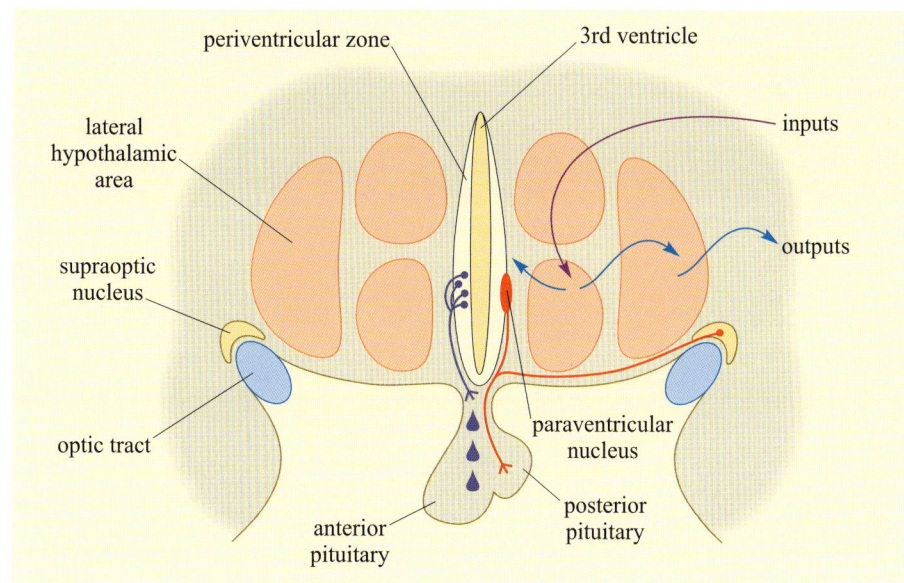

Figure 4.33 Map of control centres in the ventromedial area of the hypothalamus.

calorimeter, whilst the temperature of the hypothalamus, T_{hy}, is being both adjusted and monitored in the conscious animal using a water-filled catheter and a thermocouple. When T_{hy} is lowered below 36.5 °C, metabolic rate increases with an intensity proportional to the difference between T_{hy} and the threshold temperature. Figure 4.34b is the response in another individual, which is in hibernation at a T_a of 5 °C.

■ What do you deduce from Figure 4.34b?

The shape of the curve is identical to that in Figure 4.34a, suggesting that metabolic rate changes at a threshold temperature during hibernation as well. However, there are two striking differences. First, the threshold temperature in the hibernating marmot is about 7 °C, compared with 36.5 °C in the euthermic marmot. Secondly, the metabolic rate is much lower (10% of that in the euthermic marmot) though the response to lowered T_{hy} (a fourfold increase) is about the same. These experiments show that the hypothalamus is a thermosensitive centre which is linked to effector systems that raise the metabolic rate (and produce heat). The temperature set-point has been lowered from the normal level of 37 °C in the hibernating animal but the proportional metabolic response to lowering the temperature below the threshold is the same in euthermic and hibernating animals.

In the dormouse (*Muscardinus avellanarius*), which hibernates on its back with hind legs exposed, it is possible to cool the feet in the same way using double-layered, jacketed half-boots in which a temperature-controlled liquid can be pumped between the two layers of the boot. With this device, the temperature of the hind feet could be altered without affecting deep-body or brain temperature. It was found that, irrespective of T_{hy}, cooling the hind feet stimulated an increase in the metabolic heat response. This observation implies that the mechanism governing the set-point may not always reside in the pre-optic area of the anterior hypothalamus (POAH). Recent studies in rats, in which BAT thermogenesis was measured in response to direct electrical stimulation of the central nervous system (CNS), suggest that there are groups of nerve cells forming thermoregulatory centres at many sites in the brain and spinal cord.

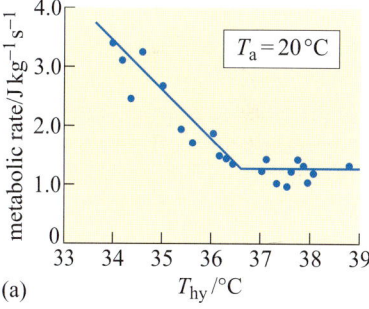

(a)

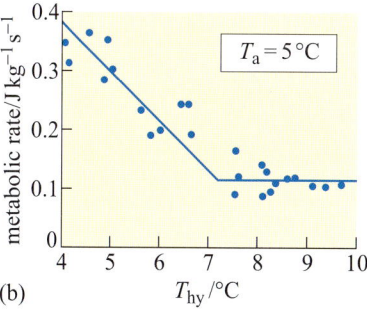

(b)

Figure 4.34 Metabolic responses to the lowering of hypothalamic temperature (T_{hy}) in (a) a euthermic marmot and (b) a hibernating marmot (Florant and Heller, 1977).

How is the progressive reduction of temperature accomplished during entry into hibernation? It might be that the set-point remains in operation, but that its value is steadily shifted downwards. The following experiments, performed by Florant and Heller (1977), explore these possibilities. By manipulating T_{hy} of marmots (*Marmota flaviventris*) entering a bout of torpor, he was able to determine absolute values for the T_{hy} threshold (T_{set}), and see how these indices changed with time. Figure 4.35 shows such an experiment in two individuals. T_{hy} is varied, using a catheter arrangement and T_s (the skin temperature) is measured. The metabolic response is indicated on the lower traces. The marmot in Figure 4.35a is entering hibernation fairly smoothly, whereas, as judged by the frequent bursts of high metabolic rate (see arrows), the marmot in Figure 4.35b is progressing into hibernation rather irregularly. The cream and blue bands on each figure indicate the periods during which T_{hy} was being manipulated (lowered (cream) and raised (blue), respectively). Take the left-hand trace (a) first. Depressing T_{hy} in the early stages stimulates an increase in metabolic response, but 30 minutes later, the same decrease in T_{hy} has no effect on metabolic rate. This experiment suggests that the threshold temperature has fallen from 22–27 °C to 23–24 °C in 30 minutes. At 3 hours into hibernation, lowering the T_{hy} even below 23 °C has little effect initially on metabolic rate. In the right-hand trace (b), T_s also declines with time but is consistently above the manipulated T_{hy}, as indicated by the bursts of metabolic activity. Therefore, at any one time, the threshold may be above or below the actual T_{hy}. If the threshold is above it, the entrance is irregular, interrupted by bursts of metabolic heat production which slow the fall in T_b. Florant and Heller concluded that the rate of entry into hibernation is limited by the rate at which threshold T_{hy} falls. It seems, therefore, that entrance into hibernation and the lowering of T_b are controlled events. Using experimental data such as those gained from Figure 4.35, it is possible to plot the declining threshold temperature. Figure 4.36 (Heller et al., 1977) shows such a plot for a golden-mantled ground squirrel entering hibernation.

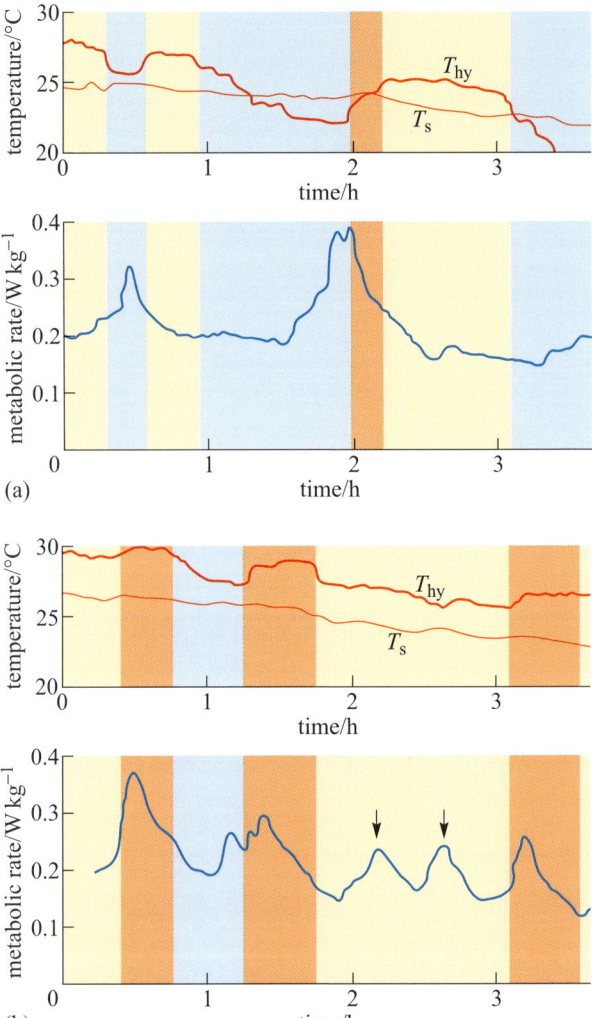

Figure 4.35 Entrance into hibernation of two marmots: (a) a smooth entry, (b) an irregular entry. The blue and orange bars indicate periods of hypothalamic cooling (blue) and heating (orange). T_{hy} is the hypothalamic temperature and T_s is the skin temperature. The arrows in (b) indicate irregular bursts of raised rate of metabolism.

The filled circles represent actual hypothalamic temperatures at specific times during entry. Manipulation of T_{hy} reveals that the threshold for metabolic heat production is somewhere in the range indicated by the vertical line below each dot. This process of continuously varying the controlled level of a feedback system has been called *rheostasis*, to distinguish it from homeostasis, which has implications of a fixed set-point.

The model exemplified by the ground squirrel entering hibernation may not, however, represent a universal phenomenon. In the eastern chipmunk (*Tamias striatus*), a facultative hibernator, a gradual, controlled decline in T_s on entering hibernation is often absent. Manipulating the hypothalamic temperature may have no effect in chipmunks hibernating with a T_b of about 4 °C, until about 2.1 °C is reached (which is T_{alarm} for this animal), when there is full arousal (Figure 4.37) (Wang and Hudson, 1971). Thus, the animal has the ability to increase

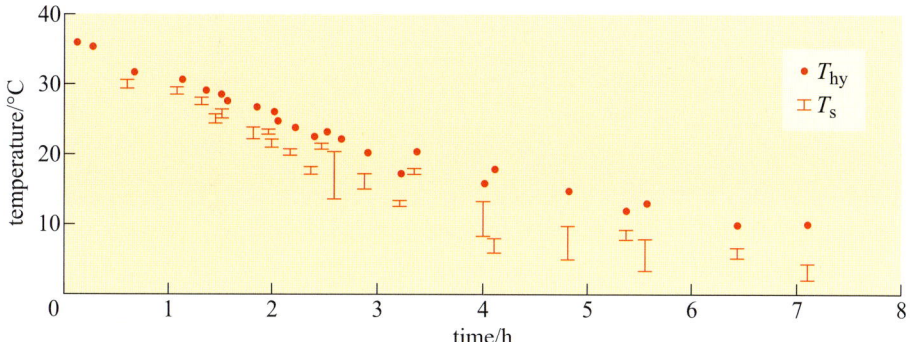

Figure 4.36 T_{hy} (●) and T_s (bars indicate range of values) during entrance into hibernation of the golden-mantled ground squirrel.

thermogenesis when cold as a result of T_{hy} falling too low, but does not normally control and defend T_b above the T_{alarm} level – at least not as a result of changes in hypothalamic temperature.

4.6.2 Metabolic regulation and the midbrain

As you found in the last section, the physiological evidence points to the likelihood that different components of regulation may be regulated separately. The hypothalamus, which appears to be central to the depression and recovery of body temperature during entry to torpor and arousal, is not the only player in the control of metabolic processes underlying non-behavioural thermogenesis. In many respects, the initiation of thermogenesis is the prime event in the reactivation of a cold body: the control mechanism which stands at the point of energy balance between dormancy and coma or even death. Centres of the brain involved in this process are likely to be efficiently protected from cooling to a critical temperature, beyond which electrical conduction and synaptic trans-mission are impossible. We do not yet know how such protection is possible in animals in hibernation, but it is clear that there is no single region of the brain with a monopoly on control of NST. The ventromedial hypothalamic nucleus is the major centre of the forebrain which stimulates the oxidation of fat in BAT as indicated by thermography studies on interscapular temperature (see Figure 4.20). But there is evidence that the operation of central activation mechanisms is balanced by regions that inhibit thermogenesis. Figure 4.38 (Hashimoto et al., 2002) shows that in a small region of the midbrain connected to the hypothalamus, electrical stimulation suppresses, whereas an injection of an anaesthetic increases T_b.

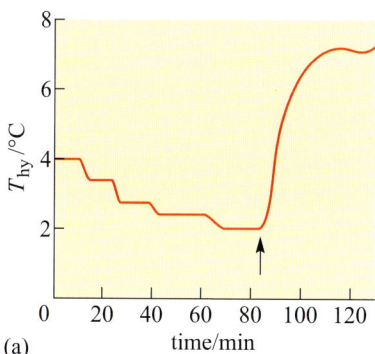

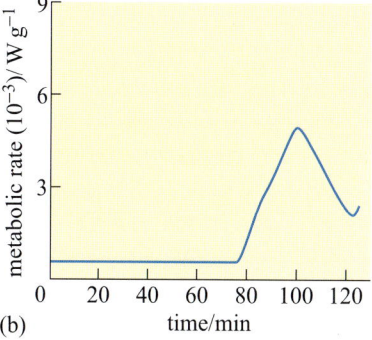

Figure 4.37 (a) Changes in hypothalamic temperature and (b) metabolic responses of a hibernating eastern chipmunk (*Tamias striatus*), with a T_b of 4.2 °C, to stepwise cooling of the POAH (T_{hy}). Cooling of the POAH to about 2.1 °C (T_{alarm}) precipitated arousal, but earlier steps produced no effect. Warming the POAH (indicated by the arrow in (a)) at the start of arousal suppressed it within 20 minutes or so.

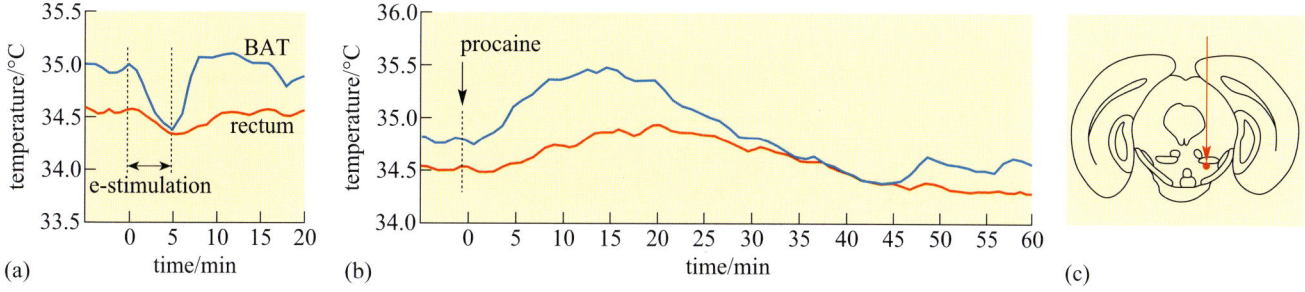

Figure 4.38 Role of the midbrain in thermogenesis during hibernation and on arousal. Effects of (a) artificial electrical stimulation (e-stimulation) and (b) local anaesthetic (procaine) micro-injection, on temperatures of interscapular BAT and rectum in anaesthetized golden hamster. (c) Brain section showing site of e-stimulation and procaine micro-injection.

It was once believed that arousal from torpor was initiated by an increase in T_b, a process which was 'paid for' by a period of intense feeding to generate metabolic energy. The presence of an elaborate interplay between activation and inhibition of NST underlines the fact that considerable accuracy in thermogenic controls can exist for an animal *within* its existing limited energy budget during torpor. Such controls could not only make possible the kind of very rapid adjustments seen in Figure 4.26, but also contribute to the mechanisms that maintain a safe minimum BMR within the key regions of the brain.

4.6.3 Rapid-response genes and rhythmic neuronal activity

Reactive changes in the brain are usually marked by changes in neuronal electrical activity. If these changes are to be of long duration, adjustments in neuronal electrical behaviour may be made through changes in gene expression. Rapid-response genes (sometimes called 'immediate–early' genes) are activated within minutes of the onset of such sustained electrical activity. These genes are master controls, acting as a gateway to a series of linked events: alteration of electrical firing patterns, repetitive activity and even the structure of neurons and their synapses, all of which modify the output of specific functional groups of nerve cells. Two such genes, called *c-fos* and *junB*, whose expression pattern in the SON follows the circadian rhythm in euthermic jerboas (*Jaculus orientalis*), cease cyclical fluctuations and remain at constant, elevated levels in the hypothalamus during bouts of hibernation. The region containing the largest number of c-fos-positive neurons in the hibernating jerboa is called the arcuate nucleus (Figure 4.39a). As hibernation progresses, and then on arousal, the distribution of c-fos-positive cells shifts to the ventromedial hypothalamus (Figures 4.39b and c). The significance of this observation is that separate regions of the brain are involved in maintenance of the hibernating state and initiation of locomotor activity on arousal (Ouezzani et al., 1999).

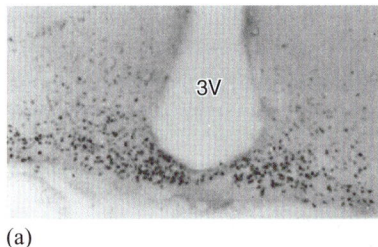
(a)

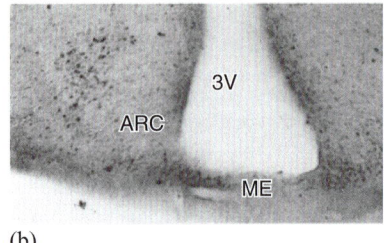
(b)

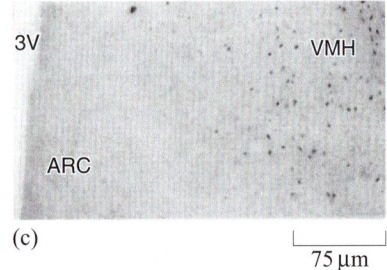
(c)

■ Where else might you expect to see *c-fos*-positive cells in the brain on emergence from hibernation?

In regions involved in initiating motor activity to the muscles.

Since the expression of both *c-fos* and *junB* genes changes in advance of alterations in neuronal behaviour, they are likely to act as signals for the re-emergence of euthermic electrical activity patterns on arousal. It is significant that the high levels of expression are located in the arcuate nucleus, separate from the SON, which could therefore be an independent regulator of locomotor activity. During hibernation, the lowest temperature for electrical activity of hypothalamic neurons is depressed from around 16 °C to 12.3 °C and there is an increase in the number of cold-sensitive neurons. In the European hamster (*Cricetus cricetus*), neurons in isolated slices of the hypothalamus can fire quite normally down to temperatures as low as 5 °C as a result of changes in the molecular configuration of ion channels in the cell membrane. Such an adaptation is particularly important because neural control systems must be intact at brain temperatures that prevail on entry to, and during arousal from, hibernation. The structure of neuronal dendrites (see Figure 4.40a) is also modified to reduce incoming synaptic contacts. All three changes result from subtle alterations in the expression of genes specific to neurons.

Figure 4.39 Neural adaptations found in hibernation and arousal. Photomicrographs representing the pattern of distribution of fos-immunopositive neurons in mediobasal hypothalamus of jerboas hibernating for (a) 2 days, (b) 10 days, and (c) after arousal from hibernation. *Abbreviations*: ARC, arcuate nucleus; 3V, third ventricle; ME, Median eminence; VMH, ventromedial hypothalamus.

4.6.4 The neurotransmitters histamine and serotonin: a role for chemical signalling between neurons of the hypothalamus

As in all other regions of the brain, the integration of physiological change in the hypothalamus, conducted by the dialogue between many thousands of nerve cells, is the result of transmissions across chemical synapses. The functions of the hypothalamus are therefore dependent upon neurotransmission by a number of different chemical mediators and are critically dependent on the balance between their respective activities. Monoamine neurotransmitters (histamine and serotonin) operate in subgroups of hypothalamic nerve cells (as well as neurons in other parts of the brain) and appear to be particularly important to the regulation of hibernation.

The possibility that histamine is a natural inducer of hibernation is supported by the observation that the neurotransmitter becomes detectable in complex networks of neuronal axons during hibernation, whereas it is absent or present at undetectable levels in euthermic animals (Figure 4.40b). The evidence for an increase in synaptic

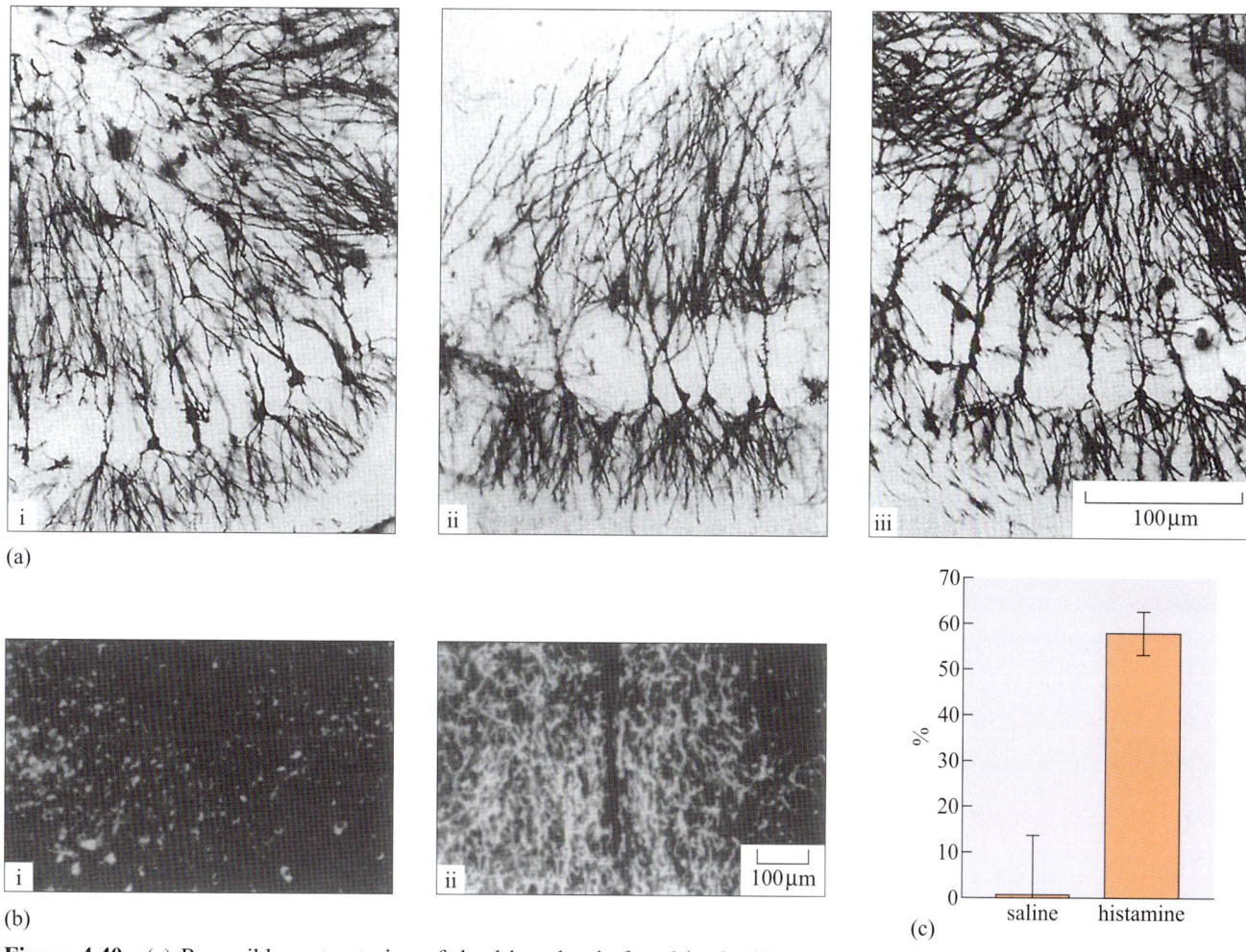

Figure 4.40 (a) Reversible restructuring of dendrites that is found in the hippocampus of a ground squirrel when (i) active, (ii) torpid and (iii) on arousal. The reduction of branching, and in the number of 'spines' which receive incoming synapses, indicate that processing of activity at many chemical synapses is drastically reduced during torpor (Popov et al., 1992). (b) Distribution of histamine-containing nerve fibres in the hippocampus of a euthermic (i) and hibernating (ii) ground squirrel, as visualized for fluorescence microscopy with an antibody recognizing histamine (Panula, 2002). (c) Effect of the infusion of saline and histamine through a cannula into the dorsal hippocampus of each animal in a group of ground squirrels on the hibernation bout length. The bout length is expressed as a percentage of the expected remaining time in the bout at the time of the start of the infusion. Data are expressed as means ±SE (Sallmen et al., 2003).

transmission by this specific neurotransmitter is seen in the light of the anatomical evidence from Figure 4.40a. The reduction of synaptic branching in torpor suggests that normal synapse activity is lost during torpor. The hippocampus is an area of the brain that controls arousal levels and in which histamine neurons are more prominent during hibernation. Injection of histamine into the hippocampus of ground squirrels (*Spermophilus lateralis*) increased the duration of hibernation bouts by over 50% (Figure 4.40c). Histamine is also believed to affect the response of the SCN to environmental stimuli, modifying circadian rhythm phase. Changes in phase responses to light can be increased by injecting histamine into the SCN and reduced by depleting the brain of histamine. These findings may help to explain the ability of histamine to prolong hibernation and show that histamine in different areas of the brain affects different aspects of the hibernation process.

The activity of tryptophan hydroxylase (TPH), a key enzyme in the biosynthesis of another monoamine transmitter, serotonin, undergoes marked changes in the brain during entry into hibernation, and arousal in *Spermophilus erythrogenys*. An increase in TPH activity was found in several regions of the brain during the pre-hibernation period in euthermic ground squirrels. A further increase in TPH activity to 150% was observed during the entry into hibernation. Significant elevation was found not only in potential TPH activity measured at the incubation temperature of 37 °C but also at incubation temperature of 7 °C, approximating the body temperature in hibernation. Serotonin may also contribute to the chemical induction and maintenance of hibernation.

4.6.5 Hormones and hibernation

Melatonin

Syrian hamsters, which display pronounced circadian temperature fluctuations before hibernation, lose these circadian cycles on entry to hibernation, and start to regain them shortly before arousal. Cycles are distorted during the early recovery period, suggesting that the SON oscillator has either been switched off or de-synchronized in hibernation. Another monoamine, melatonin, is involved in making these adjustments. A hormone rather than a neurotransmitter, melatonin is secreted by the pineal gland on the dorsal side of the brain. Removal of this gland can cause disturbances in both the 'clock' and the 'calendar' settings of mammals. Melatonin travels both to the pre-optic area of the hypothalamus, where it can reset the set-point for T_b, and to the SON where it can adjust circadian and seasonal rhythms. Continuous infusion of melatonin inhibitor into the circulation from subcutaneous implants can lead to a decrease in the duration of bouts of torpor in hibernating ground squirrels (Figure 4.41) (Pitrosky et al., 2003). A second important function of melatonin is to inhibit sexual activity and the production of gonadal steroid hormones through the mediation of the pituitary gland, which is itself under the control of neurons within the hypothalamus. Both events appear to be under the principal control of the SON, since its destruction by electrolytic lesions leads to an impairment of circadian rhythms and the ability to hibernate. More recently it was shown that an inhibitor of melatonin, on the other hand, can decrease the duration of hibernation by reducing both the number of hypothermic bouts and the amount of lipid present in BAT.

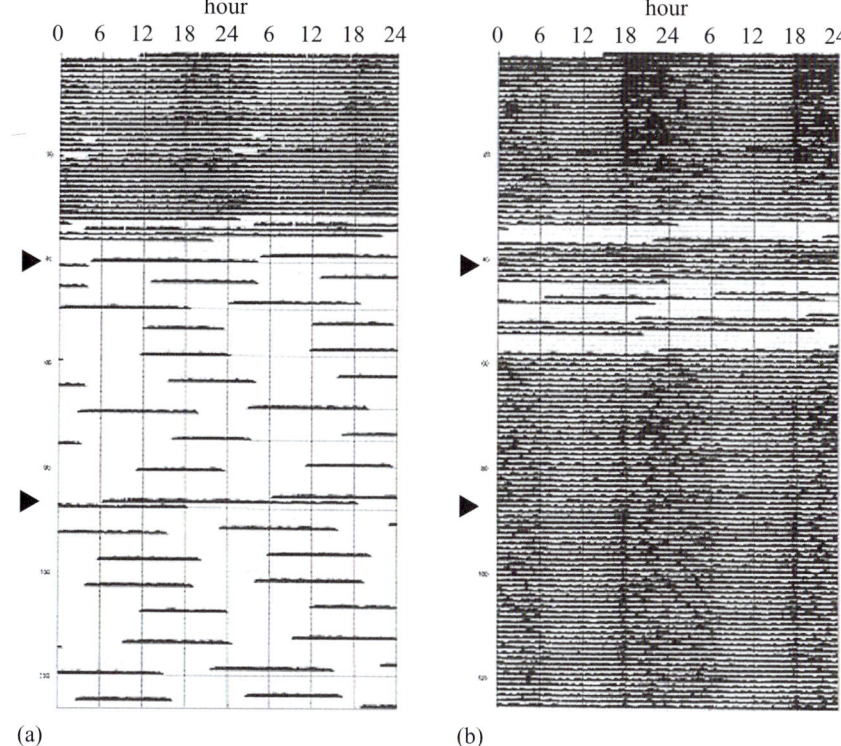

Figure 4.41 Hypothermic bouts, induced by short day laboratory regimes (10 h light and 14 h dark) with T_a of 5 °C. The trace is T_b recorded by telemetry every 5 min for 18 weeks in a Syrian hamster. Trace (b) is from an animal with a slow-release implant containing the highly selective melatonin inhibitor S22153. Trace (a) is a control animal. Each row is temperature data for 48 h. Where there is a dark band, T_b is 38 °C. The clear areas show periods where T_b is around 32 °C. Note the prolonged periods of hypothermia (torpor) in the control animal, compared with trace (b) in which torpor bouts from an animal exposed to melatonin inhibitor for the duration of the experiment are drastically curtailed. These data provide evidence that torpor bouts are induced by melatonin.

Hibernation-induction trigger

Researchers have devoted much effort to the search for a possible blood-borne chemical messenger that might communicate a signal within the brain and to other body tissues, causing entry to hibernation. Serum from hibernating animals such as the woodchuck (*Marmota monax*; Figure 4.7), when injected into active animals, can induce torpor. Partly purified serum extracts are also able to induce hibernation-like behavioural changes in a variety of mammalian species. Chemical analysis of the serum extracts reveals two components, one of high molecular mass ($M_r = 88\,000$) which has a structure closely resembling a natural protease inhibitor, and the other of low molecular mass ($M_r < 5000$) which is a member of the family of natural morphine-like or 'opioid' compounds. Pure samples of an opioid called enkephalin also induce hibernation in Colombian ground squirrels (*Spermophilus columbianus*). Dynorphin A is another opioid present in the ground squirrel brain. It acts at so-called delta receptor sites similar to those for enkephalin and rises during hibernation to a level 15 times higher than in non-hibernating euthermia, reaching an intermediate level in euthermia between bouts of hibernation. The origin of the different hibernation-inducing trigger components is not certain, but these observations point to the possibility that the blood circulation does contribute to the transmission of the signal to enter hibernation from one tissue to another. Opioids also have remarkable properties in inducing survival of cells in the brain and other organs under anoxic conditions. This discovery has led to an interest in the chemical induction of hibernation amongst medical researchers seeking ways in which to prolong tissue survival after trauma, or during and after transplantion.

4.6.6 Sleep, the brain and hibernation

There has been a popular misconception that hibernating animals are asleep when dormant, and that arousal during or at the end of hibernation involves waking analogous to that following deep sleep. Sleep in homeothermic animals can be divided into several phases, each with distinct patterns of electrical activity in the brain, as measured by an electroencephalogram (EEG). The passage into sleep is a transition from wakefulness into the stage called *slow-wave sleep* (SWS). SWS, and its characteristic electrical pattern of brain waves in the frequency range 0.7–4.0 Hz, is interrupted by periods of sleep characterized by, among other things, rapid eye movements, loss of muscle tone in the head and neck, and loss of a shivering response. These periods are known as rapid eye movement (REM) sleep. Dreaming occurs mainly at this time, and blood flow to the brain is markedly increased. The electrical patterns shown in REM sleep are very close to those of wakefulness.

Sleep deprivation and hibernation in Djungarian hamsters both lead to the suppression of slow-wave sleep activity (SWA); probably caused in the latter case by the drop in core body temperature, and a burst of SWA occurs on arousal from hibernation, further supporting the idea that the animal shows the signs of lack of sleep rather than the opposite. Stronger evidence for this theory comes from research showing that hibernation in Richardson's ground squirrel (*Spermophilus richardsonii*) leads to an accumulation of nearly three times the basal level of oleamide, a derivative of the fatty acid oleic acid, which builds up in the brain during sleep deprivation.

SWA normally follows changes in sleep activity and core body temperature during the REM sleep of euthermic animals, but still exhibits a daily cycle during non-REM sleep. In animals that display bouts of torpor, daily temperature fluctuations continue but sleep and SWA cycles shorten to 25% of their normal length. This change may be required to maintain torpor bouts lasting several days that are typically observed in hibernating mammals. Circadian rhythms are maintained by the SON of the hypothalamus, and thus we would expect that the modification of sleep patterns in hibernators is under neural control from the same brain centre as that which determines cycles in other body functions.

Summary of Section 4.6

Hibernation shares physiological mechanisms with aestivation. These adaptations are fundamental properties of animal cells, but they come under central control and coordination mediated by the vascular, endocrine and central nervous systems.

Neuronal nuclei of the hypothalamus at the base of the forebrain combine the ability to integrate internal and external signals, set biological rhythms and control metabolism and body core temperature, and are indispensable to the process of hibernation. Hibernating animals are characterized by their ability to adjust their core temperature set-point. Experiments on ground squirrels suggest a progressive and smooth decline in the T_b set-point (rheostasis), but this is not seen in all species that have been investigated. An alternative strategy is to practice thermogenesis at the point of reaching a T_{alarm} resulting in an explosive increase in metabolism leading to arousal. Displays of apparently spontaneous metabolic thermogenesis, independent of T_b, in marmots, is additional evidence that does not support the rheostasis theory for all mammals.

The ability of parts of the hypothalamus to engage in seasonal regulation of physiological functions is indicated by changes in genetic, molecular and structural changes in neurons. Rapid-response genes move from a cyclical to a sustained pattern of expression in parts of the hypothalamus implicated in circadian rhythm generation and locomotor activity. Changes in the numbers of cold-sensitive cells, the organization of neuronal dendrites and the properties of voltage-sensitive ion channels cause long-term changes in the ability of the hypothalamus to process sensory information into appropriate behavioural resonses at low T_b. The hypothalamic neurotransmitters histamine and serotonin and the pineal gland hormone melatonin appear to have an important role in synaptic integration during the onset and maintenance of hibernation. A mixture of blood-borne factors have been shown to induce hibernation behaviour in a variety of birds and mammals in plasma transfusion experiments. As well as a large protein component, HIT contains at least one member of the opioid peptide family of neuromodulators.

Hibernation leads to the suppression of slow-wave sleep activity similar to that seen in sleep deprivation. In ground squirrels, the region of the hypothalamus, the SON, which controls the day-night sleep/waking cycle also regulates cyclical changes in T_b. The modification of circadian cycles is therefore linked to the overriding need to govern the duration of 'torpor bouts.

Learning Outcomes for Chapter 4

When you have completed this chapter you should be able to:

4.1 Define and use, or recognize definitions and applications of, each of the **bold** terms.

4.2 Give definitions of the terms 'hibernation', 'torpor' and 'adaptive hyperthermia', and the three physiological processes that underlie them.

4.3 Give examples of the diversity of the major groups of mammals and birds that contain hibernating species.

4.4 Describe the physiological changes occurring during entry to hibernation and at least three of the cues that may trigger entry.

4.5 Present evidence to show that hibernating mammals and birds retain physiological control of their T_b.

4.6 Explain the role of brown adipose tissue and mitochondrial uncoupling of respiration from metabolic energy release in heat generation in mammals.

4.7 Describe the analytical and targeted experimental approaches to the identification of genes and proteins implicated in hibernation and arousal, and give examples of them.

4.8 Explain the importance of the selection of appropriate metabolic fuel sources in hibernators.

4.9 Describe the changes needed to maintain hibernation and survival at cellular level.

4.10 Critically describe experiments designed to evaluate the energy cost of hibernation as compared with euthermia, and discuss the importance of three factors that influence whether animals use hibernation as an energy-conserving strategy.

4.11 Suggest why periodic arousals occur and offer a mechanism for them.

4.12 Present experimental evidence for the view that control of T_b depends upon temperature-sensitive neurons and suggest where they may be located.

4.13 Give examples of systems of chemical control for the onset and maintenance of hibernation that operate in the brain and blood circulation.

4.14 Describe the relationship between circadian controls of sleep–waking cycles and the maintenance of torpor.

4.15 Use diagrams and flow-charts to illustrate physiological and biochemical principles.

Questions for Chapter 4

(Answers to questions are at the end of the book.)

Question 4.1 (LO 4.2)

Describe three measures of physiological regulation central to hibernation. Using these measures as definitive criteria state why the Svalbard reindeer is not a hibernator.

Question 4.2 (LO 4.3)

What reasons might you advance to support the argument that the ability to hibernate might have arisen several times in the evolution of warm-blooded vertebrates?

Question 4.3 (LO 4.4)

From what you have read in Section 4.3 indicate whether the following statements are true or false. Briefly explain your answer.

(a) The laying down of fat deposits is a criterion for identifying an animal as a hibernator.

(b) The decline in heart rate on entry to hibernation is due to an increase in the number of skipped beats and a lengthening of the period between beats.

(c) The heart and brain are the warmest tissues during arousal.

(d) Although blood pressure is lowered during entry to hibernation, there is evidence for vasoconstriction.

(e) A drop in temperature below a critical level can lead to an increase in heart and respiratory rates but no change in BMR.

Question 4.4 (LO 4.4, 4.5 and 4.7)

List the factors that determine (a) entry to hibernation, and (b) length of periods between arousals.

Question 4.5 (LO 4.7 and 4.15)

Draw a flow chart indicating the design and outcomes of an experiment which might identify a new enzyme that controls metabolism in hibernating liver cells.

Question 4.6 (LO 4.9)

Summarize the main changes occurring prior to and during hibernation at the cellular level.

Question 4.7 (LO 4.5 and 4.7)

From Sections 4.3 and 4.5, describe in one sentence each the changes that occur in the heart rate, breathing patterns, blood P_{CO_2} and BMR in hibernating mammals.

Question 4.8 (LO 4.8 and 4.10)

What factors determine the choice of carbohydrate, lipid and protein as respiratory fuel sources in mammals during torpor and hibernation.

Question 4.9 (LO 4.10)

What factors determine whether a species gains an advantage in using torpor as an energy-saving measure? Give explanations for the energy-saving figures for July and November in Table 4.5.

Question 4.10 (LO 4.12)

List two pieces of evidence each for and against the rheostasis theory of thermoregulation on entry to hibernation.

Question 4.11 (LO 4.12 and 4.13)

What adaptive changes occur in hibernating neurons? What is the evidence that neurons are 'prepared' for hibernation as conditions change?

Question 4.12 (LO 4.4)

From the evidence of brain sleep activity patterns, why is the popular concept that hibernating mammals 'are asleep' incorrect?

CHAPTER 5   POLAR BIOLOGY
Prepared for the Course Team by Caroline Pond

5.1   Introduction

This chapter is about animals' structural and physiological adaptations to living permanently in cold climates; hibernation, a special response to transient or seasonal cold, was described in Chapter 4. Living in a polar climate involves adaptations of many physiological systems: appetite, diet, energy storage and reproductive habits as well as thermoregulation. In many cases, such changes involve 'ordinary' physiological mechanisms being pushed to extremes. The study of such physiological adaptations can help us to understand how humans and domestic animals could cope with similar conditions that arise under artificial or pathological conditions. For example, obesity is rare among wild animals, even when food is very plentiful, but in humans, the condition is common and often leads to numerous physiological complications, ranging from susceptibility to diabetes to mechanical damage to legs and feet. Most naturally obese animals occur in cold climates, and there is no evidence that they suffer from the complications of the condition that are observed in people and their domestic livestock. Perhaps we have something to learn about the natural regulation of appetite and the organization and metabolic control of fat from these cold-adapted species that have evolved ways of combining fatness with fitness.

On the evolutionary time-scale, modern polar environments, and hence living species of polar organisms, evolved relatively recently. The study of polar organisms provides the opportunity to study physiological adaptations of quite recent origin that evolved in organisms which were already complex and well-integrated. Such changes are comparable to artificial evolution in domestic animals, whether by manipulation of the genome (i.e. intensive artificial selection, gene transfer, etc.), or by drastically altering the diet and husbandry conditions. Polar organisms may help us to under-stand the physiological and psychological implications of the rapid, often drastic changes that we impose upon our own lives and those of our domestic animals.

Antarctica has been isolated from other continents since the Mesozoic supercontinent Gondwanaland broke up and the fragments that became India, Australia and New Zealand drifted away. The rich fossil record in Antarctica shows that a diverse tropical fauna, including early eutherian and metatherian mammals, once lived there. As the continent became colder, many species disappeared and adaptations to the climate evolved *in situ* in surviving lineages over many millions of years. Consequently, many of the organisms of Antarctica and the surrounding oceans are endemic.

In contrast, much of the Arctic* is a large ocean, connected to the Pacific Ocean by the Bering Strait (that became a land bridge several times during the last million years) and through wider channels to the north Atlantic Ocean.

Prevailing winds and deep currents bring plenty of mineral nutrients to the Southern Ocean but the Arctic Ocean, particularly the areas north of Siberia, Alaska and Canada, is nutrient poor. Consequently, the Southern Ocean supports

---

*When used as an adjective, 'arctic' generally refers to the regions around both Poles and does not have a capital letter. 'Arctic' and 'Antarctic' are the northern and southern arctic regions, respectively, and do have capital letters. To avoid confusion, the term 'polar' is used to mean both arctic and antarctic.

a much greater abundance of marine life than is found in most of the Arctic, except in a few areas such as the Barents Sea around northern Norway and northwest Russia.

Biological evolution in the Arctic has been much affected by the Pleistocene ice age, which produced several periods of glaciation over much of the Northern Hemisphere that began about a million years ago and continued until as recently as 10 000 years ago. There were ice ages in the Palaeozoic and early Mesozoic, but until the Quaternary ice age began about 1 Ma ago, the climate had been mild, often warm, over the whole globe for the previous 250 Ma. The climate became colder and drier, promoting rapid evolution in many different lineages of animals and plants. Many species became extinct, but others, particularly descendants of cold-adapted organisms that lived on high mountains, adapted to the new conditions: numerous modifications of the skin and fur, endocrine mechanisms and behaviour and circulatory, respiratory, digestive and excretory systems evolved in many different species over a comparatively short period. Among them was an almost hairless primate, *Homo*, which adapted successfully to the cold climate in Europe and northern Asia after several million years of evolution in tropical Africa. Many such cold-adapted species ranged over much of the Northern Hemisphere until the climate became warmer during the interglacial period of the last 10 000 years, since when most have been confined to the Arctic.

5.1.1 The polar environment

At high latitudes, the Sun's rays always strike the Earth at a large angle from the vertical so they travel through a thicker layer of atmosphere and are attenuated by the time they reach the ground. Because the Earth's axis of rotation is inclined to its path around the Sun, there are large seasonal changes in daylength and the Sun is continuously below the horizon for a period in winter and continuously above the horizon for an equivalent period in summer. The annual changes in daylength and average temperature recorded just inside the Arctic Circle (at Tromsø, Norway) and far into the Circle* (at Longyearbyen on the island of Spitsbergen, Svalbard Archipelago) are summarized in Figure 5.1. The range of annual temperature change is much greater at the higher latitude, and in mid-winter (January and February), the range about the mean is more than 12 °C. In polar climates, the temperature can change abruptly and often unpredictably. In fact, both the localities featured on Figure 5.1 are on coasts, where the sea keeps the climate much more equable. Further inland, fluctuations in temperature are even greater. Polar organisms are thus adapted both to the extreme cold and to abrupt fluctuations in temperature.

As explained in the previous section, terrestrial environments in the Arctic are, by geological standards, relatively new, most of the land having been completely covered with a thick layer of ice as recently as 10 000 years ago. Consequently, the soil is thin and fragile, and poor in organic nutrients. The optimum temperatures for plant growth do not coincide exactly with peak sunshine. At Longyearbyen, continuous daylight begins in late April, but the mean temperature does not rise above 0 °C (and so the snow and ice do not melt) for another 2 months (Figure 5.1).

---

* The Arctic Circle (66° 30′ N), and the equivalent latitude in the Southern Hemisphere, are defined as the latitude above which the Sun is continuously below the horizon for at least 1 day each year. Warm, moist air from the temperate zone rarely reaches high latitudes, so in most polar areas precipitation is low. Much of the water is locked away as ice, which has a low vapour pressure, and the air is very dry (often as dry as a tropical desert) and ground water is inaccessible to plants as well as to animals.

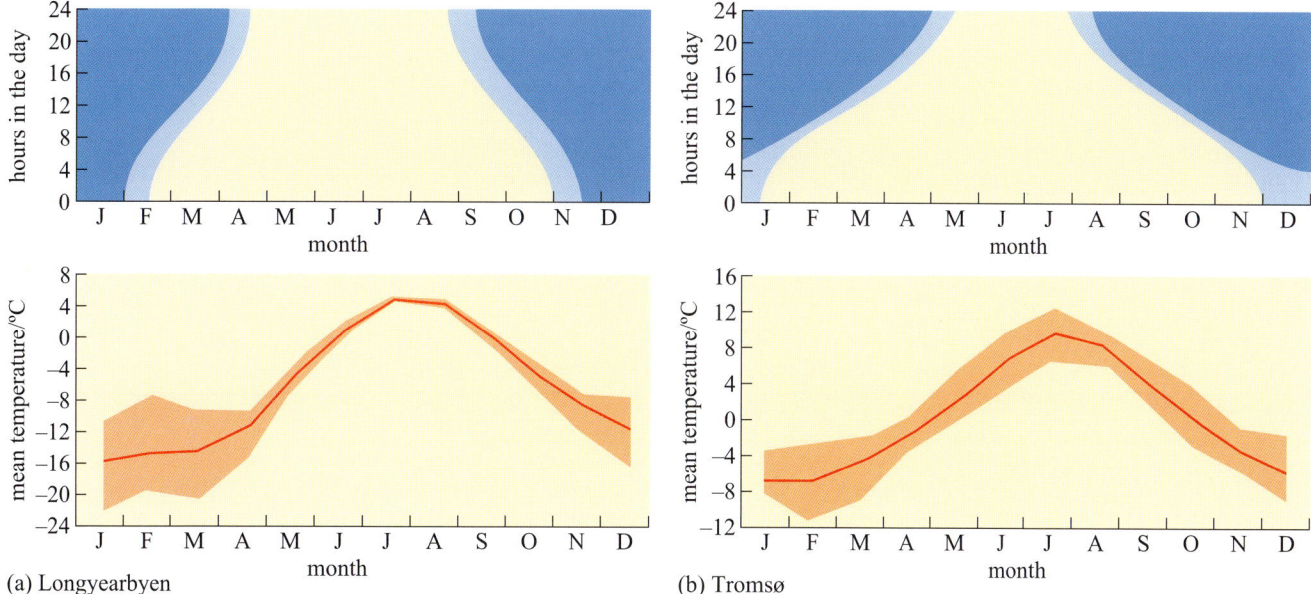

(a) Longyearbyen

(b) Tromsø

These circumstances, combined with the severe climate, mean that the growing season for plants is short but intensive, and total productivity on land is low, producing little food and still less shelter for animals. Consequently, relatively few species of terrestrial organisms live permanently at high latitudes. For example, although the land area of Svalbard is about 62 000 km^2, almost half that of England, there are only a few hundred species of insects and other invertebrates, two resident terrestrial mammals, the arctic fox (Figure 5.2a) and reindeer (Figure 5.2b), one bird (an endemic species of ptarmigan) and no reptiles, amphibians or completely freshwater fish. However, many other species spend part of the year on or near the land, often while breeding or moulting: seasonal visitors include more than 30 species of migratory birds (various kinds

Figure 5.1 The number of hours of daylight (cream), twilight (light blue) and darkness (dark blue), and the mean temperature (red line) from January to December at (a) Longyearbyen, Svalbard (78° N), and (b) Tromsø, Norway (70° N). The pale red shading shows the range about the mean temperature.

(a) (b)

Figure 5.2 Only two species of terrestrial mammal occur naturally* throughout the year on Svalbard. (a) The arctic fox (*Alopex lagopus*) also occurs throughout the Arctic, and in mountains at lower latitudes. This picture, taken in late autumn, shows an adult in its long, dense winter coat. The summer coat is usually greyish brown, often with white markings.** (b) The subspecies of reindeer (*Rangifer tarandus platyrhynchus*) that is endemic to Svalbard. This picture was taken in July, when the vegetation is at its highest, and these young males are growing antlers for the mating season in September.

* A few others have been introduced by humans during the past century.

** *Alopex* is bred in captivity for its fur, which can vary in colour from grey to bluish in winter, and chocolate brown to fawn in summer, hence the common names, silver fox or blue fox.

of geese, auks, puffins, skuas, terns, gulls, and eider ducks and snow buntings), and mammals that feed in the sea, such as polar bears, walruses and several species of seal. The simple ecosystem on land and the severe, erratic climate tend to produce 'cycles' of population abundance followed by mass mortality or migration (e.g. lemmings in Scandinavia and Russia). Interesting physiological and behavioural adaptations to these fluctuations in food supply have evolved in some of the larger animals. The vast continent of Antarctica has no indigenous terrestrial vertebrates, although many birds, including penguins, skuas, terns and gulls, and six species of seal spend time on or near land.

The situation in the sea is very different. Seawater freezes at $-1.9\ °C$, but because of the anomalous relationship between the density and temperature of water, ice floats, insulating the water underneath from the cold air above. Except in very shallow areas, the sea-ice does not extend to the sea-bed, even at the North Pole. Storms and currents sometimes break up the ice, creating many temporary, and some permanent, areas of open water even at high latitudes in mid-winter. Such turbulence also oxygenates the water and admits more light, making the environment much more hospitable to larger organisms.

The movements of ocean currents are complex (and may change erratically from year to year), often resulting in an upwelling of deep water rich in nutrients and promoting high primary productivity in the sea. In most arctic regions, the sea is both warmer and more productive than the land, so at high latitudes there are many more organisms in the sea than on land, at least during the brief summer, and, as in the case of the baleen and sperm whales, some are very large. Some groups of animals, such as bears, that are terrestrial in the temperate zone, have evolved adaptations that enable them to feed from the sea in the Arctic.

Sea-ice is less compact than freshwater ice, and contains many tiny channels containing liquid water as well as cracks caused by weather and currents. Hence sea-ice appears opaque rather than transparent like freshwater ice. The pores harbour a variety of single-celled algae, bacteria and other microbes that form the basis of surprisingly productive food chains. Most of those living on or near the surface are photosynthetic, and during the summer, such microbes are dense enough to confer a brown colour on the underside of the sea-ice. These organisms, and similar ones living on snow and in cold, dry terrestrial habitats, are collectively known as **psychrophiles** (ψυχροσ, *psychros* = cold, φιλοσ, *philos* = friend). The continent of Antarctica is generally much colder at comparable seasons and latitudes than most of the Arctic, with the possible exception of large landmasses such as Siberia, Alaska and some of the bigger islands off the north coast of Canada. With its harsher climate, and longer period of biological isolation, Antarctica has a wider variety of endemic, impressively adapted psychrophiles than most of the Arctic.

5.2 Environmental regulation of physiological processes

All plants and animals respond to environmental changes such as the light–dark cycle and temperature, but the impact of the environment on essential physiological processes such as eating, fattening and breeding is more evident and often more finely controlled in polar species than in those that are native to warmer and more equable habitats. Large effects are nearly always easier to quantify and to investigate experimentally, so arctic species offer an excellent opportunity to study the subtle but often important action of environmental changes on physiological processes.

5.2.1 Nutrient budgeting

Energy is expended in the search for food, and in ingesting and digesting it. If food is so scarce that searching is inefficient, or its nutrient content so low that little nourishment is obtained from it, animals may be able to save energy by suppressing appetite and fasting. In polar environments, food is widely scattered both in space and in time. Consequently, the physiological mechanisms that regulate appetite and energy storage are sophisticated and effective in arctic species. Herbivorous animals such as reindeer are directly dependent upon plant productivity and synchronize their foraging and other energetically expensive activities, such as mating and breeding, with it. Daylength (photoperiod) is a more reliable indicator of season than temperature (see Figure 5.1) and is often an important regulator of physiological mechanisms.

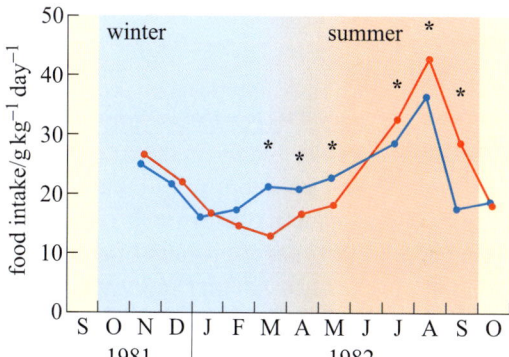

Figure 5.3 Seasonal changes in the voluntary food intake (in grams per kg body mass per day) of Norwegian reindeer (blue) and Svalbard reindeer (red) with unrestricted access to food. Asterisks mark significant differences ($P < 0.05$) between subspecies.

To investigate seasonal changes in the behaviour and metabolism of species native to the high Arctic, a few adults of the subspecies of reindeer that is endemic to Svalbard (*Rangifer tarandus platyrhynchus*, see Figure 5.2b) were transported to northern Norway and kept in small outdoor pens there, alongside similar individuals of the native subspecies, *Rangifer tarandus tarandus* (Larsen et al., 1985). All the animals had continuous, unrestricted access to forage but, as shown on Figure 5.3, the Svalbard reindeer ate three times as much food in August as in March.

■ Are these seasonal changes in the appetite of Svalbard reindeer simply a direct response to the environment?

No. There were seasonal changes in the food eaten by local Norwegian reindeer as well, but they were less pronounced than those of the animals native to high latitudes. In addition, the largest differences between the two subspecies were observed in mid-March and mid-September, around the equinoxes when day and night are equal in length over the whole globe.

■ Do seasonal differences in energy expenditure explain these data?

No. Being confined in small pens, the reindeer took little exercise all the time. Energy expended on thermoregulation should be greater in cold weather, so if thermogenesis was important, one would expect them to eat more, not less, in the winter.

Reindeer (Figure 5.2b) grow thick coats of long, hollow hair that insulates the warm skin so effectively that snow accumulates on their backs without melting. Energy expenditure on shivering or other forms of thermogenesis seems to be minimal even in the coldest weather. Foraging is slower and less efficient in winter, and the lower total daily intake is supplemented by utilization of the fat reserves built up during the brief summer, when they eat almost continuously.

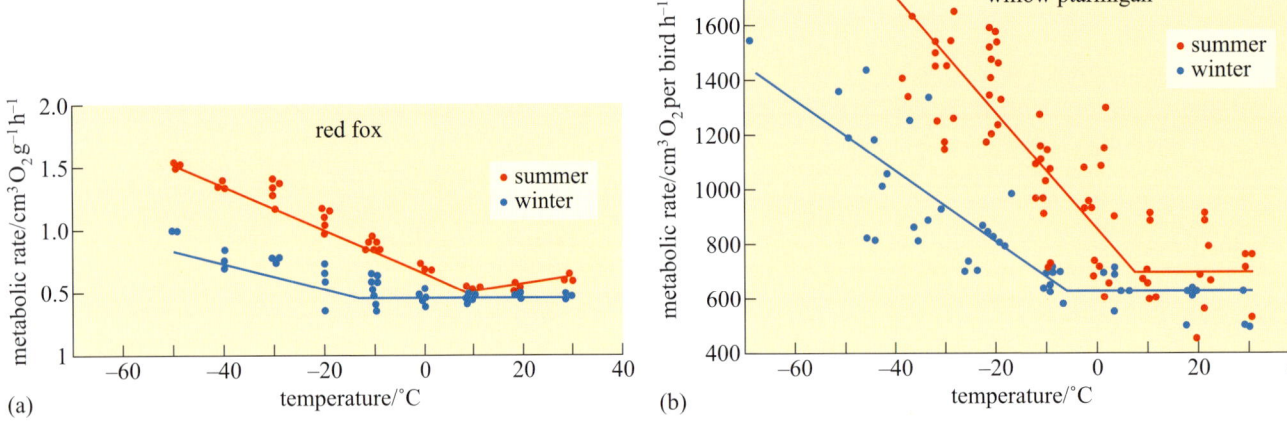

However, as these experiments show, the seasonal changes in food intake arise primarily from the endogenous control of appetite, and are not imposed upon the animals by food availability. The fine control of appetite is slightly different in subspecies adapted to different climates. The investigators also found small but significant differences at certain times of year between Norwegian and Svalbard reindeer in the rates of lipogenesis measured in adipocytes *in vitro*, and in the responses of adipose tissue to hormones such as adrenalin.

Metabolic rate, food intake and other aspects of energy balance also change seasonally in birds and mammals that are native to high latitudes. The red, or common, fox (*Vulpes vulpes*) occurs throughout Europe and northern Asia except in high mountains and arctic regions, where it is replaced by the smaller arctic fox, shown in Figure 5.2a. As shown on Figure 5.4a, at above 10 °C, the fox's BMR is about the same in summer and winter, but as the temperature falls, the rise in BMR is delayed and is slower in winter-adapted animals than in those caught in summer.

Such phenomena have been intensively investigated in ptarmigan (Figure 5.5) which are non-migratory, mainly ground-dwelling grouse-like birds that eat twigs, shoots and other plant material. There are two species in Scandinavia and Russia: the willow ptarmigan (*Lagopus lagopus lagopus*; Figure 5.5) and the rock ptarmigan (*L. mutus mutus*). ('*Lagopus*' means 'foot of a hare' and refers to the feather-covered or fur-covered feet of the ptarmigan and arctic fox, see Figures 5.2a and 5.17.)

A subspecies of rock ptarmigan occurs only on Svalbard; it is larger than the mainland forms, and has almost pure white plumage during the 8 months of winter. As shown on Figure 5.4b, the metabolic rate of willow ptarmigan measured at a wide range of temperatures is lower in winter than in summer. The seasonal differences are even greater in Svalbard ptarmigan (*L. mutus hyperboreus*). Svalbard ptarmigan also eat much more in the late summer than in winter and accumulate fat in the autumn. The experiments summarized in Figure 5.6 reveal some of the physiological mechanisms that control these changes in appetite and energy storage (Lindgård and Stokkan, 1989).

When exposure to continuous light was started in July (Figure 5.6a), the birds' usual autumnal fattening proceeded as normal, but their body mass remained high and food intake fairly low, right through to the following September.

Figure 5.4 The resting metabolic rates at different temperatures of (a) red fox and (b) willow ptarmigan acclimatized in captivity to summer and winter conditions. The adult body mass of foxes is 3–7 kg.

Figure 5.5 A willow ptarmigan (*Lagopus lagopus lagopus*) in winter plumage, with the arctic willow bushes on which they feed. This photograph was taken near Churchill, Manitoba, on the western shore of Hudson Bay, Canada in late October.

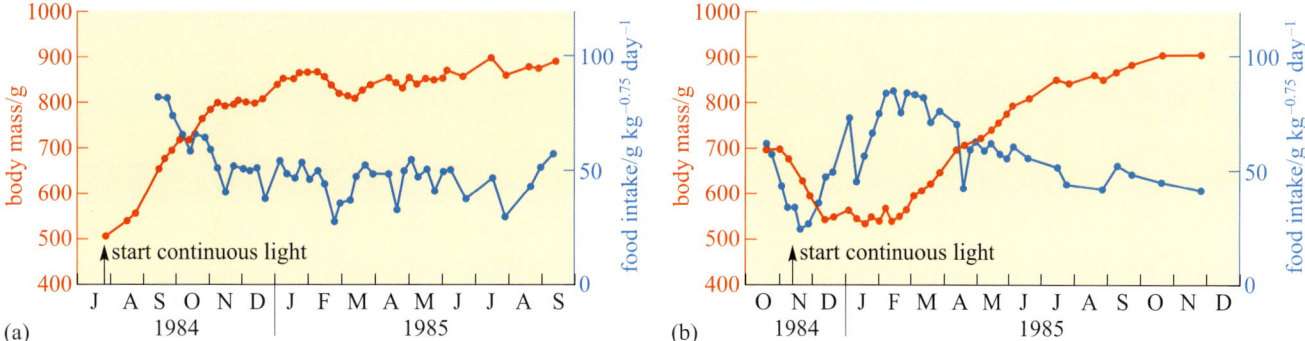

Throughout this period, their plumage remained white and they failed to breed. It was as though the continuous light held them indefinitely in their autumnal condition. However, when exposure to continuous light was started in November (Figure 5.6b), the birds underwent a complete cycle of changes in body mass and food intake (and began to develop speckled summer plumage) before settling into continuous high body mass and low appetite.

■ What do these experiments show about how seasonal changes in appetite and body mass are controlled?

They are not simply a response to environmental conditions but are at least partly controlled endogenously.

Exactly how such control mechanisms evolved and what happens when animals (or people) are abruptly transported into environments in which their endogenous controls of appetite and energy expenditure are inappropriate are not known.

Although several ruminant mammal species live in mountains and arctic regions (e.g. mountain goats and reindeer, respectively), none is known to hibernate in the strict sense of the term (see Chapter 4). There are no living species of the family Bovidae smaller than sheep or goats, but some deer (family Cervidae) are less than a tenth of that size as adults. Some tropical deer, notably species of mouse deer *Tragulus*, weigh only 1–2 kg, well within the range of size of mammals that can become torpid, but none is known to do so.

One reason might be that substantial changes in body temperature would kill the microbes in the rumen that are essential to digestion. Another possibility is that in ruminants, both storage and membrane lipids contain mostly saturated lipids, which have a higher melting point than unsaturated lipids. Laboratory experiments in which animals were fed diets rich in saturated or unsaturated lipids just before hibernation showed that, at least in small rodents, a larger proportion of unsaturated lipids in cell membranes and adipose tissue is essential to successful hibernation (see Book1, Section 1.5). Finally, pregnancy, which lasts a relatively long time in ruminants and usually takes place during the winter, could not be sustained at very low body temperatures.

Figure 5.6 Changes in food intake (blue lines) and body mass (red lines) in Svalbard ptarmigan during 15 consecutive months kept in captivity with unlimited access to food at Tromsø. (a) Birds kept outside and then indoors in continuous light from July onwards. (b) Birds kept outside and then indoors in continuous light from November onwards.

5.2.2 Migration for breeding

Birds also do not hibernate, but like reindeer, many species undergo daily or seasonal changes in energy expenditure and appetite, and many of the endocrine changes that are an integral part of true hibernation in other groups. The fact that the preliminary stages of hibernation are widespread among vertebrates may help to explain why true hibernation has evolved several times in distantly related taxa (see Chapter 4). Instead of hibernating, some species of birds migrate to and from breeding areas, where they are able to exploit transient gluts of vegetation or, more often, of the insects and other arthropods that feed on them. Long-distance migratory birds belong a wide variety of taxa, including cranes (order Gruiformes), swifts (order Apodiformes), some swans, ducks and geese (order Anseriformes), cuckoos (order Cuculiformes) and many different kinds of passeriform birds including swallows and martins (family Hirundinidae).

Figure 5.7 A red knot (*Calidris canutus*) feeding on the north coast of Norfolk near The Wash in early September. This adult's rust-red breeding plumage is fading into the inconspicuous white and grey winter plumage. It has probably recently returned from breeding in the Arctic.

Some birds travel to the Arctic to breed during the polar summer, which can be both cool and short at very high latitudes (e.g. Svalbard), or in regions such as Siberia that have particularly severe climates.* Red knots (*Calidris canutus*, order Charadriiformes) are 'waders', eating worms, shellfish and other invertebrates collected from beaches, mudflats and estuaries. These small birds (adult body mass about 0.1 kg) form large, dense flocks near sandy or muddy coasts of northern Britain and northwest Europe during the winter (Figure 5.7). Like many birds, the juveniles eat insects and other small arthropods. Some populations breed between June and August on the Taimyr Peninsula, the most northerly region of Siberia that extends into the Arctic Ocean. The area became free from permanent ice following the end of the last ice age only a few thousand years ago, and is flat and marshy, with several large slow-flowing rivers that support huge populations of mosquitoes and other insects in summer.

■ What would be (a) the advantages and (b) the disadvantages of breeding in such places?

(a) Advantages: fewer predators (though arctic foxes, snowy owls and large gulls such as skuas are present); foods suitable for the chicks and adults are available in large quantities in adjacent habitats; continuous daylight (Figure 5.1) permits continuous foraging. (b) Disadvantages: the weather is often cold and stormy, and the terrain offers little shelter, so keeping the eggs and chicks warm may pose problems. The breeding season is very short, necessitating rapid growth of the chicks. The journey between Siberia and northwest Europe is tens of thousands of kilometres.

Dutch ornithologists used doubly-labelled water (see Sections 1.6 and 2.4.2) and other techniques to study the growth and metabolism of chicks there, and compared their data to similar observations on other species of the order Charadriiformes with similar habits (sandpipers, dunlins, turnstones, godwits, plovers and oystercatchers) that breed in the temperate climates of northwest Europe (Schekkerman et al., 2003). They found that chicks of the arctic-breeding species both grew faster and generated more body heat, mainly by shivering, than similar birds breeding in temperate climates. The increased thermogenesis was necessary not only because of the severe climate, but also because in Siberia, the parents actually spent less time brooding even very young hatchlings. Red knot chicks are precocious and can forage for themselves at a few days old. They apparently also manage with very little sleep (in sharp contrast to most neonatal birds and mammals, which require many hours of

---

* Until these remote regions were explored in the 18th and 19th centuries, the breeding sites of many migratory birds were a complete mystery, giving rise to wild speculations about where, if at all, the birds bred. For example, barnacle geese (*Branta leucopsis*) derive their name from the medieval belief that they arose spontaneously from barnacles. In fact, they breed in Greenland, Svalbard, remote parts of Sweden and northern Russia, with almost all the Svalbard population spending the winter around the Solway Firth and Dumfries and Galloway, Scotland.

sleep), enabling them and their parents to forage for up to 20 h per day. The total energy expenditure from hatching to fledging was found to be up to 89% higher in the arctic-breeding knots, but the chicks were dependent on the parents for only 17–20 days, a shorter period than related species of similar size.

This study demonstrates a range of far-reaching adaptations of thermoregulation, growth rate and sleep requirements in birds that breed in polar regions. The opportunity to exploit the temporary abundance of food apparently outweighs any disadvantages associated with these adaptations of growth rate and thermogenesis and the energetic costs of migration.

The journey itself requires further metabolic specialization. The birds break their journey at several places where food is abundant and easily obtained. However, since time is short, some stopovers last as little as 1–4 days, during which time they must take on enough fuel for the next stage of the journey. The closely related sandpiper (*Calidris mauri*) that also breeds in the Arctic can fatten at 0.4 g day^{-1} (4.5 times the normal rate) during brief stopovers. This remarkably high rate of deposition of fat stores is possible due to a temporary increase in the activity of lipogenic enzymes such as fatty acid synthase. There is more about the mechanics of locomotion and the energetics of long-distance travel in Books 3 and 4.

5.2.3 Environmental regulation of breeding

As pointed out in Section 5.1.1, primary plant productivity occurs for only a few months in the summer, so the reproductive physiology of most arctic animals, particularly herbivorous species, is tightly synchronized with the seasons. On Svalbard (Figure 5.2b), more than 90% of the reindeer fawns are born in the first week of June. The mothers of those born too soon or too late are often unable to find enough food to support lactation and the fawn fails to thrive. As shown on Figure 5.1, the onset of continuous daylight and that of the conditions that support plant growth are several months out of phase. This situation poses little problem for reindeer, because the duration of pregnancy is almost constant and they mate only during a brief rutting period in September, when the daylength is changing rapidly. But this environmental cue alone would not be an accurate control on the timing of breeding of resident herbivorous birds such as ptarmigan that breed in mid-summer.

The physiological mechanisms that control the timing of several aspects of mating and breeding in the Svalbard ptarmigan (*Lagopus mutus hyperboreus*) have been investigated in detail (Stokkan et al., 1986). Their plumage is almost pure white in winter but speckled brown feathers appear in summer and the adult males have a red fleshy 'comb' over each eye. Figure 5.8 shows the seasonal changes in these secondary sexual characters, the maturation of the gonads, and the concentration of luteinizing hormone (LH) in blood plasma in ptarmigan shot on Svalbard. LH levels (Figure 5.8a) are low from August until February, when the Sun reappears (see Figure 5.1). The blood plasma LH levels and body mass (see Figure 5.6) start to increase slowly, and in March first primary, then secondary, spermatocytes appear in the testes (Figure 5.8b) and the combs begin to grow (Figure 5.8c). However, there are no mature spermatozoa until the end of May, so the gonads mature much more slowly than in most other seasonally breeding birds. Pigmented feathers also do not appear until June, just before the snow melts (Figure 5.8c).

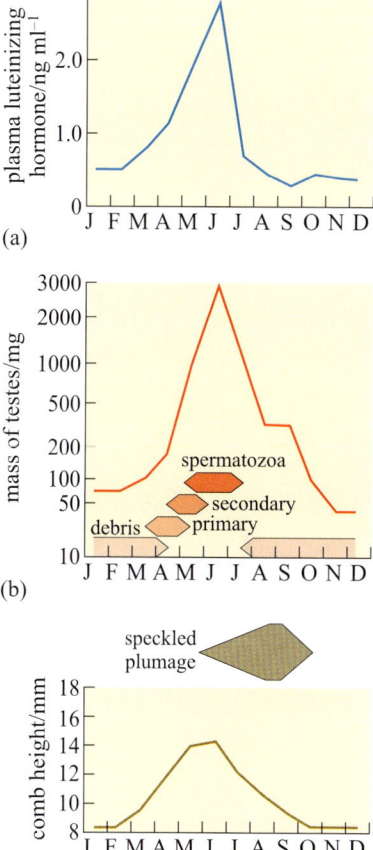

Figure 5.8 The annual cycle of maturation of the gonads and formation of external secondary sexual characters in male Svalbard ptarmigan. (a) Concentration of luteinizing hormone (LH) in blood plasma. (b) Mass of testes and appearance of cells in various stages of spermatogenesis. (c) Height of combs and appearance of pigmented feathers. The vertical axis of (b) is on a logarithmic scale.

■ Are there any advantages in delaying the development of pigmented feathers?

Speckled plumage is probably much more conspicuous to potential predators (i.e. arctic foxes) against a background of snow than pure white feathers, so it would be advantageous not to produce the breeding plumage until it is essential for courtship and mating.

Throughout the year, LH levels are lower in female (Figure 5.9a) than in male (Figure 5.8a) ptarmigan, but, as in males, there is a sharp peak in June that coincides with maximum mass of the ovary and the period during which eggs are laid (Figure 5.9b). However, LH is also fairly high in March (Figure 5.9a), several months before the gonads become active. Some other factor, perhaps non-photoperiodic inhibitory input from the environment (e.g. cold weather), must be delaying the maturation of the ovary.

5.2.4 Variable fecundity

The food supply for most polar species depends on several unpredictable factors so successful breeding is far from certain, even if births are tightly synchronized with the seasons. Maintaining pregnancy and feeding the offspring after birth (or hatching in birds) are energetically expensive. The death of the offspring before its maturity represents an irredeemable loss of 'reproductive investment' for the parents, particularly the mother, although the earlier in parental nurturing that the death occurs, the smaller the loss to the parents. Various mechanisms of environmental determination of fecundity have evolved among large birds and mammals and are particularly evident in arctic species.

Like most large ungulates, reindeer produce only one offspring a year and suckle it for more than 6 months, by which time the next pregnancy may be well underway. Observations on Svalbard reindeer show that in December, nearly all adult females are pregnant, having conceived during the mating season in the previous September. But as winter progresses, the proportion that are pregnant falls, and by June the following year, any fraction from over 90% to less than 10% of the adult females give birth to a fawn. The other pregnancies must have ended in abortion or reabsorption of the fetus. In each year, the proportion giving birth is approximately the same in all areas of Svalbard that can be studied, suggesting that it is related to the climate. Exactly how the reindeer 'knows' when to terminate a pregnancy which she is unlikely to be able to complete successfully is currently under investigation, but the quality or quantity of the food available during the winter is the most likely factor.

The fecundity of arctic foxes is also very variable: in years when prey and carrion are abundant, some litters consist of as many as 20 pups (with an average of 10–12 in Canada and 6.4 on Svalbard), a very large number for a canid (dog-like) mammal, but very few breed at all in years when food is scarce. A similar pattern is found in predatory birds such as the snowy owl, which also feeds on rodents and hares that undergo population cycles. When prey are abundant, the fox or owl parents can raise a large number of pups or chicks but if food availability suddenly falls (due to mass mortality or migration of prey or a change in the weather), most or all of the offspring may starve in

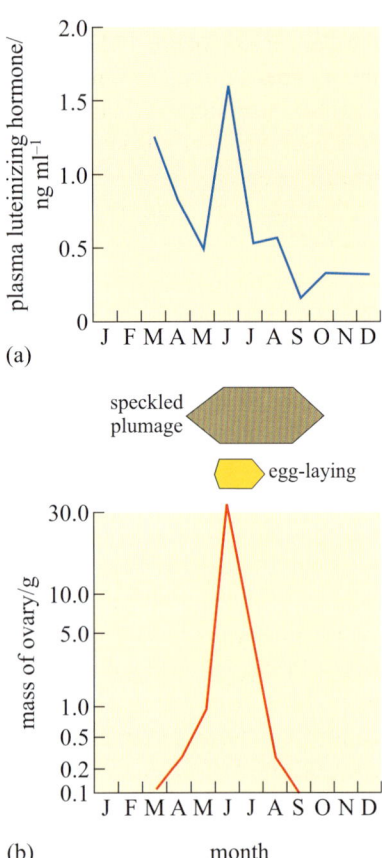

Figure 5.9 The annual cycle of maturation of the gonads and the formation of external secondary sexual characters in female Svalbard ptarmigan.
(a) Concentration of luteinizing hormone in blood plasma.
(b) Appearance of pigmented feathers, mass of ovary, and period of egg-laying. The vertical axis of (b) is on a logarithmic scale.

the nest. Food intake and/or energy stores somehow regulate ovulation and/or the number of ova that implant successfully and develop, so that as far as possible, fecundity is adjusted to food supply in a fluctuating environment. However, very little is known about the mechanisms involved: the formation and maintenance of the placenta depend upon several different hormones, some of them secreted from the pituitary and brain. The body may respond to stress or insufficient nutrition by terminating the pregnancy prematurely, thereby enabling the mother to build up reserves that could support a pregnancy in the next breeding season at which the prospects of a more successful outcome appear brighter.

Summary of Sections 5.1 and 5.2

Large seasonal changes in temperature and sunlight dominate primary plant production and hence the food supply. Food intake is regulated by the endogenous seasonal control of appetite, fattening and activity, as well as by food availability. Energetically demanding activities such as breeding and migration are only feasible during a brief period and must be tightly synchronized to season. Greater accessibility of food suitable for chicks makes long-distance migration to and from high arctic regions worthwhile for some birds. Adaptations of thermoregulation and growth rate enable breeding to be completed during the brief summer. Arctic ecosystems involve relatively few species, some of which are prone to abrupt, cyclic changes in population abundance, so food supplies change erratically from year to year and from place to place. Most physiological adaptations to these features of the polar environment probably arise from modification and refinement of mechanisms that occur in temperate-zone species.

5.3 Natural feasting and fasting

It is clear from Sections 5.1 and 5.2 that seasonal or irregular periods of fasting are an integral part of living at high latitudes, especially for large animals. When people (and many tropical and temperate-zone mammals) lose weight, either because they are eating less or because they are suffering from a digestive or metabolic disorder, protein is broken down in substantial quantities long before the lipid stores are exhausted. Even frequent and vigorous exercise cannot prevent the breakdown of lean tissue, although it can often reduce or delay the process, particularly in young people. The loss of protein causes muscles to become weak and wasted, and the skin and hair to appear shabby. Immune function is also impaired, weakening resistance to parasites and infectious diseases. These undesirable side-effects of fasting do not normally afflict mammals and birds that naturally go without food while remaining active for long periods.

5.3.1 Penguins

Penguins (order Sphenisciformes) are an ancient and distinctive group of flightless, short-legged birds that evolved in the Southern Hemisphere, probably around New Zealand, about 65 Ma ago in the late Cretaceous, although the oldest known fossils date from about 45 Ma ago.

At a maximum body mass of more than 40 kg, the emperor penguin (*Aptenodytes forsteri*; Figure 5.10a) is the largest living penguin (some fossil species were much bigger) and is found further south than any other vertebrate. Like other penguins, emperors feed on fish, squid and large crustaceans that they catch by diving and chasing the prey underwater. The main predator of adult penguins is the leopard seals (*Hydrurga leptonyx*; Figure 5.10b), the largest and most agile antarctic seal, that has a varied diet including other seabirds and smaller seals, as well as fish, squid and crustaceans.

Emperor penguins breed on the iceshelf, away from predators such as skuas that take eggs and chicks, on breeding grounds that may be as far as several hundred kilometres from the open water. The males leave the feeding areas in early April (autumn in Antarctica) and fast during 6 weeks of courtship and for a further 2 months while brooding. Only one egg is laid, and the male carries it on his feet and broods it in a special flap of feathered skin that extends from his abdomen. Brooding penguins are inactive, keeping close together in large groups and walking an average of only 30 metres per day, thereby minimizing energy expenditure to near BMR. If his mate has not returned by the time the chick hatches, the male feeds his offspring on 'curds' formed from deciduous tissue in the oesophagus and broods it as he did the egg (Figure 5.10a). As soon as he is relieved by his mate, he walks back to the open water in what is by then mid-winter, continuously dark and very cold.

The female also fasts during courtship, but she returns to the sea after presenting her mate with a single egg that is large relative to her own size. The female fattens quickly while at sea, eating 6–8 kg per day and increasing her body mass by about one-third, before returning to the breeding grounds to take her turn to feed the chick on curds and partially digested food regurgitated from her stomach.

René Groscolas and other French biologists from Strasbourg spent many months in Antarctica studying the physiological mechanisms behind these habits (Groscolas, 1982, 1986). Figure 5.11 shows the measurements that they made on wild penguins during the breeding season and in the following 3 weeks, while the birds were artificially prevented from returning to the sea to feed at the end of the natural fasting period. Every few days, marked penguins were caught, weighed, their rectal temperature measured, and a sample of venous blood taken.

While fasting at the breeding colony, the mean body mass of the males fell by 40.5%, from 38.2 kg to 22.75 kg, at an average rate of 35 g per day. After falling slightly during the first few days, the body temperature, and levels of glucose and fatty acids in the blood plasma were constant (Figure 5.11a), and well within the ranges of values measured in penguins that were feeding regularly. The ketone β-hydroxybutyrate is produced by partial oxidation of fatty acids and can substitute for glucose in some energy-producing pathways in some tissues. Its concentration increased steadily, reaching a peak when the fathers began to feed their chicks. The smaller females lost only about 22% of their initial body mass during their shorter fast. Except during the period of egg-laying, the pattern of changes is similar to that of the males.

■ Why should egg-laying affect body temperature and metabolism?

Egg production involves the synthesis of large quantities of protein and lipid (for yolk), and the withdrawal of calcium stores (for shell formation), which generate

(a)

(b)

Figure 5.10 (a) Emperor penguin (*Aptenodytes forsteri*) with a chick that is already well grown but still has its downy juvenile plumage. The chick may be heavier than its parents before they discontinue feeding it. (b) The penguins' main predator, the leopard seal (*Hydrurga leptonyx*) is awkward on land, but very agile in water. Its exceptionally long flexible neck enables it to grab swimming birds and shake them to death.

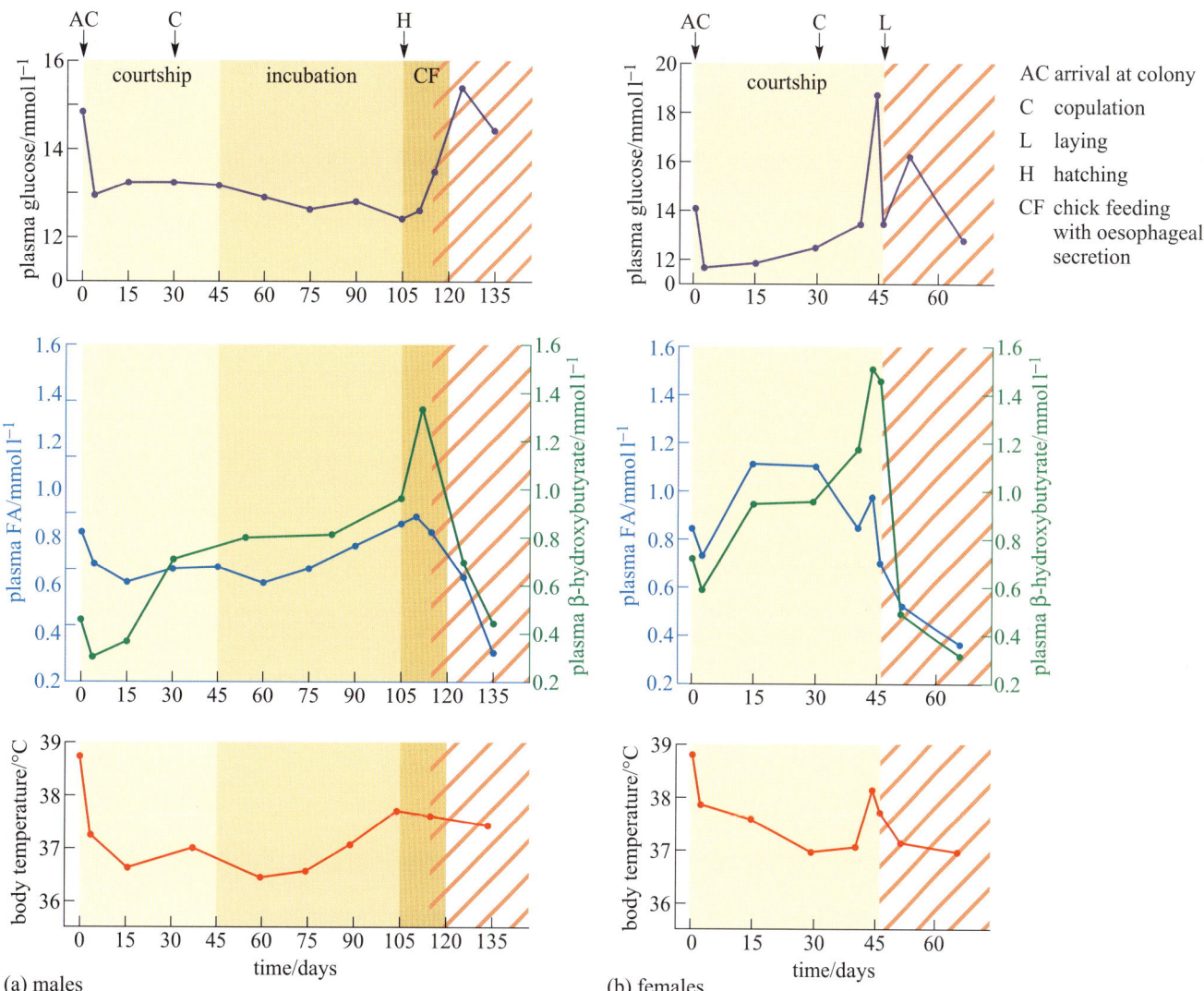

Figure 5.11 Changes in the concentrations in the blood plasma of glucose (top), non-esterified fatty acids (FA) and β-hydroxybutyrate (middle), and body temperature (bottom) in (a) male and (b) female breeding emperor penguins. The areas in yellow show natural habits, the start of the red striped area marks the end of the fast in the colony, and the red striped areas themselves show measurements during the period in which the penguins were penned, preventing them from returning to the sea and feeding (after day 115 in the males, and day 45 in females). The observation period began at the end of March and continued until the end of August (early in the Antarctic winter) for the males (a) and until the end of June for the females (b).

heat and require levels of circulating glucose similar to those that support strenuous activity.

The tenfold increase in the concentrations of β-hydroxybutyrate is small compared to changes of up to 40-fold observed in the blood of pigeons, poultry and humans after just a few days of starvation. When artificially prevented from returning to the sea at the end of their normal fast, plasma fatty acid and β-hydroxybutyrate concentrations decreased sharply in penguins of both sexes. Their rate of weight loss also increased abruptly, reaching a mean of 542 g per day for the lightest penguins that weighed only 17.5 kg.

■ What can you deduce from these observations about the penguins' fat stores and energy metabolism?

Production and utilization of free fatty acids decrease, probably because stores of triacylglycerols are almost exhausted. So the penguins start utilizing protein at a much higher rate. Because much less energy is produced from the breakdown of each gram of protein, a much higher rate of weight loss is necessary to meet the energy requirements of maintaining an almost constant body temperature.

This conclusion is confirmed by increased excretion of uric acid during enforced starvation. Other measurements indicate that during natural fasting, 93% of the penguins' energy comes from oxidation of fatty acids released from adipose tissue (Figure 5.12). The small quantity of glucose needed to support glucose-dependent tissues (e.g. the brain) is formed mainly from the glycerol in triacylglycerols, and only small quantities of protein are utilized.

■ Are the reserves replenished in the same way as they are depleted?

No. As shown on Figure 5.12, protein is withdrawn last during fasting but replenished more rapidly than lipid when the penguins start feeding again.

These observations suggest that, as in other animals, loss of protein has serious disadvantages and is only a 'last resort' used when other energy reserves are exhausted.

The mean body mass of male penguins leaving the colony is around 23 kg, which, from calculations based upon the data in Figures 5.11 and 5.12, indicates triacylglycerol reserves of about 2 kg. This amount is just sufficient to sustain the penguin as it walks, using energy at 2.8–4.5 times BMR, as far as 100 km back to the open sea. Under normal circumstances, the birds begin feeding just before exhausting completely the lipid in their adipose tissue (Figure 5.12). Utilization of protein reserves involves drastic alterations in metabolism and they do not last long, so if the weather is unusually severe, or the sea-ice is exceptionally extensive, or stocks of fish at the feeding grounds are low, penguins that were even slightly underweight at the start of the breeding season may not survive. Indeed, Groscolas suggested that the decrease in the plasma concentration of fatty acids and/or of β-hydroxybutyrate may be the metabolic signal (the black arrow on Figure 5.12) that prompts the parent to abandon its chick and return to the sea, even if its mate has not yet come back. Each year around 30% of eggs and chicks are abandoned for various reasons, and without parental care they always die. However, mortality among adult penguins is quite low, and each bird may breed many times during a long lifetime.

Comparison between different species of penguin shows that, in general, larger species can fast for longer, suggesting that the very large extinct penguins may have undergone fasts lasting many months.

5.3.2 Bears

Brown or grizzly bears (*Ursus arctos*), and black bears (*U. americanus*) feed throughout the summer on grass, fruit, nuts, fish, small mammalian prey and carrion. In autumn, all brown and black bears fatten rapidly before entering caves or hollow trees where they become dormant for weeks or months. The terms 'hibernation' and 'torpor' are sometimes used to describe this state in bears. To avoid confusion with true hibernation (Chapter 4), this phenomenon is here called '**dormancy**'. Much of the research on the metabolic basis of this physiological state has been carried out in the USA on the black bear, which

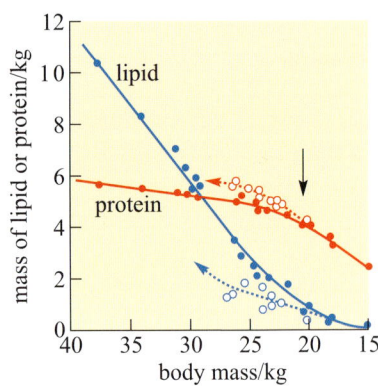

Figure 5.12 Summary of the depletion and replenishment of lipid (blue) and protein reserves (red) in breeding emperor penguins. The black arrow indicates the body mass at which most penguins abandon the egg or chick and return to the sea to feed, even if their mate has not yet come back. Closed symbols show measurements from fasting penguins; open symbols, refeeding.

occurs over most of USA and southern Canada and is smaller and easier to maintain in captivity than brown bears or polar bears.

Dormancy in black and brown bears

The dormant state of bears differs from true hibernation in that the body temperature does not fall below 31–35 °C and a major disturbance (such as an intruding biologist) can arouse them to full activity in a few minutes. Dormant bears do not eat, drink, urinate or defaecate, the heart rate drops from 50–60 beats min^{-1} to 8–12 beats min^{-1}, and oxygen consumption is only 32% of that of actively foraging bears. Nonetheless, the rate of protein turnover, as measured by the rate of dilution of ^{14}C-labelled amino acids injected into the blood, is three to five times higher during dormancy than in normal activity. Protein synthesis, particularly of enzymes involved in lipid and protein metabolism, also continues unabated during dormancy. The degradation of proteins to urea, however, is greatly slowed in dormancy. In these respects, the reciprocal changes in protein metabolism of the hibernating bear resemble those of humans and other mammals of tropical origin living on protein-deficient diets: essential amino acids are incorporated into proteins in the liver, but oxidation of amino acids and excretion of nitrogen are greatly reduced.

A small but significant quantity of urea is produced throughout dormancy but it is not excreted. Instead, it passes across the gut epithelium and into the lumen, where it is further degraded to ammonia (NH_3) and carbon dioxide by the gut bacteria. The carbon dioxide is excreted with the respiratory gases, but the fate of the highly soluble, and in high concentrations toxic, ammonium ions (NH_4^+) is more interesting. In dormant bears, the blood concentrations of amino acids, total protein, urea and uric acid during dormancy are similar to those of active bears that are feeding regularly. Since there is almost no net elimination of the nitrogen, it must be re-incorporated into amino acids. The most important source of carbon for this process is glycerol. If ^{14}C-labelled glycerol is injected into a dormant bear, the label quickly appears in alanine, then in other amino acids, and finally in plasma proteins.

■ Where would the glycerol come from normally in a dormant bear?

Glycerol is produced from lipolysis of triacylglycerols.

The fatty acids released by lipolysis are used in energy metabolism, but much of the glycerol (that in other mammals is mainly oxidized) is combined with ammonia to form amino acids, which are incorporated into proteins in the normal way. This mechanism recycles the nitrogen so efficiently that the concentration of urea in the blood actually decreases slightly after several weeks of dormancy.

The rate of excretion of nitrogen can be estimated as the ratio of the concentrations of urea (U) to creatinine (C) in the blood (U:C ratio). Malcolm Ramsay and colleagues measured the U:C ratio in blood samples collected from wild polar bears in northern Canada (Ramsay et al., 1991). Creatinine is formed from the breakdown in muscle of phosphocreatine, a high energy phosphate compound, and is a minor but constant source of excreted nitrogen. In bears, the concentration of creatinine in blood plasma increased about threefold during the first 1–2 days in dormancy and then remained constant. The U:C ratio is around 50 in most mammals, especially carnivores that are eating regularly, and does not normally fall lower than 25, even during prolonged fasting. But U:C ratios of less than 10 are frequently measured in dormant black bears, indicating that during dormancy a high proportion of the urea is re-incorporated into proteins instead of being excreted (Nelson et al., 1984). Consequently, the bears' lean body mass is hardly diminished even after months of dormancy and their muscle strength is unimpaired.

In starving humans and most other fasting animals, β-hydroxybutyrate and acetoacetate (ketone bodies) are formed by partial oxidation of fatty acids (see Figure 5.11). They are normally eliminated by further oxidation, but sometimes the presence of a high concentration of ketones disturbs the acid–base buffering of the blood and a comatose state called ketosis develops. In many hibernatory mammals, very high concentrations of ketone bodies trigger arousal. Ketone bodies increase in dormant bears as well but only to a maximum of ninefold between normal activity and dormancy and the toxic effects of ketosis have never been observed. Experiments in which labelled glycerol is injected into the blood of dormant bears show that, as well as being incorporated into amino acids, substantial amounts of labelled glycerol also appear in triacylglycerols.

■ What does this observation show?

It shows that, as well as lipolysis of lipids stored in adipose tissue, resynthesis of triacylglycerols from fatty acids and glycerol is also occurring at a significant rate. The rate of triacylglycerol turnover may be higher during dormancy than during normal activity, and may limit the rate at which free fatty acids can enter the pathways that produce β-hydroxybutyrate and acetoacetate, thereby preventing ketosis and enabling the bears to sleep undisturbed for long periods.

■ Are there any other metabolic advantages of utilizing fat during dormancy?

Oxidation of fat produces water. Since the bears do not drink during dormancy (except perhaps the occasional mouthful of snow), and the surrounding air is very dry, such metabolic water probably makes a significant contribution to water balance. Total body water, blood volume and the water content of red cells and plasma remain normal during dormancy, indicating that the water generated by such metabolism is indeed sufficient to offset the small losses due to respiration of the dry, cold air. Thus the large quantities of adipose tissue triacylglycerols in bears are much more than just an energy store: they are central to the bears' metabolic adaptations to dormancy.

Measurements of composition of the respiratory gases reveal that the respiratory exchange ratio (RER) falls from 0.78 when the bears are fully active to 0.62–0.69 during dormancy. Such values are exceptionally low: the normal minimum RER for mammals, representing oxidation of lipid only, is 0.71. The low RER shows that some of the carbon dioxide that would normally be excreted through the lungs fails to appear. Carbon dioxide cannot be stored in significant quantities because as hydrogen carbonate (HCO_3^-), it alters the acid–base balance of body fluids, so it must be converted into non-volatile compounds, possibly by the microbes in the gut or by enzymes in the bear's liver.

Like other metabolic processes, the urea cycle and protein synthesis generate quite a lot of heat. The high rate of these processes during dormancy, together with the bears' large size and thick, insulating fur, combine to maintain a much higher body temperature than that of small mammals in deep hibernation. Fully functional brown adipose tissue has not been demonstrated in bears, even in neonates, although small areas of white adipose tissue have some structural features that resemble those of BAT. Nonetheless, at a body mass of less than 1 kg, bears are smaller at birth, relative to the size of their parents, than any other eutherian mammal, and they are born in mid-winter or early spring.

Bears become fully active, eating and able to deal with predators within hours of leaving the den after weeks of dormancy. Astronauts after long periods in space and people recovering from illness or injury cannot do the same: especially in the elderly, the skeleton is weakened by more than a few days of immobilization as bone is reabsorbed. Vertebrae or limb bones may fracture under very weak forces; sometimes just standing is sufficient to cause injury. Measurements on people resuming activity after a period of bedrest show that the rate of reformation of bone is 2–3 times slower than its loss during immobilization. How to bears manage to avoid similar problems during and immediately after dormancy?

Carboxy-terminal propeptide of type 1 collagen (PICP) is the remnant of nascent type 1 collagen (the principal protein in bone – see Book 3) that is cleaved off by proteases as the protein is incorporated into bone. It can be measured in the blood serum, and thus serves as a convenient marker of bone formation. Another serum protein, carboxy-terminal cross-linked telopeptide of type 1 procollagen (ICTP), acts as a marker of bone resorption. Figure 5.13 shows some measurements of these proteins in breeding and non-breeding black bears (*Ursus americanus*) during dormancy and when actively feeding in the wild in the mountainous region of Virginia, USA (Donahue et al., 2003).

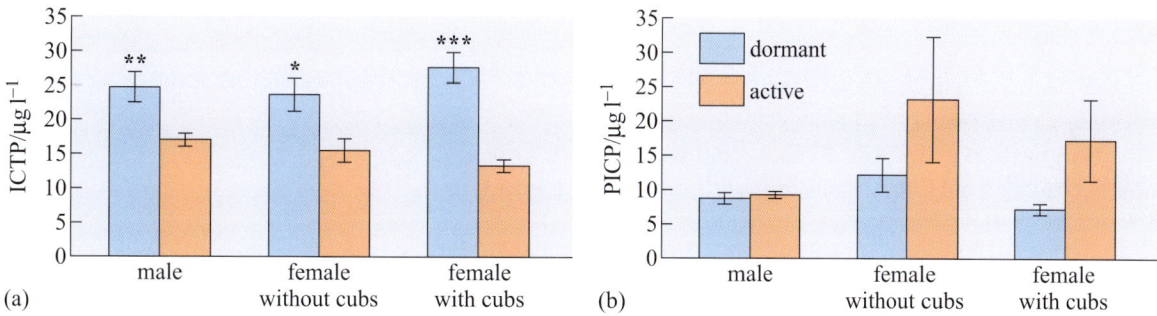

Figure 5.13 The quantities of (a) ICTP and (b) PICP measured in blood samples taken from various black bears. The asterisks indicate where the differences between dormancy and activity for the same animals are statistically significant: * significant at $P < 0.05$, ** significant at $P < 0.01$, *** significant at $P < 0.001$.

■ What conclusions about bone loss and reformation can you draw from Figure 5.13?

No statistically significant differences between dormant and active bears were found in the measurements of PICP, though there was a trend towards higher values for active females. In contrast, ICTP, especially in females who were feeding cubs, was higher while they were in the den.

■ Do the data in Figure 5.13 explain how bears avoid increased risk of bone fractures just after emerging from the den after a period of dormancy?

No, not fully. Bone loss (Figure 5.13a) is still higher during dormancy (though the differences between measurements from bears found in dens and full activity are not as great as between sedentary and exercising people or rats). But the data provide no evidence that bone formation is consistently greater while the bears are active.

- Could the investigation have failed to reveal important aspects of the time course of bone turnover during dormancy and activity?

Yes. This investigation would fail to detect any brief period of acceleration of bone formation just before or just after emergence from the den. It is important to study the markers of bone turnover during the transitions between dormancy and activity.

The scientists were able to obtain measurements of the bone formation marker (PICP) before, during and after emergence from a few bears (Figure 5.14). These measurements show that PICP is four- to fivefold higher early in the remobilization period compared with mid-summer. This brief peak in the concentration of PICP may indicate that bone formation increases just when it is most needed.

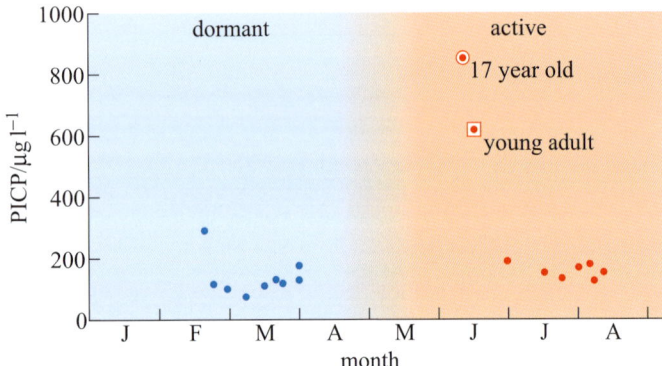

Figure 5.14 The bone formation marker (PICP) in the blood of female black bears. Bears entered dens in December and left in mid-April. Each point represents measurements from a blood sample taken from an individual bear. Each of nine bears was sampled once while in the den and once while active, but the exact day was determined by when they could be found and conveniently tranquilized.

- Do the data in Figure 5.14 indicate exactly how long the peak concentration of PICP in bears emerging from dormancy lasts?

No. Unfortunately, no bears were found and tranquilized between the end of March and the beginning of June, so we have no information about this crucial period. Such are the problems of studying wild animals under natural conditions.

This more efficient compensatory mechanism for recovering from immobilization-induced bone loss, combined with low bone loss during disuse, enables bears to switch abruptly between dormancy and full activity. The control mechanisms that prompt the rapid increase in bone formation remain to be explored. Its elucidation might enable physicians to induce similar effects in people recovering from hip-replacement operations and other situations in which long periods of immobilization are unavoidable.

- Why are the differences between the values for dormancy and activity greatest for females that are suckling cubs (Figure 5.13)?

Lactating females are transferring some of the minerals (especially calcium) in the bones to the milk (where it is incorporated into the cubs' skeletons), further weakening their own skeleton. Breeding females thus need to maximize bone reformation after dormancy.

Fasting in polar bears

Polar bears (*U. maritimus*) are almost entirely carnivorous and are the only living species of bear to obtain almost all their food from the sea. Their main prey is ringed seals (*Phoca hispida*), both adults and suckling pups; they catch the adults when they come up to breathe through holes in the ice, though detailed observations show that as few as one in five attempts results in a kill. The seals are born in holes in the snow, and their mothers leave them hidden there while they go to feed, returning to suckle them at least once a day. The young seals fatten and grow very rapidly on their mothers' exceptionally rich milk. Until their fluffy white fur is replaced by the sleek, more waterproof adult coat, the young seals cannot enter the water even if it is accessible because they quickly get too cold. The seal breeding season in spring and early summer thus provides polar bears with plentiful prey.

■ What features of the arctic environment enable the seals to maintain their population in spite of predation from bears?

The seals travel long distances and ice conditions suitable for breeding are not necessarily in the same place each year, so the bears may not find many of the seal breeding colonies.

How often are polar bears successful in finding food? Do they fast when out on the ice, as well as when in dens? Polar bears range over such a wide area of inhospitable terrain that such questions, though vital to the management of the species in the wild, are not easily answered by direct observation. The study of nitrogen metabolism in black bears suggests an indirect way of investigating such topics.

Figure 5.15 shows some measurement of the ratio of urea to creatinine in blood samples collected from polar bears in northern Canada that were temporarily sedated with drugs injected from a dart gun.

■ What do these data suggest about food sources and hunting success in bears?

More than 75% of bears in dens (Figure 5.15a) had U:C ratios of 10 or less so they were obviously not eating, but the U:C ratios were 19.9 or lower in 70% of those caught on land in summer and autumn, showing that they were also fasting (Figures 5.15b and c). Out on the sea-ice in spring (Figure 5.15d), more than half the bears sampled had U:C ratios of 30 or more, indicating that they were feeding frequently. At least 10% of the bears in this sample were fasting: either they were inexperienced, inefficient or unlucky hunters of seals or (as quite frequently happens) they were forced to give up their kills to larger bears that threatened them. Alternatively, they may be 'voluntarily' anorexic while mating: during the spring, large males attend oestrous females closely and may fight with rivals, leaving little time for hunting.

Ice conditions that favour catching adult seals are strongly dependent upon weather and water currents, and so are widely scattered in space and time. Food supply is probably erratic even for the most proficient bears. Seal hunting is almost impossible for several months in the summer and autumn when the sea is unfrozen, and the only food available to polar bears is the odd bit of carrion and a very small quantity of plant food. The data in Figure 5.15 show that nearly all polar bears fast for long periods in the summer and that, for many,

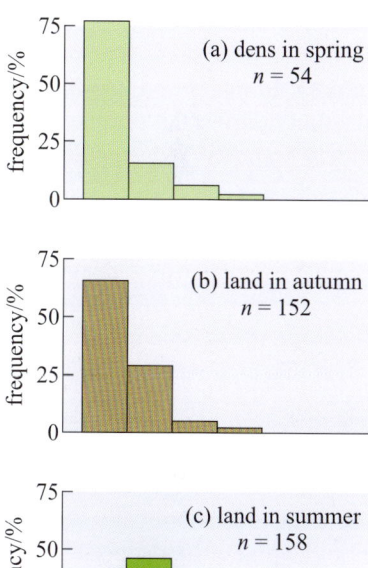
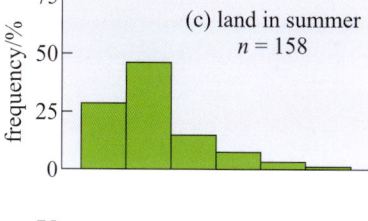
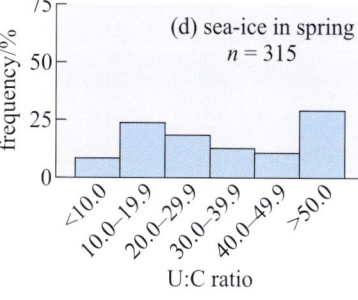

Figure 5.15 The frequency (as a percentage of the total) of U:C ratios measured in free-ranging polar bears caught in different areas of northern Canada at different seasons.

the food supply is unreliable even in winter and spring. Thus polar bears seem to adopt many of the metabolic features of winter dormancy in the omnivorous brown and black bears, while remaining active enough to be able to travel long distances between seal-hunting grounds. They become lethargic and remain inactive for long periods when weather conditions or terrain make hunting impossible, suggesting that they have become 'dormant' without actually being asleep in a den.

This theory is confirmed by observations on polar bears held temporarily in captivity. Bears caught in autumn were starved for 5–7 weeks, fed for 3 days, and then fasted again. Blood samples were taken just before and for several days after feeding, and the concentration of urea and creatinine measured. During the imposed fast, the U:C ratios averaged only 11.0, but rose abruptly to 32.0 after feeding and then declined to 22.8.

Polar bears are a relatively new species, almost certainly evolving from brown bears during the last 100 000 years. They must have inherited the capacity for dormancy from their omnivorous ancestors that fed mainly in the summer and autumn. In males and non-breeding females, dormancy takes the form of inactivity during the summer months and in winter, intermittent fasting between widely scattered and irregular feeding opportunities.

Breeding females undergo periods of fasting more closely tied to the seasons. Shortly before giving birth in December or January, the pregnant females migrate to areas where deep snow drifts suitable for making dens have formed against river banks or similar obstacles. Around Alaska, polar bears find denning places on the sea-ice but those around Svalbard and Hudson Bay in Canada travel inland, sometimes substantial distances. The mothers give birth in the snow den and suckle their cubs there for up to 5 months, during which time they remain inactive and do not feed at all, but they are alert and, so far as we know, the body temperature is normal.

Lactation is metabolically very demanding, especially during fasting. The mobilization of lipids, proteins and minerals from storage tissues and the skeleton and the synthesis of milk would produce more than enough heat as a by-product of metabolism to maintain normal body temperature. For obvious practical reasons, there are few physiological data on breeding polar bears. However, indigenous people who traditionally hunted the animals for their meat and skins report that bears in maternity dens are no soft target, and the adult females are quick to put up a fight.

Polar bear cubs are weaned over many months, and while food is scarce, mothers sometimes suckle offspring that are almost as large as themselves. Females never breed more frequently than every other year, and in some areas, the interbirth interval may average more than four years. Although polar bears mate in the late winter, the early embryo undergoes delayed implantation (a phenomenon that seems to be widespread among mammals, especially in carnivores whose food supply is irregular) and gestation does not start until the autumn.

■ How could delayed implantation enable female polar bears to adjust reproduction to food supply? How is this adaptation similar to that of reindeer?

Like the reindeer, female polar bears can adjust their reproduction to the food supply for the particular year and location: fortunate bears may find large, fecund colonies of breeding seals and be able to fatten rapidly by eating several pups a day. Such animals may lay down reserves of nutrients sufficient to raise triplets. Others may produce only one cub, or, if food is really scarce, the pregnancy may be terminated, and the female becomes receptive in the following mating season.

5.3.3 The structure of adipose tissue

Since food is only available seasonally or intermittently at high latitudes, many arctic birds and mammals, including polar bears, Svalbard reindeer, arctic foxes, seals and walruses, naturally accumulate large stores of fat. The quantity of energy stored and the metabolic control of its use are finely adjusted to the habits and habitat of the species. This section is concerned with the cellular structure and anatomical organization of adipose tissue in such naturally obese species. Most laboratory mammals do not naturally become obese, and must be induced to do so by drastic measures such as changes in diet, drugs or surgery. Although it is impossible to carry out as detailed measurements or carefully controlled experiments on wild animals as it is in the laboratory, arctic species provide a rare opportunity to study fattening and obesity as natural phenomena, rather than as pathological or artificial conditions. Observations on these naturally obese animals can help resolve discrepancies between mechanisms that can be demonstrated experimentally in rats and those that seem to occur in people.

One important aspect of obesity is the contribution of adipocyte enlargement and the formation of additional adipocytes to animals' increased capacities for storing lipid. In adult rats and mice, fattening is achieved almost entirely by enlargement of adipocytes: the number of cells does not change. The matter is not easy to investigate in humans because there is no really accurate, non-destructive way of measuring total adipocyte complement, but indirect estimates suggest that the accumulation of more adipocytes makes a significant, in some people the dominant, contribution to obesity. In order to establish whether adipocyte proliferation is also essential to expansion of the lipid storage capacity in naturally obese arctic animals, we have to find a way of calculating how many adipocytes would be expected in an animal of any particular body mass. Figure 5.16 shows some measurements of the numbers of adipocytes in some temperate-zone and tropical mammals (Pond and Mattacks, 1985). The equations for the regression lines drawn on Figure 5.16 can be used to calculate the number of adipocytes expected in an animal from its body mass. The predicted adipocyte complement can then be compared with the measured adipocyte complement.

Such comparison shows that naturally obese arctic mammals such as polar bears, arctic foxes, wolverines* and reindeer have more adipocytes than expected, usually between twice and four times as many, although a few specimens have almost exactly the predicted number of adipocytes. Such proliferation of adipocytes is modest compared to that of humans: some obese people have more than ten times as many adipocytes as expected from comparison with the data in Figure 5.16.

---

* Wolverines (*Gulo gulo*) are large mustelid carnivores, related to otters, stoats, weasels, mink, ferrets and badgers.

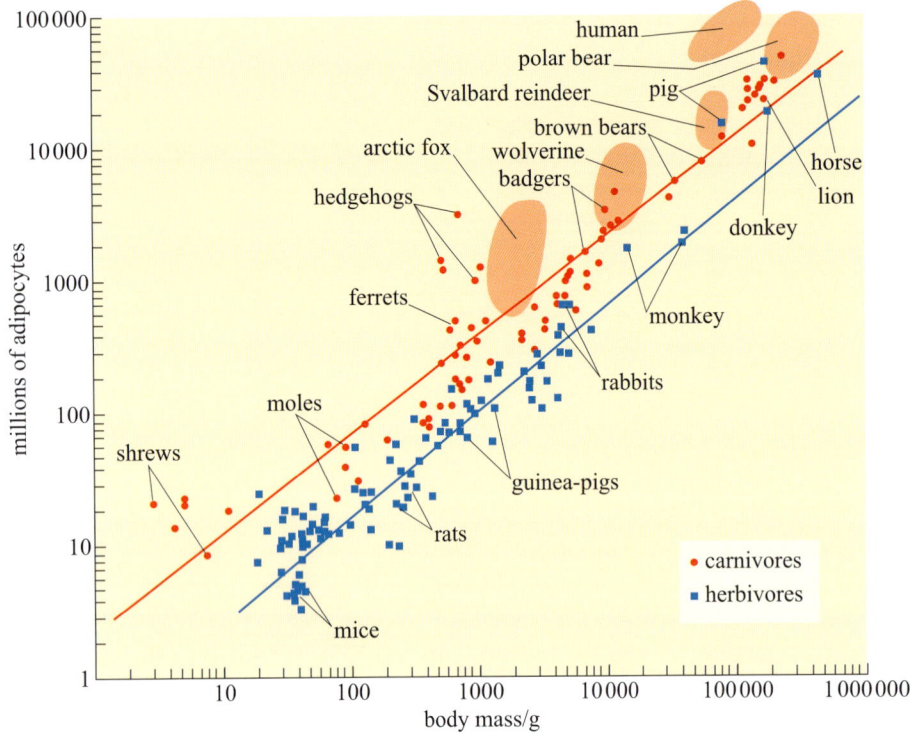

Figure 5.16 The numbers of adipocytes in some temperate-zone and tropical carnivorous and non-ruminant herbivorous mammals, compared with similar measurements on some naturally obese arctic species: polar bears, arctic foxes, wolverines and Svalbard reindeer. The large shaded circles enclose numerous measurements from humans and wild specimens of the four arctic species.

However, the adipocyte complement of the wild mammals was found to be quite variable in otherwise similar specimens collected from the same area at the same time. Many factors such as hunting ability and appetite determine individual differences in fatness, but among the carnivores there was no evidence that individuals with more adipocytes were normally any fatter than those with fewer adipocytes: the adipose tissue of the former simply consisted of numerous, relatively smaller adipocytes. Individual variation in adipocyte complement is also observed in humans, with some people having relatively few, large adipocytes and others more numerous smaller ones, but it is not as conspicuous in laboratory rats, all of which seem to have about the same adipocyte complement in relation to their body mass unless artificially manipulated to make them unnaturally obese.

■ Do these observations suggest that people who have large adipocyte complements are, or will inevitably become, obese?

No. In the wild carnivores, fatness does not correlate with adipocyte complement. The same may be true of other mammals including people.

We know very little about the origin of such individual differences in adipocyte complement: adipocyte proliferation takes place mainly during the suckling period, and the exact course of growth at this age may differ from one individual bear, arctic fox, wolverine and reindeer to another, depending upon the number of littermates and the amount of food available to its mother.

■ Would it be possible to determine the fatness of a particular bear or reindeer by measuring the volume of a sample of its adipocytes?

No. Adipocyte volume would not be an accurate measure of fatness because the relationship between the total mass of adipose tissue and the volume of its adipocytes would be different in specimens that have large or small adipocyte complements.

Unfortunately, assessment of fatness from biopsy samples of adipose tissue is much more satisfactory in rats (because their adipocyte complement is more constant) than it is in either naturally obese arctic mammals or in humans.

Summary of Section 5.3

Penguins and many other large polar animals fast for long periods while remaining active and at near-normal body temperature. Emperor penguins fatten before the breeding season and fast for weeks during courtship and reproduction. Very little protein is broken down until lipid stores are nearly exhausted. Energy reserves determine an individual's behaviour such as feeding or abandoning the chick. Omnivorous brown and black bears feed in summer and become dormant in winter: they stop feeding and enter dens, where their metabolism slows and is supported almost entirely by lipids released from adipose tissue. Urea is recycled and very little nitrogenous waste is excreted, so the protein in muscle, liver and other lean tissues is not depleted, as normally happens in prolonged fasting. Bone may be withdrawn from the skeleton during long periods of inactivity, but the tissue is restored to normal strength by rapid deposition of new bone in spring. Similar physiological processes occur in carnivorous polar bears when food is scarce but, except for breeding females, there is no regular, prolonged period of dormancy in a den.

5.4 Thermal insulation

The principles of thermal insulation in birds and mammals are described in Chapter 2. For organisms of similar size and shape in a similar thermal gradient, the rate of heat loss from convection is up to 90 times as fast in water as in air, so in temperate climates, aquatic endotherms need much more efficient insulation than terrestrial species. Since seawater freezes at −1.9 °C, but the temperature of the air around the Poles can fall below −50 °C, the insulation requirements of aquatic and terrestrial polar animals are not very different. Nonetheless, there are important differences in the tissues involved and in their responses to different environments.

5.4.1 Insulation in terrestrial endotherms

Relatively minor changes in body shape can contribute much to reducing heat loss. Polar bears have relatively small, round ears, huge, shaggy feet and the tail is much reduced. Svalbard reindeer (Figure 5.2b) are smaller and stockier, and have shorter ears, legs and snout than subspecies that live further south.

■ The continental climates of northern Canada and Russia are as cold or colder than that of most of Svalbard, but the native reindeer nonetheless have long legs. What factors other than minimizing heat loss might determine their body shape?

Svalbard reindeer have no natural predators so they never run fast enough or for long enough to risk overheating. Reindeer on mainland Europe and America have long been subject to predation from wolves, which chase their prey over long distances. The need to lose excess heat while running fast may curtail the evolution of the short, stocky body form. The value of being able to see predators while the head is lowered for grazing may also account for the longer, narrow face of mainland subspecies compared to Svalbard reindeer.

The insulating properties of furs and feathers can be easily compared by wrapping pelts around heated objects such as bars and measuring their rate of cooling under various conditions. Such observations indicate that in still air, the insulation of all coats of fur or feathers is proportional to their length and thickness, but the texture and secretions from cutaneous glands produce very different properties when exposed to wind and water.

Some small arctic mammals such as lemmings and hamsters spend the winter in burrows and tunnels under the snow, where the air is effectively still all the time. Arctic foxes, and sometimes bears, shelter in snow drifts, and their young are born in dens, but in polar regions there is little shelter from plants, because trees and large shrubs are absent, and there are not many caves or other geological structures formed by flowing water. The ears of arctic foxes are reduced and thickly furred (Figure 5.2a) and the long, very bushy tail can provide extra insulation to any exposed part of the body when the animal is resting.

The effects of wind are important for large mammals, particularly if, like reindeer, they spend a large proportion of the time foraging in exposed places. In such animals, the outer guard hairs are long and relatively stiff, providing mechanical protection and support for the fine, dense underfur that traps layers of warm air near the skin. Stiff outer feathers and fluffy down combine to insulate birds in much the same way.

In polar homeotherms, fur or feathers often extends over parts of the body that are usually naked in temperate-zone species: the feathers extend along the legs and over the feet of ptarmigan (Figure 5.5) and snowy owls, and the pawpads of arctic foxes and arctic hares are covered in short, tough fur (Figure 5.17). The fur of Svalbard reindeer is longer and denser than that of Norwegian reindeer and it covers the ears, eyelids, snout, lips and feet much more extensively.

Figure 5.17 The underside of a hind paw of an arctic fox in winter coat, showing the paw pads covered with fur.

■ Could there be any disadvantages in fur covering all parts of the body?

The animal's ability to dissipate heat during strenuous exercise or in warm weather is reduced and it risks overheating. When overheated, seals hold their flippers up in the wind or try to get back into water. Reindeer, bears and other terrestrial mammals pant vigorously, but hyperthermia is a real risk, especially for very large or pregnant specimens. The need to dissipate heat during prolonged, strenuous exercise may be one reason why the large hunters, such as wolves, which occur throughout the Russian and Canadian Arctic, are not completely covered in thick fur. In husky dogs (and their wolf ancestors), counter-current blood flow in the legs and nose results in much lower temperatures of the peripheral parts of these organs (Figure 5.18a), greatly reducing heat loss from them. When the animals are asleep, they tuck their feet and nose into their coat or cover them with the thickly furred tail, but they avoid overheating during long chases by retaining some exposed surfaces through which heat can be lost rapidly. All animals with wettable fur lose heat faster when wet, and most species, including polar bears, shake themselves vigorously immediately after swimming (as dogs do).

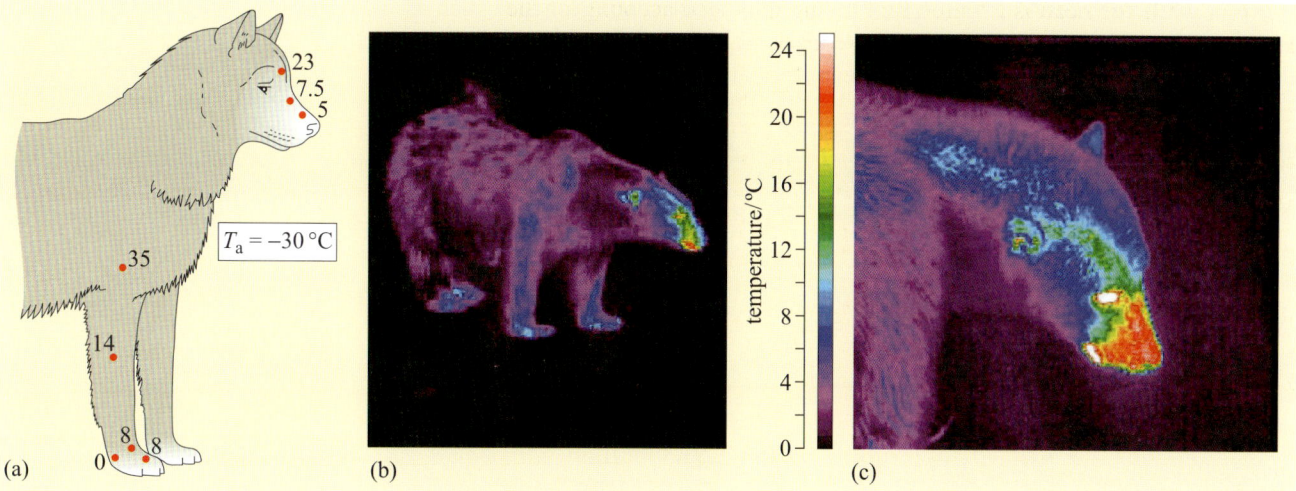

Figure 5.18 (a) The temperatures at various areas of the skin of a husky dog measured with a thermometer. (b) and (c) The heat emitted from a wild polar bear, photographed by infrared thermography. The scale indicates the surface temperatures inferred from the amount of heat radiated. The background snow radiates so little heat compared to the bear that it appears black.

Figures 5.18b and 5.18c show similar information obtained by infrared thermography of a free-living polar bear standing on snow in the Canadian Arctic. The nose, mouth, eyes and ears are warm relative to the rest of the body, and are thus the sites of the greatest heat loss per unit area.

■ Do the measurements in Figure 5.18b suggest that this bear was under thermal stress (i.e. too hot or too cold)?

No. Its feet and lower legs are emitting a moderate amount of heat. Heat lost from the feet would be much curtailed if the animal was too cold, or greatly increased if it was too hot, e.g. after running, as happens in many mammals, including dogs (Figure 5.18a) and to some extent in ourselves.

■ In which of the warm structures revealed by Figure 5.18c is the high temperature due to (a) rich perfusion with blood or (b) high intrinsic metabolic rate?

(a) The external ear, eyelids, nose and mouth receive a rich blood supply (even very small wounds bleed profusely), bringing warm blood from the body core to the surface. (b) In contrast, the outer surface of the eye is not perfused with blood (except transiently at the site of injury or infection, when the damaged eye becomes 'bloodshot'). Its heat is produced locally by the very high metabolic rate of the transparent tissues (lens and cornea) and the numerous tiny neurons that comprise the retina (light-sensitive surface) and the optic nerve.

■ Why do bears (and people) keep their eyes almost closed in strong winds and cold weather?

Figure 5.18c shows that the surface of the eyeball was above 26 °C, the highest surface temperatures recorded from this bear. Even a thin covering such as eyelids would significantly reduce heat loss (as well as protecting the eye from freezing and drying). So arctic animals, especially large species standing out in the wind, view the world through narrow, slit-like eyelids, which confers an aloof, imperious expression.

The surface temperatures of red foxes, kit foxes (another temperate-zone species native to USA) and arctic foxes at air temperatures from −25 to +30 °C have also been compared using infrared thermography, producing images such as those shown in Figures 5.18b and c. There were surprisingly few differences: all foxes lost heat through their legs, paws, ears and snout, but while the two temperate-zone species also lost heat through their thinly furred forehead, this area of the arctic fox is efficiently insulated. In the winter, the forehead of arctic foxes is covered in long, dense fur (Figure 5.2a), making the animal look like a pet dog, but far from being a trivial character, this tuft of fur is an integral part of the species' adaptation to extreme cold. It is greatly reduced in the greyish-brown summer pelt.

People's breath freezes on beards and eyelashes but ice does not accumulate on the fur of many arctic mammals, notably that of arctic foxes, wolves and wolverines, probably because the microscopic structure of the hair surface and/or oily secretions from the skin prevent the formation of ice-crystals. Arctic people value such fur, particularly for trimming hoods and mufflers, although it is less soft and silky than mink.

For reasons associated with the erratic food supply (Section 5.2), many arctic mammals are obese, particularly in the winter, a situation that has led to the notion that thick subcutaneous adipose tissue makes an important contribution to insulation. As long ago as the 1950s, measurements on people swimming the English Channel and on men on polar expeditions failed to reveal any firm association between thickness of superficial adipose tissue and the capacity to withstand exposure to cold. Nonetheless, statements that subcutaneous adipose tissue is superficial because it is essential to insulation still appear in many recent textbooks.

- How could you test the hypothesis that subcutaneous adipose tissue is important for thermal insulation?

The hypothesis predicts that there is normally a thermal gradient across the adipose tissue. In practice, it is quite difficult to measure such a thermal gradient over a long period, and demonstrating it would not prove that adipose tissue (rather than any other superficial tissue such as muscle) is essential to insulation.

Another approach is to compare the partitioning of adipose tissue between internal and superficial depots in arctic and tropical species and look for evidence of selective expansion of the superficial fat in homeothermic animals adapted to cold climates. Polar bears eat seals, and occasionally swim long distances in ice-cold water between hunting grounds. Ice conditions often make their prey inaccessible, forcing them to fast for long periods (Section 5.3.2), so bears fatten when prey are readily available and are usually obese for large parts of the year. These facts have given rise to the idea that, as in fully marine mammals such as walruses and polar cetaceans (that are often found in the same habitats), adipose tissue makes an important contribution to insulation in polar bears. The data in Figure 5.19 (Pond and Ramsay, 1992) enable us to test this hypothesis directly. The order Carnivora includes species that share a common ancestry and many habits and range in size from weasels (with a body mass of about 0.1 kg) to bears (with a body mass of up to 700 kg) and are adapted to live in very hot (e.g. Rüppell's fox) and very cold (e.g. arctic fox, polar bear) climates.

- Is the partitioning of adipose tissue between internal and superficial depots in semi-aquatic arctic polar bears different from that in fully terrestrial temperate-zone carnivores?

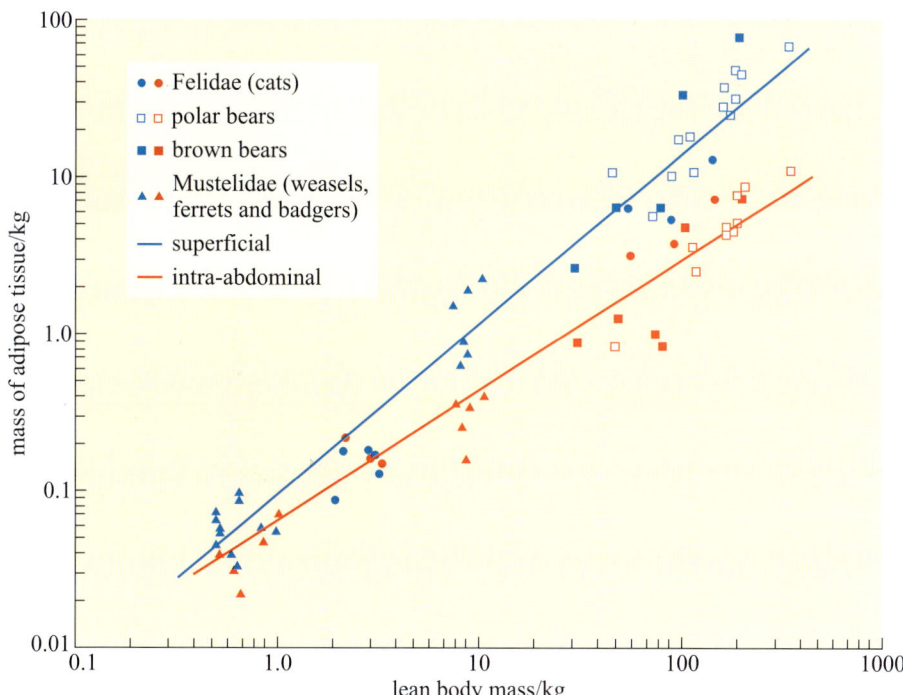

Figure 5.19 The mass of superficial and intra-abdominal adipose tissue in some relatively obese carnivores. There was no significant correlation between fatness and lean body mass in this sample of 44 moderately obese carnivores. To compare the arctic species with temperate-zone and tropical Carnivora, the regression lines are calculated from all the data except those from polar bears.

No. The mass of the intra-abdominal depots becomes proportionately smaller and that of the superficial adipose tissue larger with increasing body mass in all species studied. The data for polar bears lie close to the regression lines fitted to the data from the temperate-zone species. There is no evidence for adaptive redistribution of adipose tissue in polar bears. Their skin is warm to the touch and the coarse guard hairs and woolly underfur are probably the principal insulating tissues. The distribution of adipose tissue in mammals native to cold freshwater, such as otters, beavers and muskrats, is also not different from that of related terrestrial species (e.g. badgers, squirrels and lemmings, respectively), providing no evidence that their adipose tissue is adapted to function as an insulator.

As much as 50% of the body mass of large adult bears can be lipid, and selective expansion of the superficial adipose depots relative to the internal depots seems to arise mainly because 'there is nowhere else for so much fat to go'. The surface area also declines with size in animals of similar shape, so the superficial layer of adipose tissue becomes thicker with increasing body mass, even if it does not become proportionately more massive. There is also proportionately more superficial adipose tissue in large, naturally obese birds: the subcutaneous depots amount to more than 80% of the total adipose tissue in emperor penguins.

5.4.2 Insulation in aquatic endotherms

Most seals and sealions (order Pinnipedia) are furred. The adult fur usually consists of short, dense stiff guard hairs that are oily from profuse secretions of the sebaceous glands. The hair probably acts like a wet suit of a human diver: a layer of water is trapped around the hair, where it is warmed by body heat and prevents much colder water from coming into direct contact with the skin. Short, oily fur dries quickly when the seals come onto land. Two genera, the northern fur seal (*Callorhinus ursinus*), in the north Pacific (Figure 5.20), and eight species of southern fur seals, *Arctocephalus* spp., in the Pacific and Southern Oceans, have a dense layer of underfur throughout their lives that traps small bubbles of air and keeps the skin dry.

■ Are there any disadvantages of such insulation for aquatic mammals?

During deep dives, the pressure of the water would compress the air bubbles, greatly reducing the insulating efficiency of the fur and increasing the possibility that the skin is wetted. The air also makes the mammal or bird more buoyant, thus hindering diving and swimming underwater.

This kind of insulation is most common in semi-aquatic mammals such as beavers, muskrats and otters, and in the feathers of ducks and penguins, which live mainly in shallow water and spend long periods exposed to very cold air. Newborn seals have a fluffy coat, quite different in both colour and texture from that of the adults, that provides good insulation in air but is ineffective in water. Seal pups cannot survive in very cold water for longer than a few minutes until they moult the neonatal coat and grow the firm, darker adult fur that provides better protection in water. Because the fur is an efficient insulator in air, adult fur seals and the juveniles of many other seal species have long been hunted for their pelts.

Figure 5.20 The northern fur seal (*Callorhinus ursinus*), on the coast of Alaska. Note the thick fur and long, stiff whiskers (vibrissae) that are important sense organs when foraging in dimly lit waters.

In view of the effectiveness of insulating fur among seals, it may seem surprising that one of the most northerly pinnipeds, the walrus (*Odobenus rosmarus*), and all the whales and dolphins (order Cetacea) are almost hairless, even as neonates. In these marine mammals, the skin and a specialized form of fibrous adipose tissue called blubber are the major insulators. Walrus skin is up to 5 cm thick and the blubber, although minimal or absent over the tail, flippers and parts of the head, is up to 15 cm thick over some areas of the neck and trunk (Figure 5.21). Measurements on dead tissues indicate that conductance of heat is about half as efficient through adipose tissue as through aqueous tissues such as muscle. However, such passive properties are probably much less important to thermal insulation than counter-current systems of blood vessels (see Section 3.3.2), and control of the rate of flow of blood through the tissue, that brings heat from the warm core to the surface. Blood flow through the superficial adipose tissue can be reduced almost to zero for hours without damaging the adipocytes.

■ Why cannot the blood flow to muscle be similarly reduced?

Muscle is much more metabolically active than adipose tissue and cannot remain functional unless supplied with sufficient blood-borne nutrients and oxygen. Deprivation of blood for longer than a few minutes causes permanent damage to most muscles.

When basking out of the water, the skin of walruses is so flushed with blood that it appears pink (Figure 5.21), particularly in warm weather, which is when they are most often seen by hunters, photographers and biologists. As soon as walruses enter cold water, blood vessels in the skin and outer layers of the blubber

Figure 5.21 A large adult walrus (*Odobenus rosmarus*) hauled out on ice in the Bering Strait near Alaska. The blue water in the background suggests that it is summer, which accounts for the pink coloration of much of the hairless skin. Compared to typical seals (Figure 5.20), walruses have a massive snout with thick muscular lips covered with short, stout vibrissae.

constrict, shutting off the circulation almost completely. The animals become dull grey in colour, and much less heat is lost at the surface. During strenuous exercise, or when in warmer water, perfusion can be increased, thereby adjusting accurately the rate of heat loss to internal heat production, as happens in counter-current mechanisms.

■ Why is the largest area of pale pink skin in Figure 5.21 a symmetrical patch between the walrus's eyes? Hint: compare this photograph with Figure 5.18a and c, and think about where you sweat most when hot!

The pale patch covers the walrus's forehead, which overlies the forebrain, a metabolically active and functionally important region of the brain that is impaired by even small changes in temperature. Blubber under this area of skin is much thinner than that on the neck and body, perhaps absent altogether. All higher mammals, including dogs (Figure 5.18a) and bears (Figure 5.18b and c) lose a lot of heat from this area of the head, protecting the brain from overheating. The human brow sweats profusely and is a good site from which to measure minor changes in overall body temperature, as in fever.

Vigorous exercise is rarely necessary for feeding, because walruses eat mainly bottom-dwelling invertebrates, especially clams and similar burrowing molluscs, which they locate with their sensitive vibrissae (Figure 5.21), and dig out by squirting jets of water. Being so large (up to 1.5 tonnes), adult walruses have few predators except killer whales and, occasionally, polar bears. Adipose tissue has the advantage of being almost incompressible and, although fat is less dense than water, it contributes less to buoyancy than air trapped in the pelt.

However, restricted blood flow is incompatible with certain other functions of the superficial tissues. Walruses shed the outer layers of the skin each year, possibly as a means of getting rid of external parasites. To support regrowth of the skin, blood perfusion of the superficial tissues is plentiful throughout the moult, and walruses normally spend almost the entire period basking on beaches or ice-floes. Although moulting takes place in mid-summer, walruses can die of cold if forced to spend too much time in the water during this period. Cetaceans spend their entire life in water and moult less efficiently, enabling barnacles and ectoparasites to colonize their skin.

Thick layers of superficial adipose tissue also contribute to insulation in other species of seal, although the relative importance of fur and fat probably differs greatly between the seasons and in different species. The distribution of adipose tissue in most adult seals, dolphins and the small toothed whales suggests that it is adapted to contribute to thermal insulation: the superficial blubber forms an almost continuous layer, albeit of very variable thickness, and adipose tissue is almost absent from inside the abdomen and muscles.

All kinds of seals that have been investigated have surprisingly little superficial adipose tissue at birth, the superficial depots being only 2–4 mm thick in northern fur seal pups that weigh 5–6 kg at birth. Their thick natal coats keep them warm in dry weather on land, but, although their BMR can increase to as much as 18 W kg^{-1}, seal pups quickly become hypothermic if immersed in water or during heavy rain. Furthermore, the distribution of adipose tissue of neonates resembles that of typical terrestrial mammals: as well as several superficial depots, there are significant quantities of adipose tissue inside the abdomen, around the kidneys, and in the pericardium. Some of these internal depots contain adipocytes which appear under the electron microscope to have features in common with BAT (see Chapter 4, Box 4.1): mitochondria are quite numerous but they lack cristae. However, although it may be thermogenic to some extent, the tissue is not true BAT in either structure or metabolism.

Birds replace old, worn feathers with new ones, usually one or twice a year, often just before breeding or migration. Moulting and replacement of the plumage impose heavy demands on the nutrient reserves because large quantities of energy and protein are used in the synthesis of new feathers. Foraging is also difficult or impossible: most large birds cannot fly (in the absence of primary wing feathers) and polar species do not swim because their insulation is so severely impaired that they would become too cold in water. The moult takes 2–5 weeks in emperor penguins and king penguins (*Aptenodytes patagonica*), during which time they remain on land (or on ice-floes) and fast, losing up to 45% of their body mass and up to 50% of their protein reserves.

5.4.3 Humans in polar regions

Humans evolved in tropical Africa and gradually colonized colder climates during the Pleistocene ice ages. There have been permanent populations in the Arctic for several thousand years, mostly Inuit (Eskimos) in what are now Canada, Alaska and Greenland, and several groups in northern Europe and Russia, such as the Saami (Lapp) in Scandinavia and the Chukchi in Siberia. Such people do not grow crops and keep only a few domestic animals, mostly for transport (e.g. husky dogs or reindeer), not for food. Until very recently, they lived by hunting seals, walruses, polar bears, fish and wild and semi-domesticated reindeer and, during the brief summer, gathering wild berries.

Adaptation to living in the Arctic has been more technological and cultural than physiological: Inuit (Figure 5.22) are shorter and stockier than Canadians of European ancestry, but comparisons of the distribution and abundance of their adipose tissue revealed that the native people have less, rather than more, superficial fat.

■ What does this comparison suggest about the function of superficial adipose tissue in humans?

It is not adapted to a role as thermal insulation. Frost damage to exposed parts such as the face and hands is prevented by efficient perfusion with warm blood (see Figure 5.18), rather than by any form of insulation. Human colonization of the Arctic was made possible by the effective use of animal skins as clothing.

Adaptations of digestion and metabolism have evolved among Inuit: although their diet was very rich in fat and protein, and for 9 months of the year included almost no fruit or vegetables, diseases such as obesity, diabetes,

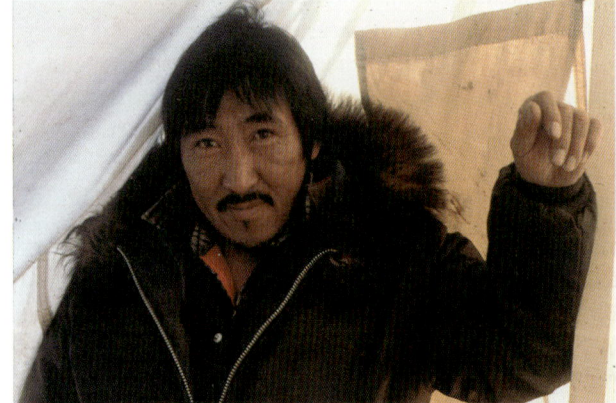

Figure 5.22 Inuit in Arviat, Nunavut (formerly known as Eskimo Point, Northwest Territories), Canada. Genetic, archaeological and linguistic evidence indicate that Inuit ancestors migrated westward from northern Japan and northeastern Siberia, crossing the Bering Strait to Alaska and Canada.

scurvy, rickets, dental caries, constipation and colon cancer were rare. However, obesity, diabetes and dental caries have become much more common during the last 40–50 years since they adopted a western diet. Inuit have never grown or stored crops, and so alcoholic drinks produced from fermented carbohydrates (i.e. grain, potatoes, fruit or sugar) were never part of their diet: alcohol dehydrogenase, the enzyme that detoxifies alcohol, is present in very small quantities in their livers and it is not as readily induced as it is in people whose ancestors have a long tradition of drinking alcoholic beverages. Consequently, grown men are easily intoxicated by as little as 0.25 l (half a pint) of beer.

The capacity of the human nose to conserve moisture by warming and hydrating inhaled air and reclaiming the heat and moisture of exhaled air (shown in the dog and bear in Figure 5.18), is much less efficient than the long nasal turbinals of native arctic mammals such as bears, reindeer and wolves. The ability to breathe steadily through the nose rather than through the mouth improves with practice, but most inexperienced visitors to polar regions are bothered as much by thirst as by cold.

Living in such a severe climate is very tough: archaeological studies suggest that human habitation of arctic regions was often transient, with many settlements being abandoned when the climate worsened or food became scarce. Until very recently, resources were never abundant enough to support the development of large, dense cities or towns.

People from the temperate zone have only recently explored the high arctic regions, attracted by opportunities for hunting fur-bearing animals (seals, bears, beaver, lynx, wolves, arctic fox, musk rat, otters), and whales and other marine mammals for their meat and oil, and the search for gold, crude oil and other minerals. European expeditions, such those led by the Dutch sea captain, Willem Barents, in 1596–1597 and by the Russian-financed German explorer, Vitus Bering, in 1741, visited the Arctic Ocean and many of its islands, including the Svalbard Archipelago (see Section 5.1.1). Sir John Franklin led several British expeditions to northern Canada and the islands off the north coast, starting in 1819. Although equipped with the latest ships and extensive provisions, and assisted by the Inuit, Franklin and almost all his crew died on the final journey. Their primary objective, to find and map the North-West Passage from Europe to Asia, was never realized. No permanent settlements of Europeans were established in the high Arctic until the 20th century.

There is no pre-historic evidence for humans on Antarctica. The voyage of Captain James Cook in 1772–1773 is the first known exploration of the Southern Ocean. Fisherman and hunters of whales and seals landed on many of the islands during the 18th and 19th centuries, but Antarctica itself was not explored until the first decade of the 20th century. Although research in and around Antarctica has been much expanded since the 1960s, there is still no permanent, breeding human population.

Summary of Section 5.4

Many polar mammals and birds are obese because their food supply is highly seasonal or erratic. In large species, proportionately more adipose tissue accumulates in the superficial depots and less in the internal depots in tropical, temperate-zone and polar animals. There is evidence for redistribution of adipose tissue as an adaptation to thermal insulation only in pinnipeds and smaller cetaceans. Unlike fur, adipose tissue is incompressible: its effectiveness as insulation depends upon rapid, efficient control of blood perfusion through it and the skin. Humans are basically tropical and have colonized arctic regions only very recently in evolutionary terms, so they have minimal anatomical and physiological adaptations to the environment and are capable of only limited acclimatization.

5.5 Polar ectotherms

The land and shallow water experience at least a brief summer at high latitudes, so terrestrial and freshwater ectotherms can be active during warm periods and hibernate when the temperature is below freezing (see Chapter 4). In contrast, the polar seas are not warmed significantly by the Sun: the mass of water is too large and sea-ice covered in snow reflects sunlight very well. The seawater beneath the permanent sea-ice is continuously at between −1.9 and +6 °C, so its inhabitants complete their entire life cycle at temperatures at which tropical ectotherms would die and most temperate-zone species would become torpid. There is much less mixing between warm and cold currents around Antarctica, so in the southern oceans, temperature zones are sharply delimited and have distinctive faunas. However, movements of water currents in the North Atlantic and North Pacific cause quite large seasonal changes in water temperature around the Arctic and hence less clearly defined faunal zones.

The common arctic fish are closely related to species in north temperate-zone waters. They include two salmonids, the capelin (*Mallotus villosus*) and the arctic char (*Salvelinus alpinus*), which breeds in rivers but spends part of its adult life in the sea, sculpins (family Cottidae), various flatfish such as polar halibut (*Reinhardtius hippoglossoides*) and flounder (*Pleuronectes americanus*) and members of the cod family, such as arctic cod (*Arctogadus glacialis*) and haddock. Many of the most abundant and widespread fish around Antarctica belong to a suborder Notothenioides of the order Perciformes (perches). Notothenids (Figure 5.23) probably evolved in the oceans around Antarctica during the last 20–30 million years, and the living species are almost confined to that region. In contrast to the Arctic Ocean, there are very few species of the cod (Gadiformes), herring (Clupeiformes) and salmon (Salmoniformes) families in the Southern Ocean. Only a few chondrichthyan fish live in polar waters, among them the Greenland shark (*Somniosus microcephalus*).

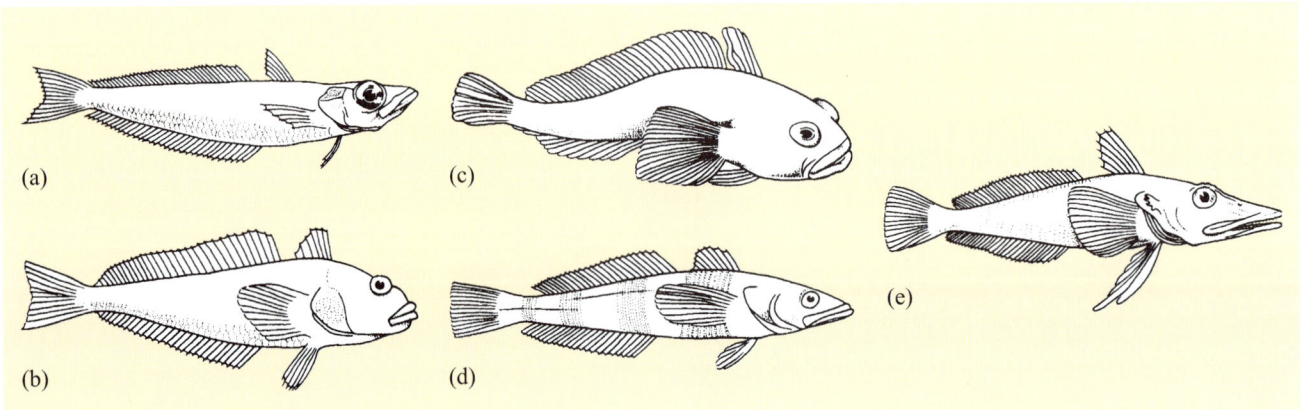

Figure 5.23 Some nototheniid fish native to the Southern Ocean that have been studied in the laboratory. (a) The antarctic silver fish (*Pleuragramma antarcticum*) 12–17 cm long; (b) *Notothenia neglecta* 25–40 cm long. (c) The emerald rockcod (*Trematomus bernacchii*) about 20 cm long. (d) The toothfish (*Dissostichus mawsoni*) up to 1.25 m long, is the largest fish in antarctic coastal waters. (e) The icefish (*Chaenocephalus aceratus*) up to 1 m long.

5.5.1 Passive properties

Freezing is nearly always harmful to living cells because the tertiary structure of hydrophilic molecules such as proteins is disrupted and the permeability of membranes is drastically altered. The concentration of solutes in the blood of teleost fish is only about half that of seawater so, while seawater freezes at −1.9 °C, fish blood would be expected to freeze at −1 to −0.6 °C. One way in which fish living in very cold seawater avoid what seems like inevitable disaster is by **supercooling**: the body fluids can remain indefinitely below −0.6 °C, provided they do not come into contact with any ice-crystals. The consequences of so doing were first demonstrated over 50 years ago by the Norwegian physiologist, Per Scholander (Figure 5.24).

■ What tissues are the likely route for entry of ice-crystals into the fish?

The gills, which present no effective barrier to ice-crystals, the gut when food is ingested, and the flow of urine from the excretory system.

Except during urination, the urethra is closed tightly by a muscular sphincter lined with large quantities of mucus, thereby minimizing the risk of ice-crystals forming in the relatively dilute urine. Ice floats in water and the deeper layers of the oceans are usually slightly warmer, −1.8 °C, than the surface water. So one way of avoiding the fate described on Figure 5.24 is to remain in deeper water. However, some fish, notably the capelin, spawn on beaches where the air temperature can be much colder than that of the sea. The sticky secretions on the outside of the eggs both stick them to the rocks and promote supercooling down to −5 °C, but if the shells are pierced, tiny ice-crystals quickly form and the embryos freeze at −1.4 °C.

Freezing is also avoided by the presence of 'antifreezes' called **cryoprotectants** (κρυοσ, *kryos* or κρυμοσ, *krymos* = frost) in the blood. Natural cryoprotectants are usually glycopeptides or peptides of molecular weight 2 400–36 000 that bind to ice-crystals and prevent them from growing larger than tiny nuclei. These cryoprotectants are present in almost all body fluids, including the blood, the cerebrospinal fluid, the peritoneal fluid, the interstitial fluid of the muscles, and (via bile secretions) the lumen of the gut. Other body fluids, such as the ocular fluid in the eye, are protected from contact with ice-crystals by the surrounding tissues. Cryoprotectant molecules are not eliminated by the kidney because most polar fish that have them also have aglomerular kidneys: their urine is formed by secretion into the nephron (Section 3.3.1) rather than by filtration.

At least eight different cryoprotectant molecules have already been identified, and almost all antarctic fish have some kind of antifreeze in their body fluids, usually throughout the year. Similar cryoprotectants have evolved in several kinds of arctic fish, but such adaptations are less widespread and cryoprotectants are present only during the winter in many species. In the northern oceans, most of the fish (and their mammalian and avian predators, see Section 5.3.2) stay in the highly productive waters at the edge of the iceshelf and avoid direct contact with frozen seawater. Other cryoprotectant agents have evolved in terrestrial arthropods, amphibians and reptiles that hibernate in sub-arctic areas of Canada and Russia.

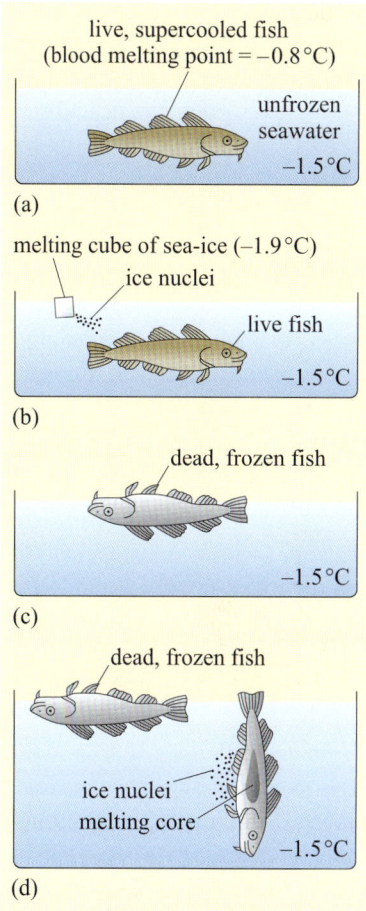

Figure 5.24 The effects of 'seeding' ice-crystals into water containing a supercooled fjord cod (*Boreogadus saida*). (a) The fish can remain indefinitely at −1.5 °C, because its blood is supercooled, but if (b) a piece of frozen seawater at −1.9 °C is put near it, the supercooling is destroyed, and the fish freezes and dies (c) in a few minutes, although the seawater itself remains liquid at −1.5 °C. Similar processes occur if the fish comes into contact with a dead conspecific that contains ice-crystals (d).

Blood pigments

The solubility of oxygen (and of many other gases) in water increases with decreasing temperature: at 0 °C, seawater holds 1.6 times as much oxygen when saturated as at 20 °C. This fact, and continual disturbance by frequent storms, mean that the surface waters of polar oceans are very well oxygenated. A family of 17 species of notothenid fish, the Channichthyidae, have no erythrocytes, no haemoglobin and almost no myoglobin at all stages of the life cycle (Section 1.5).

■ What would such fish look like?

Unless the skin is pigmented, they would be almost colourless. Collagenous tissues usually scatter light evenly, thus appearing dull white, like mammalian fascia or tendon.

■ For what tissue is being coloured essential to its function?

Visual pigments called opsins absorb light in the retina, making the back of the eye appear black. Fish adapted to function in dim light have relative large eyes. Sea-ice transmits little sunlight (see Section 5.1.1), and daylength at high latitudes is often very short (see Figure 5.1). Many fish in the Southern Ocean around Antarctica (Figure 5.23) have large eyes in which the dense, dark retinal pigments stand out against the pale bodies, making them conspicuous. Pigment cells in the skin provide camouflage for many species. *Trematomus bernacchii* (Figure 5.23c) is mottled brown, black and pink when it rests on rocks in shallow water, but its pigment cells contract when it is deep water, making the skin uniformly pale. Its common name, emerald rockcod, refers to several green spots on its broad, flexible pectoral fins. *Pleuragramma antarcticum* (Figure 5.23a) is called the silver fish because its fine scales reflect light, camouflaging its outline as it moves.

The most thoroughly studied species, the icefish (*Chaenocephalus aceratus*, Figure 5.23e), is almost transparent because the proteins in its skin and muscles are sufficiently ordered to enable light to pass through them with minimal scattering.

■ How could the absence of erythrocytes improve blood flow?

The viscosity of most fluids, including water, increases with decreasing temperature, so the heart must pump harder to maintain the circulation of the blood. Blood cells, many of which are quite large relative to the diameter of the vessels through which they pass, make a major contribution to the viscosity, so the elimination of erythrocytes would make the blood much less viscous and hence reduce the work of pumping.

As well as being large (up to 1m long), icefish are active predators that swim in the surface waters and use oxygen at about the same rate as red-blooded antarctic fish such as *Pleuragramma* or *Notothenia* (see Figures 5.23a and b). Icefish blood contains solutes and a few white blood cells, so it is a yellowish, watery fluid similar to mammalian lymph. Large volumes of it flow in wide capillaries propelled by a heart that pumps out three to four times as much blood as that of a red-blooded fish of similar size and habits. The bulbus arteriosus on the anterior side of the heart is greatly expanded and is the only muscular tissue to contain any myoglobin. Measurements on the uptake and circulation of oxygen in captive icefish suggest that, as well as the gills, the thin, scaleless skin is important as a site of gas exchange, with up to 8% of the oxygen absorbed through the tail skin alone. At low temperatures, the blood of icefish circulates faster and takes up nearly as much oxygen from the water as red blood, although transfer of oxygen to tissues is less efficient. However, in even slightly warmer water, these advantages disappear and icefish suffocate.

■ What other aspect of energy metabolism is likely to be different in icefish?

Anaerobic metabolism. Without myoglobin, their muscles are unable to store oxygen so the fish quickly become anoxic during very fast swimming or in deoxygenated water. They tire quickly after a brief burst of swimming and cannot tolerate the build up of more than a very little lactic acid in the muscles and, of course, they do not survive in warmer, less oxygenated water.

5.5.2 Metabolism

Molecules diffuse more slowly at low temperature: measurements of the rates of diffusion of small molecules such as lactic acid, Ca^{2+} and analogues of glucose and ATP through fish muscles produced Q_{10} values of 1.75–2.04 between 5 and 25 °C. Nearly all enzyme reactions are slower at low temperatures (although sometimes whole pathways can be faster if an inhibitor is more inhibited by low temperature than are the catalysts). So, in the absence of temperature compensation (see Section 2.5.2), most metabolic processes, including contraction and relaxation of muscle, digestion and growth, are slowed. The Q_{10} values (see Section 2.4.1) of most enzyme-mediated processes that have been studied directly are in the range of 1.5–3.0.

One of the fundamental cellular processes that has been most intensively studied is the maintenance of ion gradients across the cell membrane. This system is particularly relevant to cold adaptation because it is an almost universal property of cells and because ions leak into the cell passively but are actively extruded by an ATP-based pump. Inward movement of ions through ion channels is basically a physical process, for which the Q_{10} is about 1.2–1.4 in the range 0–10 °C, but, like other active, enzymatic mechanisms, the Q_{10} of ATP-producing pathways is 2–3. So as the temperature falls, extrusion cannot keep up with inflow, ions accumulate in the cells, and the membrane potential falls, with disastrous effects in neurons, muscle, kidney and many other kinds of cells. In theory, stable coupling between the two processes at low temperatures could be achieved either by increasing the capacity for active transport of ions or by decreasing membrane permeability.

■ What would be the implications of these adaptations for BMR and exercise habits?

ATP production and utilization are major components of BMR, so more active extrusion of ions would lead to higher BMR. Decreasing membrane permeability would reduce the need for ion pumping, leading to lower BMR and lower oxygen utilization, but also to diminished capacity for osmoregulation and sluggish movement.

Polar fish have recently been studied intensively both in the wild and in the laboratory. The resting metabolic rate of several antarctic fish at −2 °C proved to be at least twice as high as that expected from extrapolation of BMR data of temperate-zone or tropical fish to this temperature. However, if such warm-water fish were cooled to this temperature, they would probably not be able to swim at all and would quickly die, so the comparison is not really valid. A more relevant comparison is with temperate-zone fish that live in the deep sea, where the water temperature is always 0–4 °C. When such species are compared at their normal physiological temperatures of about 0 °C, the few antarctic fish that have been studied are found to have relatively high BMR. Their respiratory capacity is also more efficient: at 0 °C, the isolated gills of the emerald rockcod (Figure 5.23c) take up oxygen at the same rate as those of the common goldfish at 15 °C.

Muscles

The rates of muscle contraction and relaxation, and the maximum force generated, are complex enzymatic processes that determine speed of swimming. Ian Johnston of St Andrews University (Johnston, 1989) has compared the maximum tension of muscle fibres isolated from several species of antarctic, temperate-zone and tropical fish (Figure 5.25).

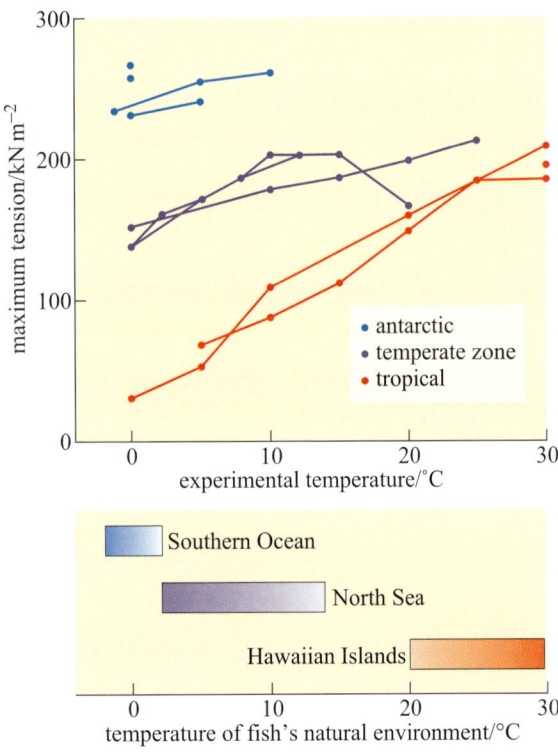

Figure 5.25 The effects of temperature on maximum tension generated by fast-contracting muscle fibres isolated from several species of antarctic, temperate-zone and tropical fish. The outer membranes were removed to permit study of the intracellular contractile mechanism and the mitochondria in isolation.

■ Do the data on Figure 5.25 provide evidence for temperature compensation of the contractile mechanism?

Yes. Between 0 and 10 °C, muscles from antarctic fish generate forces five to ten times larger than those measured from muscles of tropical species.

However, these properties of isolated muscle fibres did not match well with studies of individual molecules and whole animals. Temperature compensation could not be demonstrated in the maximum activity of some key enzymes in muscle contraction (e.g. ATPase) and many mitochondrial enzymes (e.g. cytochrome oxidases) studied *in vitro*. When temperate-zone fish such as goldfish (*Carassius auratus*), eels (*Anguilla anguilla*) and carp (*Carassius carassius*) are acclimated to low temperatures, the proportion of the volume of red muscle fibres occupied by mitochondria increases from 14% at 28 °C to 25% at 2 °C, indicating that oxidative capacity is maintained by the presence of more mitochondria, rather than by temperature compensation of the enzymes. Antarctic fish also generally have more and/or larger mitochondria but the data are not very clear cut: the proportion of the volume of red muscle fibres occupied by mitochondria ranges from 13–56% in the five antarctic species studied, compared to 4–45% in temperate-zone fish. The muscle fibres of the icefish (*Chaenocephalus aceratus*, Figure 5.26a) are more than half mitochondria, leaving little room for the contractile mechanism itself. Those of red-blooded antarctic fish, such as *Notothenia gibberifrons* (Figure 5.26b), contain a greater proportion of contractile myofibrils than the icefish, but mitochondria are still more abundant than in temperate-zone fish, particularly towards the edges of the muscle fibres near the blood vessel.

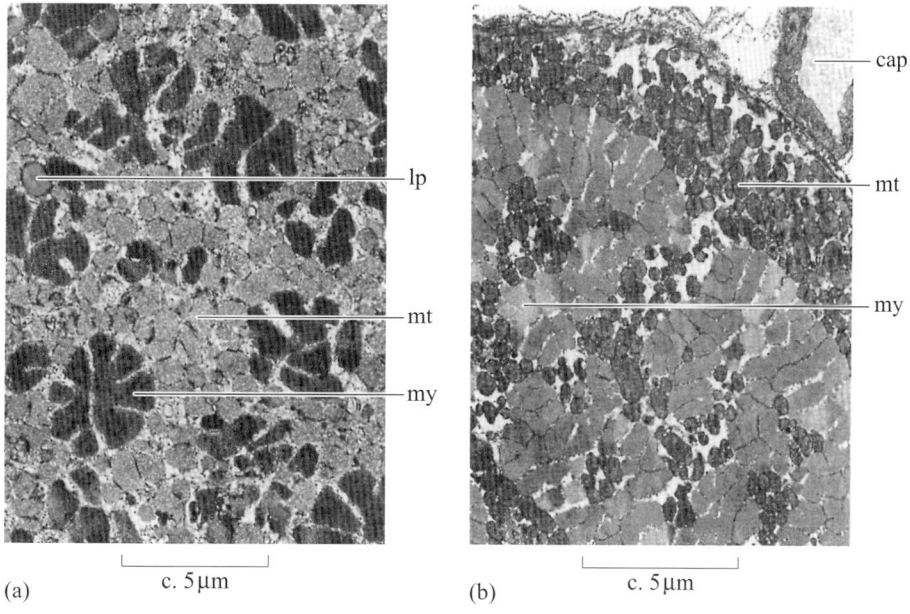

Figure 5.26 Electron micrographs of the slow swimming muscles of two antarctic fish. (a) The icefish (*Chaenocephalus aceratus*). The mitochondria are so large and numerous that they occupy more space within the muscle fibre than the contractile myofibrils. (b) *Notothenia gibberifrons*. The fibres contain fewer mitochondria and the myofibrils are more regularly arranged than in (a). *Abbreviations*: my, myofibrils; mt, mitochondria; cap, capillary; lp, lipid droplet.

■ How could mitochondria arranged as in Figure 5.26 adapt the muscles to activity at low temperature?

Diffusion is slower in the cold. Delays in metabolites such as ATP reaching the contractile proteins are minimized if numerous mitochondria are interspersed between the muscle fibres, thereby shortening the mean distance between the mitochondria and the ATP-using enzymes.

Biologists from the University of Maine compared the maximum activities at 1 °C of several enzymes in the swimming muscles and the heart of two antarctic fish, *Notothenia gibberifrons* (similar to Figure 5.23b) and *Trematomus newnesi* (similar to Figure 5.23c), with those of two species of similar size and habits caught in the western Atlantic Ocean off the coast of Delaware, USA. They found that the activities of enzymes involved in lipid catabolism and aerobic respiration, such as carnitine palmitoyltransferase and

3-hydroxyacyl CoA dehydrogenase, were 1.3–27.0 times higher in the red muscles of the antarctic species than in those of the temperate-zone species and the Q_{10} values were less than 2. However, the activities of phosphofructokinase, pyruvate kinase and lactate dehydrogenase that are essential to anaerobic utilization of carbohydrates were either not significantly different or were lower in the polar species. Thus these antarctic fish seem to be equipped to use lipid fuels more efficiently than carbohydrate fuels.

Maximum swimming speed during brief 'bursts' of activity (e.g. when escaping from a predator) has been measured accurately in only two species of antarctic fish, and was found to be at the lower end of the range found in temperate-zone fish of similar size and body shape. Clearly, the situation is complicated and further research on a greater range of species is necessary to understand adaptation to low temperatures. Nonetheless, even such limited information enables us to identify some principles of adaptation to polar conditions in ectotherms.

- With which of the two mechanisms for maintaining ionic balance at low temperature are these data most consistent?

Taken together, the observations suggest that antarctic fish living in surface waters achieve temperature compensation by increased activity of the ionic pump.

- Why would this mechanism be better for antarctic fish living in surface waters?

Slow movement is unlikely to be adaptive where fast-swimming, endothermic predators such as penguins and seals are about. There are relatively few surface-swimming fish in antarctic waters. Most fish live on or near the bottom, or in deep waters, out of reach of most air-breathing predators.

The silver fish (*Pleuragramma antarcticum*; Figure 5.23a) lives in the surface layers of coastal waters and eats pelagic invertebrates. Around McMurdo Sound in Antarctica, silver fish are known to be an important food for penguins, skuas and Weddell seals (*Leptonychotes weddellii*). Although its high density of mitochondria must increase its BMR and the energy cost of swimming, such adaptations to quick responses and fast escape are probably essential to avoiding predation. In contrast, most deep-sea fishes that have been investigated (and only a few species have been kept alive in surface laboratories for long enough to be studied) have lower BMR than expected and probably compensate for low temperature by reducing membrane permeability. Mammalian and avian predators are absent in the deep sea, and food (and probably also oxygen) is scarce, so the alternative mechanism for maintaining the potential gradient across the cell membranes is more efficient.

Another peculiar and consistent feature of nototheniid fish is that most species, including *P. antarcticum* (Figure 5.23a), *Dissostichus mawsoni* (Figure 5.23d) and species of *Notothenia* and *Trematomus*, have numerous sacs of lipid within and around their swimming muscles and between the muscles and the skin. Several functions have been suggested: energy stores, buoyancy (which could be important in fish that lack swim bladders) and more recently, oxygen diffusion. The solubility of oxygen in lipid is four times higher than in aqueous cytoplasm, and, particularly in fish in which blood pigments are reduced or absent, lipids closely associated with muscles may facilitate oxygenation of the tissues.

You may have noticed that in Figure 5.25 the contraction of muscles from tropical and temperate-zone fish were measured over a temperature range from 0 °C to 25–30 °C, but there are no data for the muscles of antarctic fish above 10 °C. Many polar fish tolerate only a very narrow range of temperatures and die within minutes if warmed more than a few degrees above the temperature of the water in which they normally live. Exactly why they die is not clear, but a likely cause is widely different Q_{10} values of enzymes in critical metabolic pathways, such as ion pumps or mitochondria: a small change of temperature puts the whole pathway 'out of kilter', causing metabolic intermediates to accumulate to toxic concentrations. In this respect, polar fish resemble non-hibernating homeotherms such as humans and rats: their metabolism is seriously, often irreversibly, disrupted by small departures from their normal body temperature. This property, of course, makes it much more difficult to transport such fish alive and to keep them in captivity for long enough to study habits such as breeding, growth and dietary preferences.

5.5.3 Fatty acids as indicators of diet

Although polar fish and invertebrates are difficult to study alive for the reasons just described, some information about their diet and habits can be obtained from analysis of the lipid composition of their tissues. At high latitudes, the supply of most kinds of marine food changes with the seasons, just as it does on land, and many fish eat little or nothing for long periods, living off their reserves of triacylglycerols. Lipids are major fuels for polar fish (see Section 5.5.2) and, although many fish have little or no adipose tissue, storage lipids are deposited in the muscles or liver, sometimes in large quantities.

The vertebrate digestive system breaks down proteins, carbohydrates and most other nutrients to amino acids, glucose or other small molecules containing just a few carbon atoms. In contrast, triacylglycerols and phospholipids from membranes are hydrolysed to glycerol and fatty acids, but in simple-stomached animals, the latter are not further broken down: medium- and long-chain fatty acids pass intact into the blood and from there into the tissues.

■ Name three different roles that long-chain fatty acids could play in vertebrate tissues.

The fatty acids may be oxidized for ATP production, or incorporated into the cell fabric, primarily as membrane phospholipids, but also combined with other molecules to form acylated proteins or glycolipids, or re-esterified into storage triacylglycerols. Certain polyunsaturated fatty acids could also become signal molecules.

In carnivores and other simple-stomached animals, almost all the fatty acids deposited in membrane phospholipids and storage triacylglycerols are unaltered, so the fatty acid composition of such components of animal tissue reflects that of the diet over the previous weeks or months. The relationship between diet and tissue composition is different in ruminants such as reindeer: microbes in the rumen convert much of the dietary carbohydrates to short-chain fatty acids, and alter the structure of the plant lipids. The reindeer's liver and adipocytes also synthesize long-chain fatty acids from short-chain fatty acids.

Hundreds of different fatty acids are known from plants, microbes and animals, but many are chemically so similar that until recently, separating and identifying them was slow and difficult. Over the past few decades, instruments such gas-liquid chromatographs have become both more accurate and easier to use, enabling biologists to identify and quantify dozens of different fatty acids in small samples of tissue. Such 'fatty acid profiles' are particularly valuable for understanding the diets of cold-water fish such as herring, sprats, capelin, mackerel and halibut, which usually store more lipid than their tropical relatives. In the North Atlantic, these commercially important fish often fluctuate greatly in numbers, and in their flavour and nutritional quality as human food. The causes of changes in their abundance are poorly understood, but may be related to fluctuations in the availability of their prey arising from instability of weather, and movements of sea-ice.

Marine mammals including most kinds of seals and many whales and porpoises also eat fish, usually swallowing their prey whole, and digesting it completely. Most of the prey species contain some storage lipids, which pass almost unmodified into the predators' tissue. The availability of the different kinds of prey species is also major cause of their breeding success. Dietary analysis also enables scientists to monitor the recovery of populations of species after legislation restricting commercial hunting is enforced.

■   What anatomical features of these marine mammals would facilitate using fatty acid profiles to obtain information about recent diet?

The adipose tissue is plentiful and forms superficial blubber that usually receives little blood perfusion. Biopsies of up to a gram of blubber can be taken with minimal injury to the animal. This quantity of tissue yields sufficient triacylglycerols that can be analysed to provide information about what the mammals have been eating. However, the outer layers of blubber, that are cool most of the time (see Section 5.4.2), contain more polyunsaturated fatty acids than the inner, warmer layers. Lower perfusion with blood probably also means that triacylglycerols are mobilized and deposited more slowly in the outer layers. To avoid sampling errors, blubber biopsies must always be taken from the same relative depth of blubber (not easy if the total thickness changes with anatomical site and with whole-body fattening).

The conclusions are often surprising, and have revised our understanding of what some kinds of seals and whales are finding to eat, especially in polar regions. Populations of the same species in different areas prove to have quite different diets, indicating that predators must be behaviourally, anatomically and metabolically highly adaptable.

Summary of Section 5.5

Several anatomical and biochemical adaptations to living in very cold water have evolved in polar fish, particularly those of the southern oceans, which have evolved in isolation for many millions of years. Cold, turbulent water is rich in oxygen. One family of fairly large fish lacks blood pigments but its blood is less viscous and it has additional respiratory surfaces. Many fish have cryoprotectants in the blood and other body fluids, and the muscles of some contain numerous mitochondria and are adapted to use lipid as fuel in preference to carbohydrate. Many polar fish tolerate only a narrow range of temperatures and quickly die if exposed to water even slightly warmer than that in which they normally live. The fatty acid composition of the tissue (the 'fatty acid profile') provides information about the animals' recent diet, which varies greatly with season and location.

5.6 Conclusion

There is much more to living in polar regions than insulation against the cold. Food may be very scattered both in space and in time and breeding must be tightly synchronized to seasons and food availability. Some of the most spectacular examples of natural obesity and efficient regulation of appetite are found among polar animals. The study of such species not only demonstrates that it is possible to remain healthy and active when very obese and during prolonged fasting, it also helps us to identify the similarities and differences between natural and pathological or artificial obesity. Until recently, people were unable to remain in the Arctic and Antarctic for long enough to study the fauna and flora in detail, but as new techniques become available we can expect to find out more about these organisms living at the extremes of climate.

Endothermic birds and mammals comprise a large part of the polar faunas and survive mainly by more efficient insulation and energy budgeting. Profound modifications of the circulation and muscles enable them to avoid excessive predation from marine mammals and birds, whose body temperature may be 40 °C warmer. Some teleost fish and numerous invertebrates (including crustaceans, molluscs and several phyla of worms) have also evolved ways of completing their entire life cycle in the cold. Reptiles and amphibians have failed to adapt to continuous cold and are absent from the arctic and antarctic fauna, although a few species occur in cold areas at lower latitudes, where they hibernate during the winter and feed and reproduce during the brief warm summer.

5.6.1 Extra-terrestrial life?

Polar biology not only enables us to understand an extensive but, until recently little explored habitat, it may also be an important guide to life elsewhere in the universe. Within the Solar System, extra-terrestrial life is now thought to be most likely on the planet Mars, and on Ganymede and Europa, the largest and smallest of the major moons of Jupiter.* This hypothesis is based mainly upon analysis of the close-up pictures and physical measurements sent back by space probes from the mid-1970s onwards. Mars and Ganymede probably have ice-covered poles, beneath which may be water; the extensive ice-covered seas of Europa may also contain large quantities of water, kept warm by volcanic activity in the rocks below. All these habits resemble the polar regions on Earth: ice, probably cracked and perforated, covers water that is only dimly illuminated by sunlight.

Many astronomers believe that extra-terrestrial organisms would probably resemble the psychrophiles found in the Arctic and Antarctic. Terrestrial psychrophiles include some eukaryotes (mostly fungi and protoctists), but the

---

* The Italian physicist and astronomer Galileo (1564–1642) described and named the four largest moons orbiting Jupiter in 1609–1610; by 2003, at least 61 satellites of Jupiter had been identified.

majority are bacteria, including a wide variety of the primitive Archaebacteria, believed to be the earliest kind of life on Earth. Their metabolism is distinctive in various fundamental ways, including unusual features of their genetic code and cell membranes. The astronomers' current interest in detecting traces of contemporary or extinct extra-terrestrial life is stimulating much new research into the structure, physiology and evolution of psychrophiles.

The search for traces of life is one of the principal objectives of the Galileo space mission to Jupiter and its moons, which has been sending back information since 1995 and ended in September 2003, and of The Open University's Beagle 2 and two American probes that are scheduled to land on Mars a few months later. So we can expect further scientific and popular interest in cold-adapted organisms.

Learning Outcomes for Chapter 5

When you have completed this chapter you should be able to:

5.1 Define and use, or recognize definitions and applications of each of the **bold** terms.

5.2 Outline the special features of the polar regions as a habitat and list some contrasts between the Arctic and the Antarctic.

5.3 Describe some effects of daylength on feeding, fat deposition and reproduction in arctic animals.

5.4 Explain why the environmental controls of appetite, activity level and fecundity are essential adaptations to living at high latitudes and describe some physiological mechanisms involved.

5.5 Describe some adaptations of fuel metabolism and bone formation to dormancy in bears.

5.6 Describe the metabolic control of prolonged fasting in breeding polar bears and penguins.

5.7 Explain the use of comparative studies to identify anatomical and physiological adaptations to thermal insulation in aquatic and terrestrial endotherms.

5.8 Describe some adaptations of the blood, respiratory system and muscles of fish to the polar environment.

5.9 Explain the role of fatty acid profiles in investigating the diet of polar vertebrates.

Questions for Chapter 5

(*Answers to questions are at the end of the book.*)

Question 5.1 (LO 5.2)

Which of the factors (a)–(g) is/are a valid reason(s) for the fact that penguins are numerous and diverse around Antarctica but absent from the Arctic?

(a) The climate in the Arctic is too severe for penguins.

(b) There are not enough fish in Arctic waters to sustain a population of penguins.

(c) Penguins are excluded from the Arctic by the presence of bears.

(d) Penguins are excluded from the Arctic by the presence of seals.

(e) Penguins evolved in the Southern Hemisphere and have never occurred naturally in or around Europe, Asia or North America.

(f) In general, there are fewer birds in the Arctic than in the Antarctic.

(g) In general, there are fewer vertebrate animals in the Arctic than in the Antarctic.

Question 5.2 (LOs 5.2 and 5.3)

What is the evidence that endogenous factors as well as daylength control food intake and breeding? Why are such dual control mechanisms adaptive for polar animals?

Question 5.3 (LO 5.4)

Which of the statements (a)–(g) about the food intake and metabolism of polar mammals and birds is/are generally true?

(a) There are many more animals relative to the food supply in polar regions than in the tropics.

(b) The food supply in polar regions is highly seasonal and/or irregular.

(c) Being obese makes animals lethargic.

(d) Very lean animals are incapable of strenuous physical activity.

(e) Polar animals have fewer predators than tropical animals so they can afford to be fat and lazy.

(f) The only food available in polar regions is less nutritious than that available in the tropics.

(g) Polar animals have a higher metabolic rate and so need more food than those in the tropics.

Question 5.4 (LO 5.5)

Explain in a few sentences why measurements of the concentration of ICTP and PICP in blood samples taken during peak activity and in the middle of the dormant period failed to explain how black bears avoided increased risk of bone fracture at emergence from dormancy.

Question 5.5 (LO 5.6)

What information relevant to the metabolism of energy and/or proteins can be obtained from measurements of:

(a) body temperature

(b) RER

(c) the composition of the blood

Question 5.6 (LO 5.7)

Which of the statements (a)–(h) about the structure and arrangement of adipose tissue is/are true?

(a) All polar mammals have thick subcutaneous adipose tissue.

(b) Birds, whether polar or not, do not have thick subcutaneous adipose tissue.

(c) Polar bears have thick subcutaneous adipose tissue because they live in very cold climates and swim in the sea.

(d) In Carnivora, the partitioning of adipose tissue between superficial and intra-abdominal depots depends upon body size.

(e) In Carnivora, the partitioning of adipose tissue between superficial and intra-abdominal depots depends upon habits and habitat.

(f) In naturally obese mammals, fatter individuals always have more adipocytes than thinner individuals.

(g) In naturally obese mammals, as an individual gets fatter, its adipocytes enlarge.

(h) The number and size of adipocytes in relation to fatness are quite variable in naturally obese mammals and in humans.

Question 5.7 (LO 5.8)

Explain in a few sentences:

(a) The effects of the absence of red blood cells on delivery of oxygen to the muscles of *Chaenocephalus*.

(b) Why most fish living in very cold water remain on or very near the bottom, but there are plenty of invertebrates in mid- and surface waters.

Question 5.8 (LO 5.9)

Explain in a few sentences why:

(a) The diet of large fish and marine mammals changes seasonally, and may differ greatly for members of the same species found at different sites.

(b) The relative abundance of the fatty acids in adipose tissue triacylglycerols of herbivores such as reindeer would reveal little about the diets of animals found in different areas.

CHAPTER 6 ALTITUDE

Prepared for the Course Team by Patricia Ash

6.1 Introduction: the high-altitude environment

As in previous chapters, this chapter looks at both phenotypic and genotypic adaptations to what may be regarded as an 'extreme' environment. People who live at sea-level and ascend to high altitude are exposed to hypoxia; there then follows a suite of physiological and biochemical changes that maintain an oxygen supply to the respiring tissues. These physiological changes are known as acclimatization, and are possible because of phenotypic flexibility. The chapter also considers long-term physiological and biochemical adaptations to life at high altitude. Increasing numbers of people are emigrating from sea-level to high altitude, e.g. from China to Tibet. Long-term physiological adaptations for life at high altitude may also be derived from phenotypic plasticity which, as you will recall from Chapter 2 (Section 2.10.2), is the process by which factors in the environment, e.g. aridity, low T_a or low P_{O_2} of a neonate or immature animal, can shape the anatomy and influence the physiology and biochemistry of the adult. We should remember too that phenotypic plasticity and flexibility (and therefore the *ability* to adapt physiologically) are themselves under genetic control.

Low atmospheric pressures at high altitude mean low P_{O_2}, and therefore hypoxia in respiring tissues. Sudden exposure to high altitude, say 4000 m, by means of a helicopter flight to the upper slopes of a mountain, may result in a set of symptoms known as mountain sickness, which is caused by the effects of acute hypoxia on the blood circulatory system and the brain. Acute exposure to even higher altitudes, and hence lower P_{O_2}, may result in loss of consciousness, followed by death (Figure 6.1).

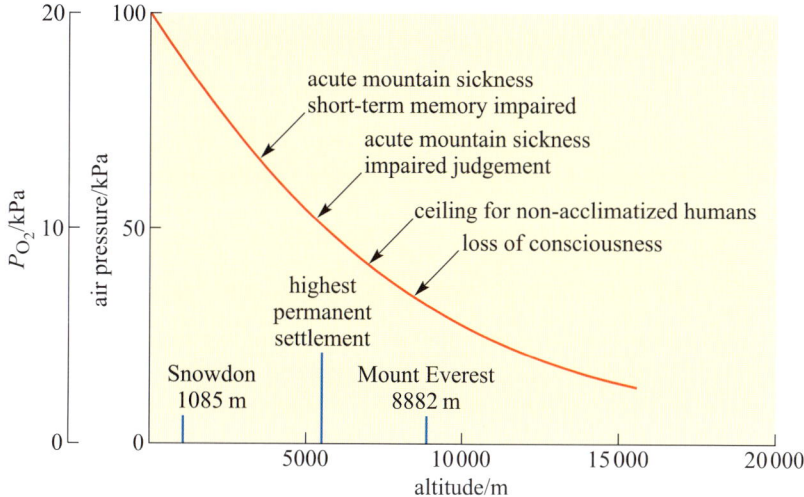

Figure 6.1 The total air pressure and the partial pressure of oxygen in the atmosphere at various altitudes with a summary of the effects usually observed in non-acclimatized people.

Nevertheless, more than 22 million people live permanently at high altitude, with domestic animals grazing the alpine pastures, where crops such as barley can also be grown. It is estimated that humans have lived on the Qinghai-Tibetan plateau for about 4000–5500 years, and on the high Andes (altitudes 3000–4000 m) for up to 12 000 years. The Bolivian village of Morococha is at an altitude of 4500 m and the capital, La Paz, is at 3800 m. Cusco in Peru is at 3400 m, and villages around Lake Titicaca at 3800 m. The ethnic Andean population living at high altitude numbers around 8 million, about 75% of whom are Quechuan people and the remainder Aymara. In Section 6.4 we explore and evaluate research studies that investigated whether Quechuan people have heritable traits that can be linked to adaptations for life at high altitude.

Animal species have lived at high altitudes for much longer than humans. In fact, fossils of pika have been found on the Tibetan plateau and dated at about 37 million years old. Camelids such as llama and vicuna (*Vicugna vicugna*; Figure 6.2) live at high altitudes in the Andes and have no apparent problems with hypoxia. The yak (*Bos gruniens*; Figure 6.3) is an important livestock species in mountainous areas in Asia. Small mammalian species such as pika (*Ochotona* sp.; Figure 6.4) and root voles (*Microetus* sp.) live in the same high-altitude alpine meadows as yak.

Reptiles including iguanids (*Liolaemus* sp.) from the Andes and amphibians such as the Andean frog (*Telmatobius peruvianus*) are also permanent residents at high altitudes. Birds fly over mountain ranges on their migratory flights. The bar-headed goose (*Anser indicus*; Figure 6.5) migrates over the Himalayas, flying over Mount Everest at 9000 m altitude whilst honking loudly, and so clearly not suffering from shortness of breath!

Tourists, trekkers and climbers who live at sea-level or at low altitudes, visit high mountains in ever-increasing numbers. Many people living at sea-level can walk up mountains, even those as high as Everest, as long as they do so gradually. Rapid transport of a lowlander to high altitude, e.g. in a car or a helicopter, may result in a suite of symptoms known as acute mountain sickness (AMS). The well-known symptoms include headache, fatigue, shortness of breath, dizziness, sleep disturbance and vomiting. The physiology of mountain sickness is covered in Sections 6.2.4 and 6.2.5.

Domestic animals are also vulnerable when transported to high altitude. In the summer, farmers transport large numbers of cattle to rich pastures at high altitudes. The cow (*Bos taurus*) is a lowland species and, like humans, suffers from symptoms linked to hypoxia. In Section 6.4 we return to the initial response to hypoxia, this time at the molecular level. Hypoxia-inducible factor (HIF-1α), a transcription factor, plays a crucial role in mediating the cellular response to hypoxia (Section 6.4.1). We end in Section 6.5 with a research study comparing the activities of three catabolic enzymes in the muscle tissue of three pika species and look at enzyme properties that could be linked to low tissue P_{O_2}.

We begin our study of high-altitude physiology with a brief survey of the physiological and biochemical responses to high altitude.

Figure 6.2 Vicuna (*Vicugna vicugna*).

Figure 6.3 Yak (*Bos gruniens*).

Figure 6.4 Pika (*Ochotona* sp.).

Figure 6.5 Bar-headed goose (*Anser indicus*).

6.2 Initial physiological and biochemical responses to high altitude

6.2.1 The control of respiration

Before we study how animals and humans cope in atmospheres with low oxygen pressure, we need to understand oxygen transfer. The 'oxygen cascade' (Figure 6.6) describes the partial pressure of oxygen as it passes from air to the deep body tissues.

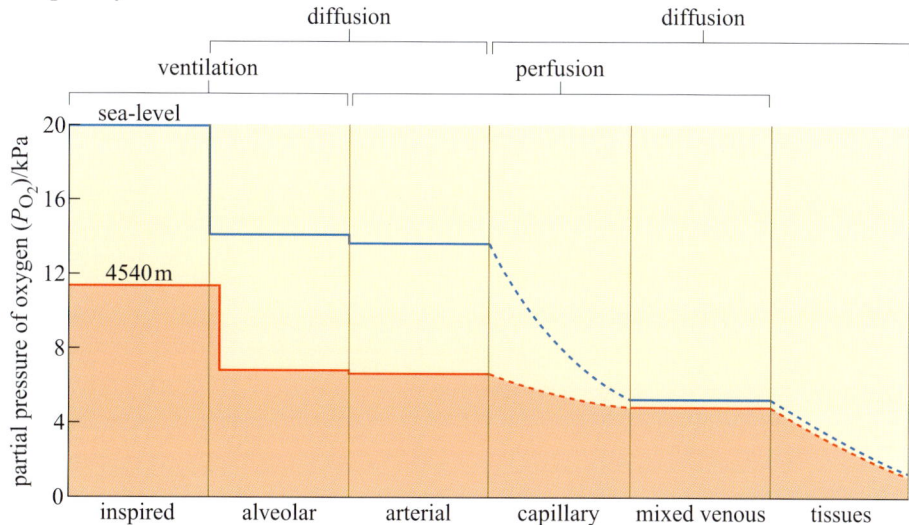

Figure 6.6 Oxygen cascade in human tissues. Mean P_{O_2} gradients from inspired air to tissues in human subjects native to sea-level and in subjects native to 4540 m.

Although the P_{O_2} in the atmosphere at high altitude is much less than that at sea-level, and the P_{O_2} gradient down the oxygen cascade is much less steep than that for sea-level residents, the venous P_{O_2} is not greatly different. Therefore, the P_{O_2} gradient for oxygen transfer to the mitochondria is maintained. The dashed portions of the lines represent estimates of the P_{O_2} changes and are not precise measurements. The labels identify the four steps in the cascade where oxygen transfer is governed by diffusive conductance (Fick's law), or by mechanical movement of fluid (ventilation or perfusion).

In large animals, it is possible to separate the oxygen cascade into steps that describe gradients of:

- **diffusive conductance**, where oxygen transfer is by physical diffusion. So, it is directly proportional to the exchange area and P_{O_2} gradient, and inversely proportional to the diffusion distance (e.g. lungs/gills to blood and blood to tissues).
- **convective conductance**, where oxygen transfer is by mechanical movements of the surrounding fluid, e.g. lung/gill ventilation with air/water and perfusion of lung/gill and tissue capillaries by blood that is carrying oxygen bound to haemoglobin.

In vertebrates, an oxygen cascade with four steps from the environment to the tissues may be identified.

- *Step 1* – delivery of water or air containing oxygen to the respiratory gas exchange surfaces of the animal. This process of external convection is termed **ventilation.**

- *Step 2* – passage of oxygen from the air/water presented to the respiratory gas exchange surfaces (gills/lungs), across the epithelium, by diffusion into the body fluids.
- *Step 3* – transport of oxygen (combined with haemoglobin) to the metabolizing tissues by internal convection, or perfusion, through the cardiovascular system.
- *Step 4* – passage of oxygen from the circulating body fluids to the metabolizing mitochondria, by diffusion through the tissues.

Hence the stages in the cascade (Figure 6.6) are alternately convection, diffusion, convection and then diffusion again. For a human ascending from sea-level to high altitude in a helicopter flight, from sea-level to say 4000 m, the immediate challenge is low atmospheric P_{O_2} and the initial failure of the lungs to increase ventilation to compensate. Why is there such a delay? In order to understand why lungs cannot increase ventilation immediately, we need to investigate the chemical control of breathing. At sea-level, as long as the alveoli in the lungs are ventilated and the cardiovascular system is working normally, respiring tissues receive blood containing sufficient oxygen and normal levels of carbon dioxide. Changes in both P_{CO_2} and P_{O_2} can have damaging effects however. A fall in P_{O_2} in the blood (known as **hypoxia**) compromises the oxygen supply to tissues when people are exposed to high altitude without the opportunity for gradual acclimatization.

Even small changes in P_{CO_2} are significant. Chapter 2 (Section 2.4.2) explained that when carbon dioxide dissolves in blood, it reacts slowly with water forming carbonic acid (H_2CO_3), which dissociates reversibly to form hydrogen ions (H^+) and hydrogen carbonate ions (HCO_3^-).

$$CO_2 + H_2O \rightleftharpoons H_2CO_3 \rightleftharpoons H^+ + HCO_3^- \qquad (6.1)$$

Like all chemical equilibria, the reactions proceed to an extent determined by the relative concentrations of the reactants and the products. Note that plasma contains hydrogen carbonate ions (HCO_3^-) derived from a reaction in red blood cells, which pushes the reaction to the left.

■ What effect would an increase in P_{CO_2} have?

An increase in P_{CO_2} would push the equilibrium to the right, ending with an increase in hydrogen ion concentration [H^+] – recall that the [ ] symbol denotes concentration. Increased [H^+] in turn causes a decrease in pH, i.e. an increase in acidity. Increase in P_{CO_2} of blood is known as **hypercapnia**, and a fall is **hypocapnia.** Equation 6.1 demonstrates that P_{CO_2} determines the pH of blood, which is normally tightly regulated. Note that an increase in [CO_2] would push the equilibrium in Equation 6.1 to the right, and so promote the formation of product, which includes H^+ ions. The higher the [H^+], the lower the pH, and the greater the acidity. The **Henderson–Hasselbalch equation** (Equation 6.2), shows mathematically how if P_{CO_2} increases, the blood becomes more acid. If P_{CO_2} decreases then the blood becomes alkaline.

$$pH = pK_a + \log_{10}\left[\frac{[\text{base}]}{[\text{acid}]}\right] \qquad (6.2)$$

pK_a = pH at which acid is 50% dissociated.

We can express the Henderson–Hasselbalch equation as:

$$pH = pK_a + \log_{10}\left[\frac{[HCO_3^-]}{P_{CO_2} \times a}\right] \qquad (6.3)$$

where pK_a is the pH of $H_2CO_3^-$ when it is 50% dissociated and *a* is the absorption coefficient for CO_2, i.e. the solubility of CO_2 in units of mmol gas litre^{-1} plasma kPa^{-1}.

For example, in normal blood, the absorption coefficient of CO_2 is 0.19 mmol l^{-1} kPa^{-1}, P_{CO_2} = 6.3 kPa, [HCO_3^-] = 25 mmol l^{-1} and the pK_a of H_2CO_3 = 6.1. So,

$$\text{pH} = 6.1 + \log_{10}\left[\frac{25 \text{ mmol l}^{-1}}{6.3 \text{ kPa} \times 0.19 \text{ mmol}^{-1} \text{ kPa}^{-1}}\right] = 7.42$$

A rise of P_{CO_2} in arterial blood produces a significant change in pH. As an example here, say P_{CO_2} changes to 7.3 kPa, pH will then fall to 7.36. This is a significant change because biochemical reactions are very sensitive to pH.

$$\text{pH} = 6.1 + \log_{10}\left[\frac{25 \text{ mmol l}^{-1}}{7.3 \text{ kPa} \times 0.19 \text{ mmol}^{-1} \text{ kPa}^{-1}}\right] = 7.36$$

If pH increases, as it does when P_{CO_2} falls, the consequences are even more serious because increased alkalinity of plasma causes calcium ions to bind to certain negatively charged blood proteins. The resulting decrease in plasma [Ca^{2+}] and extracellular fluid causes an increased excitation of nerves, which leads to uncoordinated movements and eventually convulsions.

Peripheral chemoreceptors located in the carotid and aortic bodies (Figures 6.7a and b) detect changes in P_{O_2} in the blood and send impulses via the vagus nerve and the carotid sinus nerve to the medulla oblongata of the brain. This in turn responds by signalling an increase in the rate of ventilation of the lungs if P_{O_2} falls.

Figure 6.7 (a) Location of chemoreceptors in the carotid and aortic bodies in a human. (b) The link between the central and peripheral chemoreceptors across the blood–brain barrier.

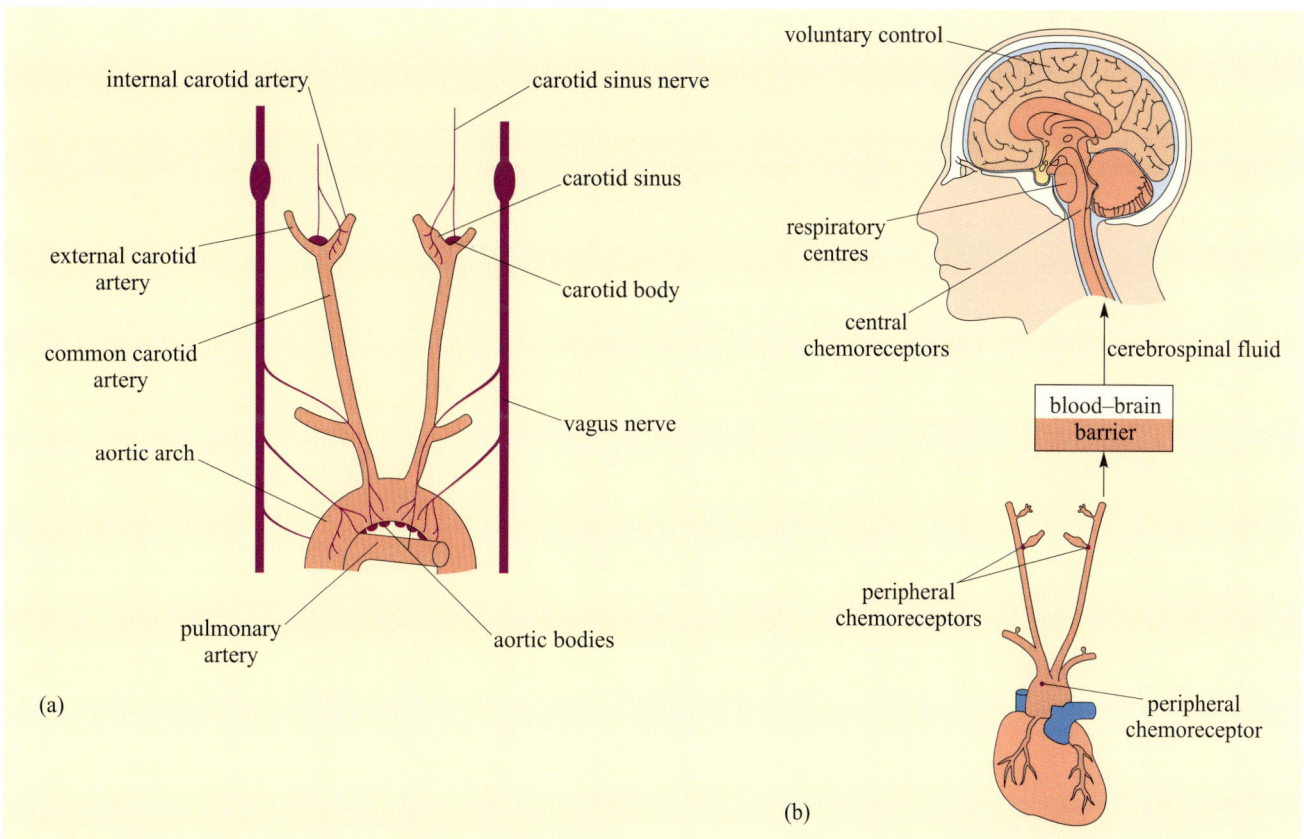

The sudden transport of a person, or any lowland mammal, to altitudes greater than 3000 m results in acute hypoxia, caused by an initial failure to increase ventilation and compensate for the low P_{O_2}. However, arterial P_{CO_2} is not affected by exposure to hypoxia and the increased rate of ventilation 'blows off' CO_2, and so reduces arterial P_{CO_2}. Equation 6.1 shows that increased removal of CO_2 from blood in the alveoli pushes the equilibrium of the two reactions to the left, so producing even more CO_2 in the blood. This CO_2 in turn diffuses out of the blood into the alveoli, and is breathed out. The loss of CO_2 results in a lower $[H^+]$, so the pH rises, and therefore the alkalinity of the blood increases. Central chemoreceptors are located in the brain, on the side of the medulla, where they are bathed by the cerebrospinal fluid (CSF), and it is this decrease in P_{CO_2} that triggers the central receptors to reduce the rate of breathing.

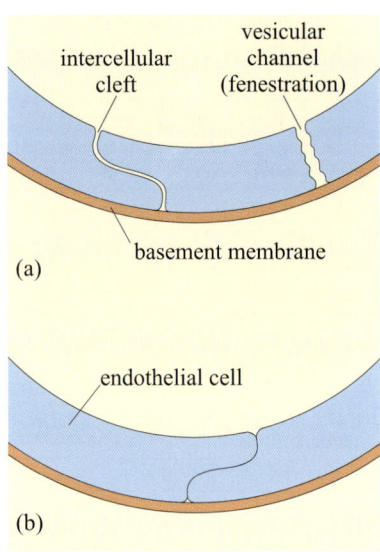

The endothelial cells of tissue capillaries have gaps, intercellular clefts, between them. Fluids, and even white blood cells, can pass through these (Figure 6.8a). When the capillary blood pressure is high, fluids can leak between the cells into tissues and cause edema. In some cases these endothelial cells have openings, fenestrations, which effectively form vesicular channels through which molecules can pass. However, specialized endothelial cells line the brain blood vessels, forming the blood–brain barrier (Figure 6.8b); isolating the CSF from the blood. These endothelial cells are joined by tight junctions, which close the intercellular space, limiting the entry of fluid, large molecules and white cells from the capillary contents.

Figure 6.8 (a) Endothelial cells of tissue capillaries. (b) Endothelial cells of brain capillaries.

Most substances cannot cross the blood–brain barrier, so the composition of CSF is controlled by the cells that secrete it, the choroid plexus cells. Carbon dioxide can cross the blood–brain barrier, so P_{CO_2} is the same in the blood as in CSF. Dissolved CO_2 in the CSF forms H^+ ions to an extent determined by P_{CO_2} and the concentration of hydrogen carbonate ions according to the Henderson–Hasselbalch equation (Equation 6.2). Hydrogen carbonate ions cannot cross the blood–brain barrier, and $[HCO_3^-]$ is controlled by the secretion of HCO_3^- by the choroid plexus cells. Secretion rates of HCO_3^- do not change immediately if P_{CO_2} changes, which means that small changes in P_{CO_2} produce large changes in the pH of the CSF. The central chemoreceptors (Figures 6.7b and 6.9) do not respond to P_{CO_2} directly, but to the effect of changes in P_{CO_2}, in this case reduction in $[H^+]$ and therefore increased pH of the CSF. The increase in pH is the signal for slowing down the breathing rate. Therefore, the central chemoreceptors are 'mini' pH meters that sense the effect that changes in P_{CO_2} have on the pH of CSF.

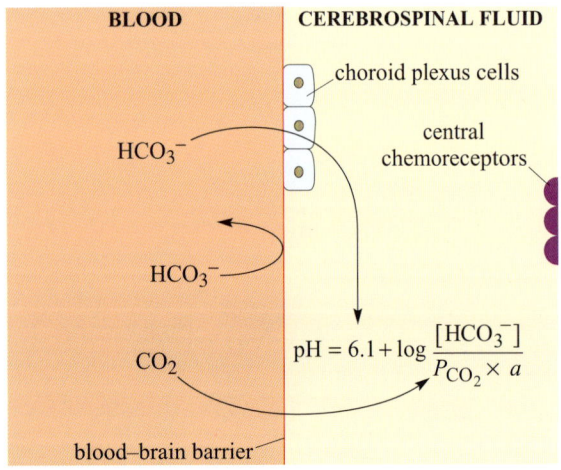

Figure 6.9 Pathways across the blood–brain barrier which lead to changes in the pH of the CSF.

- A person exposed suddenly to an altitude of 3000 m breathes rapidly because peripheral chemoreceptors have stimulated an increase in breathing rate. Explain why this increase in breathing rate is not maintained for longer than a few minutes.

As more carbon dioxide is excreted through the lungs owing to the increased breathing rate, P_{CO_2} in plasma is decreased and eventually P_{CO_2} in CSF also decreases, resulting in an increased pH, a signal to reduce the breathing rate.

The signal from the central chemoreceptors cancels out the signal from the peripheral ones, resulting in a very small net effect on the ventilation rate of the lungs. This example demonstrates the 'conflict' between peripheral and central chemoreceptors. In contrast with other receptors, peripheral chemoreceptors do not adapt but continue firing as long as there is hypoxia. Their continued generation of neural signals increases ventilation a little, so the CSF becomes slightly alkaline. The choroid plexus responds by secreting CSF that contains less HCO_3^-, which restores CSF pH to its 'normal' value. Over about 12–15 hours, this progressive compensation allows the signal from the peripheral chemoreceptors to be fully expressed, so ventilation increases, raising P_{O_2} in the alveoli. The blood remains alkaline for a few days, but the pH is eventually corrected by excretion of excess HCO_3^- by the kidneys.

6.2.2 Changes in blood and the blood circulatory system

The initial failure of the respiratory system to respond to hypoxia is to some extent ameliorated by responses in the blood and the cardiovascular system. Immediately following exposure to high altitudes, red blood cells which are stored mainly in the spleen are released. This increases the level of haematocrit*, so producing an immediate increase in the capacity of the blood to carry oxygen. Within 12 hours, the rate at which red blood cells are formed also starts to increase, mediated by the glycoprotein hormone, erythropoietin, which stimulates cell division in haematopoietic stem cells in the bone marrow. Erythropoietin is synthesized in hormone-secreting cells in the inner cortex of the kidney in response to hypoxia (Figure 6.10). The haematocrit level may not peak until after many months of exposure to high altitude.

Another physiological response to high altitude is an increase in heart rate and therefore **cardiac output**. You have seen how it takes time for respiratory control mechanisms to adapt to hypoxia at high altitude. The amount of oxygen delivered to respiring tissues depends on the rate of blood flow to the tissues as well as the value of P_{O_2} in the blood and tissues. Cardiac output is increased by up to five times the resting level at sea-level. In addition, previously constricted capillaries in the tissues open up so enhancing blood supply to the tissues and reducing blood pressure.

6.2.3 Implications of the biochemistry of haemoglobin for visitors to high altitude

For a lowlander climbing up to high altitude, the biochemical properties of haemoglobin mean that even at low P_{O_2} in the alveoli, the blood pigment can still deliver oxygen to respiring tissues. Haemoglobin molecules load up with oxygen

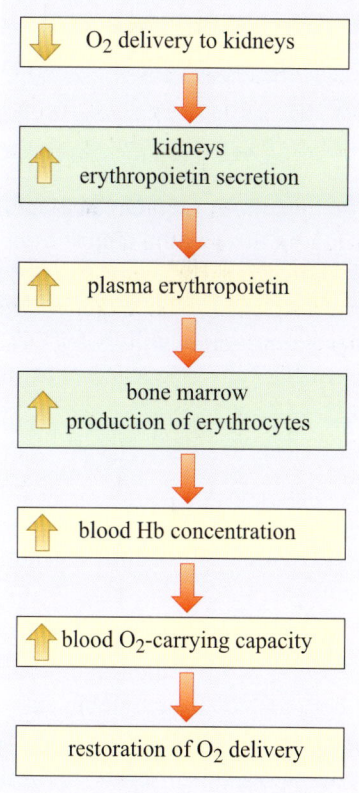

Figure 6.10 Flow chart showing how decreased oxygen delivery to the kidneys increases production of erythrocytes via increased secretion of erythropoietin.

* The haematocrit is the ratio of the volume occupied by the blood cells as a percentage of the total blood volume.

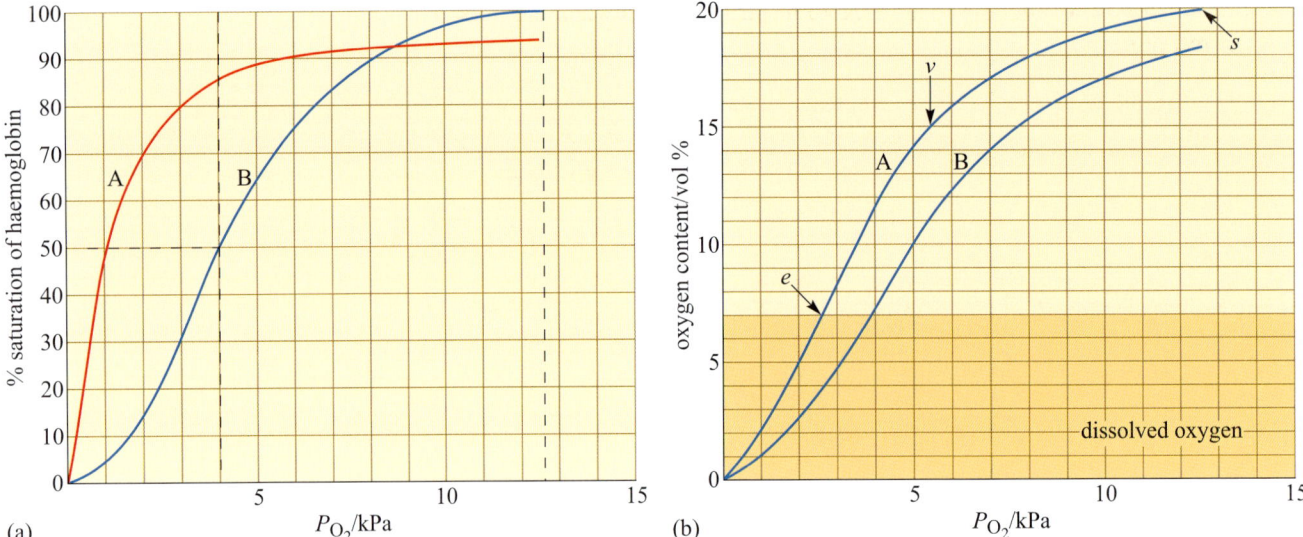

Figure 6.11 (a) Oxygen dissociation curves for myoglobin (red curve, A) and adult haemoglobin (blue curve, B). The myoglobin curve has a hyperbolic shape; the haemoglobin curve is S-shaped or sigmoid. (b) Curve A is an oxygen dissociation curve for adult human blood. $s = P_{O_2}$ at saturation; v = venous P_{O_2} when subject is at rest; e = the venous P_{O_2} during exercise. Curve B is an oxygen dissociation curve for adult human blood containing 25 µmol 2,3-diphosphoglycerate (2,3-DPG) g^{-1} haemoglobin.

in the lungs where P_{O_2} is high, and release their bound oxygen to the respiring tissues. Figure 6.11a shows oxygen dissociation curves for two different oxygen transporting pigments, myoglobin, curve A, and haemoglobin, curve B. The curve for A is hyperbolic; the curve for B is S-shaped or sigmoid.

- Which of myoglobin or haemoglobin in Figure 6.11a would in theory be the most effective oxygen transporters in vertebrate blood? Assume that these two respiratory pigments have identical oxygen capacity and that they would operate over a range of P_{O_2} between 12 kPa (at the external respiratory surface in the lung), and 4 kPa (in the tissue capillary networks).

The hyperbolic oxygen dissociation curve of myoglobin means that although it would in theory bind almost as much oxygen at the lungs as haemoglobin, it gives up only about 10% of its oxygen at the P_{O_2} prevailing in the tissues, 4 kPa. In contrast, haemoglobin can unload almost 50% of its bound oxygen at 4 kPa. Myoglobin would deliver 50% of its bound oxygen in the tissue capillaries only when P_{O_2} falls to a very low level, around 1 kPa.

The value of P_{O_2} at which a pigment releases 50% of its maximum oxygen capacity is known as the P_{50}, and this provides a measure of the affinity of the pigment for oxygen. Therefore, the higher the value for P_{50}, the lower the affinity of the pigment for oxygen. Haemoglobin has a lower affinity for oxygen than myoglobin, so it has an oxygen dissociation curve displaced to the *right* in comparison with the curve for myoglobin. Respiratory pigments with high affinity for oxygen have oxygen dissociation curves displaced well to the *left*.

However, we are interested in the amount of oxygen carried per unit volume of blood, and the haemoglobin curve in Figure 6.11a does not help us with this. The oxygen content of blood is expressed as volume percent, vol%, a term that tells us the oxygen content in cm^3 in a 100 cm^3 sample of blood. Figure 6.11b (line A) shows the oxygen dissociation curve for haemoglobin in human blood. Look at the right-hand side of the graph, at point s, where P_{O_2} is high, at 12.6 kPa. Here, the oxygen content of arterial blood is 20 vol%, or 20 cm^3 oxygen 100 cm^{-3}, a typical value for oxygen-saturated blood. The following exercise, based on Figure 6.11b demonstrates how the S-shaped oxygen dissociation curve for haemoglobin confers considerable advantage for humans at high altitude.

■ What proportion of oxygen is released to respiring tissues when the person is at rest at point v?

The proportion of oxygen released to respiring tissues when at rest can be worked out from the difference between s and v on the graph. Subtract 15 vol% from 20 vol%, which is 5 vol%. The proportion of oxygen released to the respiring tissues is:

$$5\,\text{vol\%}/20\,\text{vol\%} \times 100\% = 25\%$$

Much of the remainder of the oxygen bound to haemoglobin is 'reserve capacity' that can be used during exercise.

■ What proportion of the oxygen is released to respiring tissues during exercise?

The total proportion of oxygen released during exercise is about 65%. This value is obtained by subtracting the oxygen content during exercise (7 vol%) from the value for the total carrying capacity of the blood (20 vol%). This gives 13 vol%, which is 65% of 20 vol%.

There is always a small amount of oxygen bound to haemoglobin that is never released. Actively respiring tissues use up a great deal of oxygen so that the P_{O_2} in the capillary beds drops sharply during exercise, down to a minimum of about 2.6 kPa. Such a low P_{O_2} in the tissues causes the haemoglobin to unload its oxygen. Note how this shift in P_{O_2} is associated with a very steep portion of the oxygen dissociation curve, where a *small* drop in P_{O_2} at the capillaries results in a *relatively large* increase in oxygen available to the tissues. A drop in P_{O_2} from 12.6 kPa at point s to 5.3 kPa at point v provides about 5%, whereas a drop from 5.3 kPa at v to 2.6 kPa at e more than doubles the oxygen available to 13 vol%.

So the change in the steepness of the oxygen dissociation curve can increase the proportion of bound oxygen that is released to the respiring tissues. The arterial blood of a human subject at about 4500 m has a P_{O_2} of about 6 kPa so it will have an oxygen content of about 16 vol% (see Figure 6.11b). Contrast these figures with those for a human subject at sea-level, where arterial P_{O_2} would be about 12.6 kPa (up on the plateau of the oxygen dissociation curve), thus corresponding to an oxygen content of about 20 vol%. Next take a scenario where both the highlander and the lowlander required an oxygen turnover of 5 vol%. In order to achieve this, the lowlander would require a very low P_{O_2} in the tissue capillaries of about 5.5 kPa, which corresponds to a drop of 7 kPa in P_{O_2} between the arteries and the tissues. In the highlander, the same amount of oxygen could be delivered to tissues if P_{O_2} was only 3.7 kPa, an arterio–capillary drop of only 2.3 kPa. Therefore, because of the shape of the oxygen dissociation curve, the highlander can deliver the *same* amount of oxygen to tissues as the lowlander for a much smaller drop in P_{O_2} at the capillaries.

The characteristic shape of the oxygen dissociation curve is a result of the protein structure of the haemoglobin molecule which is a tetramer made up of four sub-units, each a polypeptide chain, of similar size and structure to myoglobin.

The polypeptide chains undergo cooperative interactions in response to oxygen. Haemoglobin with no oxygen molecules attached exists in a conformation known as the 'tense' or 'T' form, that has a low affinity for oxygen. Binding of an oxygen molecule to a sub-unit of haemoglobin causes a change in conformation to the 'R' or relaxed form, which increases the avidity of the other sub-units for oxygen. Binding of successive oxygen molecules increases its avidity for oxygen even further, resulting in the steep increase in oxygen content of haemoglobin as P_{O_2} rises.

The oxygen dissociation curve for haemoglobin is affected by the concentration of organic phosphates, especially 2,3-diphosphoglycerate (2,3-DPG). During acclimatization to high altitude, there is a gradual lowering of the affinity of haemoglobin for oxygen as the oxygen dissociation curve moves to the right (curve B in Figure 6.11b). This shift is associated with increased levels of 2,3-DPG in the red blood cells. When humans are exposed to high altitude, 2,3-DPG levels increase by up to 10% within 24 hours.

- Would you expect a decrease in the affinity of haemoglobin for oxygen to have a beneficial effect on oxygen uptake by pulmonary blood or a beneficial effect on delivery of oxygen to the tissues?

The affinity of haemoglobin for oxygen is important at two sites. In the lung alveoli, where oxygen is taken up, it would be advantageous for haemoglobin to have a high affinity for oxygen. Within the respiring tissues, where oxygen is delivered to cells, it would be more advantageous for haemoglobin to have a lower affinity for oxygen in order to give up its oxygen load more readily. However, haemoglobin with low affinity for oxygen, would not load up so much oxygen in the alveoli.

You may know from your general biology that the right shift of the haemoglobin saturation curve due to lower pH is known as a Bohr shift.

6.2.4 High altitude cerebral edema: HACE

Sections 6.2.1 to 6.2.3 suggest neat physiological and biochemical solutions to the problem of hypoxia for humans and other mammals that ascend to high altitude, but unfortunately the situation is not quite so simple. About 20% of people who ascend rapidly to 2500–3000 m, will suffer from acute mountain sickness (AMS). Most people recover quickly, but a small proportion of patients go on to develop high altitude cerebral edema (HACE) in which the brain swells. AMS and HACE are regarded as stages in a continuum. Early signs of HACE include headache and stumbling. Blood pressure is high and hallucinations may occur. HACE can be confirmed by simple tests in the field, such as checking the ability of a person to place a finger accurately on their nose. The best cure is immediate descent to low altitude but the patient's mental confusion may manifest itself by a vigorous denial that there is anything wrong. Computerized tomography scans and magnetic resonance imaging (MRI) show that fluid may be localized in certain areas of the brain, or be spread across the whole brain. Cerebral haemorrhages may accompany the edema because the brain tissue swells, squeezing the capillaries and reducing the blood supply to the brain. There is evidence that the brain swelling is caused by changes in the permeability of the capillary endothelial cell membranes. Increased permeability would promote leakage of fluid into the loose supporting tissue of the brain. The flattened endothelial cells of brain capillaries rest on a basement membrane. So why does the brain's response to hypoxia result in breaching of the blood–brain barrier?

Vascular endothelial growth factor (VEGF), is a cytokine produced by cells in response to hypoxia. VEGF stimulates angiogenesis, the development and growth of blood vessels. The initial response to VEGF is an increase in the permeability of blood vessel endothelial cells, mediated by gap formation between the cells and formation of fenestrations (Figure 6.8a), which are small openings in the capillary endothelium. Where there is a fenestration, the only barrier between the capillary lumen and the surrounding tissue is the basement membrane. Although the latter is sufficient to prevent movement of plasma proteins out of the blood, smaller molecules can diffuse through, and out of, the capillary. Formation of fenestrations is followed by dissolution of the capillary basement membrane, leading to leakage of plasma and later, red blood cells into loose connective tissue within the brain. Many researchers now support the view that VEGF is involved in the cause of HACE. For a long time, however, VEGF was not sought in the hypoxic brain, because it was considered that the blood–brain barrier would exclude these polypeptide molecules. However, Fischer et al. (1999) demonstrated that VEGF is produced by cultured brain endothelial cells in response to hypoxia. Thus it is likely that VEGF is present on the brain side of the blood–brain barrier because it is produced there, rather than being transferred across the barrier. VEGF gene expression in brain capillary endothelial cells in response to hypoxia is discussed in Section 6.4.

6.2.5 High altitude pulmonary edema: HAPE

A small proportion of people, about 2%, who ascend to altitudes above 3000 m, develop high altitude pulmonary edema (HAPE), a suite of life-threatening symptoms, in which the patient literally drowns in the fluid in the lungs. An example of how HAPE can kill is described by Houston (1998):

> A commercial pilot and his wife flew from their home to Colorado and drove to a ski area at 3000 m. He skied hard the next day, but tired easily, and on the third day he fell a lot and went home early. During the night his breathing became laboured and early in the morning his wife managed to take him to the medical clinic. In the car, he stopped breathing. Though resuscitated and flown to intensive care, he died eleven days later with pulmonary and cerebral edema and subsequent pneumonia.

As so many people visit high altitudes for business and holidays, and because the limited time schedules for such trips involves rapid ascent to high altitude, HAPE is currently a significant medical problem. To understand how HAPE develops, we need to understand the structure of the respiratory surfaces of the lungs. The total surface area for respiratory gas exchange in the lungs is huge and located within tiny air sacs, the alveoli. As you know from Chapter 2 (Figure 2.16), the very thin walls of alveoli are closely apposed to the thin capillary walls, a feature which facilitates diffusion of oxygen and carbon dioxide. Blood enters the lung in the pulmonary artery, which contains venous blood from the right ventricle of the heart, at the same pressure as it is in the heart. As the blood flows from the pulmonary artery into the arterioles in the lung tissue, its pressure falls sharply to about 0.8 kPa, less than a quarter of that in systemic capillaries (4.0 kPa). The thin barrier between blood in the capillaries and air in the alveoli means that fluid and small dissolved molecules are easily forced from the blood into the alveoli at a rate which is determined by the steepness of the gradient in hydrostatic pressure between the pulmonary blood and alveolar air. As the alveolar membrane is impermeable to proteins, the pulmonary blood exerts an osmotic pressure, which tends to pull water back into the pulmonary capillaries. There is a little leakage of fluid, so the surface of the alveolar epithelium is always wet and is thus an important source of respiratory water loss, which can amount to 300 cm^3 per day in an adult man in a temperate climate. The low pressure of the pulmonary capillaries keeps down the hydrostatic pressure gradient across the alveolar membrane, so that it does not override the osmotic pressure gradient. Blood pressure also remains low in the pulmonary veins that transport blood back to the left ventricle of the heart.

However, at high altitude, when P_{O_2} in the alveoli is low, the muscular pulmonary arterioles respond by constricting, in contrast to the arteries in the rest of the body which dilate when blood P_{O_2} is low. Constriction and dilation of these vessels is possible because there is a relatively thick layer of smooth muscle in the vessel walls (Figure 6.12).

At sea-level the vasoconstrictive response to low P_{O_2} inside a few alveoli is of physiological advantage as it promotes the diversion of blood flow to well-ventilated alveoli. When all the alveoli in the lungs have low P_{O_2}, however, the vasoconstrictive response takes place over most of the lung volume, resulting in a rise in pulmonary artery pressure (P_{pa}). In this situation, the hydrostatic pressure gradient does override the osmotic pressure gradient, and fluid leaks out of the capillaries into the alveolar air spaces. The increased pressure may damage the alveolar walls increasing leakage, or increase the permeability of the capillary walls, which exacerbates the situation. Whatever the mechanism, the result is HAPE, in which the alveoli become flooded with liquid so that the patient has difficulty in breathing and may literally drown.

Nitric oxide (NO) is synthesized in endothelial cells of blood vessels. It is a short-acting signalling molecule that relaxes smooth muscle tone, causing small blood vessels to dilate. It has been suggested that people susceptible to HAPE may have a genetic deficiency in NO synthesis. A study showed that when susceptible patients with HAPE inhaled 40 ppm NO, pulmonary artery pressure decreased and blood shifted away to less flooded regions of the lungs, increasing oxygen saturation of arterial blood.

In people long established at high altitude, problems with raised pulmonary blood pressure (hypertension) are rare. Groves et al. (1993) studied five male residents of Tibet, living at altitudes > 3600 m. They found that the subjects' resting mean pulmonary pressures were within normal values for humans at sea-level, even when they breathed a hypoxic gas mixture (resulting in arterial P_{O_2} of 48 kPa). Near maximal exercise in the subjects increased cardiac output by a factor of three but there was no associated increase in pulmonary blood pressure. The authors concluded that their small sample of high-altitude residents had no problems with pulmonary vasoconstriction even when breathing gas mixtures that enhanced hypoxia.

Nevertheless, some cases have been reported in Tibetan and Nepalese males, in which pulmonary artery pressure is high and the right side of the heart is enlarged (pulmonary hypertension). Prolonged pulmonary hypertension in humans increases the workload for the right ventricle and may lead to failure of the right ventricle of the heart, a rare disease found in humans who live at high

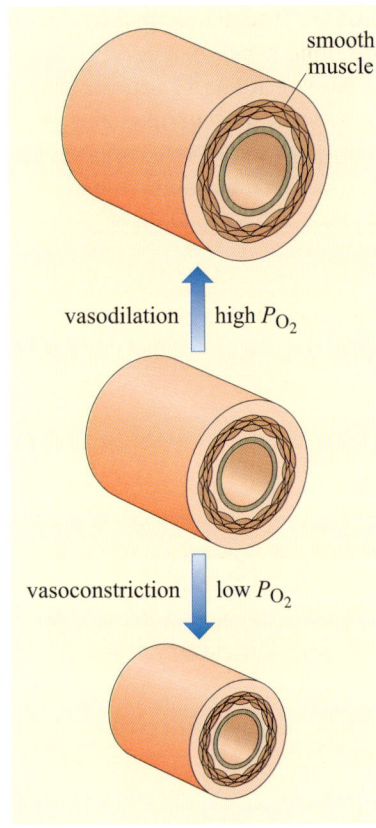

Figure 6.12 Vasodilation and vasoconstriction in small arteries in response to high and low P_{O_2}.

altitude. The patient has flushed cheeks, purple skin and blue nails, signs of systemic hypoxia. The illness is known as chronic mountain sickness (CMS), or Monge's disease. Right-side heart failure in these cases can only be prevented by a permanent move to low altitude, a difficult option for individuals who have lived their entire lives in high-altitude communities.

6.2.6 The role of phenotypic flexibility in physiological adaptation at high altitude

Once people have overcome any problems with acute mountain sickness, most individuals can live at high altitude – up to about 5500 m – long-term or even permanently. The heart rate returns to normal after acclimatization to high altitude. The shift of the oxygen dissociation curve to the right is maintained by appropriate levels of 2,3-DPG in the blood. The haematocrit level of permanent highland residents remains higher than that of resident lowlanders, which partly compensates for the reduction in percentage saturation of the blood at high altitude. In a study on 8 women who moved from the University of Missouri (213 m), to Pikes Peak, Colorado (4267 m), haematocrit increased from 43% to 48% after 30 days at high altitude (Figures 6.13a and b). Values for haemoglobin concentration tracked those for haematocrit, reaching 16 g dl^{-3} after 30 days at 4267 m.

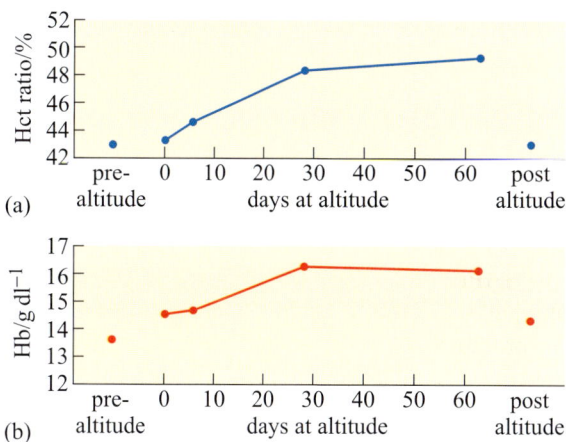

Figure 6.13 Effects of altitude in 8 young women from the University of Missouri (altitude 213 m) prior to, during and 2 weeks after exposure to 4267 m altitude at Pikes Peak, Colorado: (a) haematocrit (Hct); (b) blood haemoglobin (Hb) levels.

Research into the effects of hypoxia on humans has focussed on the effects on muscle tissue and capillary density, because of the interest in sport-related physiology. Loss of body mass and muscle volume has been reported in individuals participating in climbing expeditions to the Himalayas. Hoppeler and Vogt (2001) studied a group of 14 mountaineers who spent 8 weeks at altitudes above 5000 m. Compared with measurements taken at sea-level before the climb, measurements after 8 weeks at >5000 m demonstrated muscle atrophy, with a reduction of 20% in muscle fibre cross-sectional area in the vastus lateralis muscle – a superficial thigh muscle that extends the knee. Although capillary density was increased by 9–12%, the increase could be attributed to the reduction in muscle fibre cross-sectional area. However, Hoppeler and Vogt pointed out that oxygen diffusion to the muscle fibres is improved because the same capillary bed is supplying a lower volume of muscle tissue. Furthermore, they measured a decrease of 20% in the density of mitochondria in the mountaineers' muscle tissue.

■ What is the implication of lower mitochondrial density in terms of the capacity of tissue to carry out oxidative metabolism?

Measures of mitochondrial density provide a guide to the capacity of the tissue for aerobic metabolism. A lower density of mitochondria implies a lower capacity for oxidative breakdown of fuel molecules.

The study showed that the total length of the muscle capillary bed is retained, so with a reduced number of mitochondria, presumably the oxygen supply to the remaining mitochondria is improved. Participation in an expedition with a strict time schedule for climbing a Himalayan mountain, involves endurance of harsh living conditions and mountain sickness that in turn reduces appetite, which both contribute to loss of muscle mass and weight. So the effects observed in mountaineers may not be relevant to people moving to high altitude to live a normal, relatively sedentary life.

Hoppeler and Vogt were therefore interested in studying the skeletal muscle of long-term residents living at high altitude, living in comfortable housing, with access to a normal diet. Such people would be expected to show complete acclimatization to life at high altitude. A group of 20 students of mixed ethnic origin living at La Paz (altitude 3600–4000 m) were studied before and after 6 weeks of planned endurance exercise. The individuals taking part were living and eating well, in contrast to the members of the climbing expedition. Initially the students' muscle fibre cross-sectional area at high altitude, was similar to, or slightly less than that of, sea-level residents and muscle capillary density, at 404 mm^{-2}, was lower than that in a comparable lowland population. The capacity of muscle for aerobic metabolism (measured by the density of mitochondria, 3.94%) was 30% less than would be expected for untrained young lowlanders. After 6 weeks of endurance training, the muscle oxidative capacity and capillarity increased, but to an extent similar to that observed in lowland training studies. The content of intracellular lipid droplets in muscle biopsies after training was halved in comparison with similarly trained lowland residents. Endurance athletes training at low altitude accumulate large intracellular lipid droplets, which occupy 0.5% of total muscle fibre volume. In contrast, the lipid content of muscle fibres of the students training at La Paz was only 0.2%.

Another observation from this study is that long-term highland residents have slightly reduced muscle fibre cross-sectional area, reduced capacity for aerobic metabolism and reduced capillary density. The low intracellular lipid stores of highland residents may reflect a shift in muscle metabolism towards carbohydrate as a substrate for oxidative metabolism rather than lipid.

The increased capillary density in individuals undergoing endurance training at both low and high altitude is of crucial physiological importance. The number of muscle fibres is unchanged so that the ratio of capillaries to fibres is substantially elevated. An increase in capillary number would be associated with increased tissue diffusive conductance, derived from shortening of the diffusion path for oxygen from capillary to mitochondrion. Without the increased tissue capillary density, the gradient for oxygen diffusion from hypoxic blood in tissues to oxygenated blood in capillaries, would be insufficient to supply the most distal sites, so some muscle tissue would become anaerobic during exercise.

Summary of Section 6.2

The principles of the oxygen cascade and the chemical control of breathing must be understood before studying the physiological and biochemical responses to low P_{O_2}. The chemistry of dissolved CO_2 is especially relevant as it explains the effect of CO_2 on blood pH, which is summarized in the Henderson–Hasselbalch equation. As CO_2 can cross the blood–brain barrier, blood P_{O_2} also affects the pH of cerebrospinal fluid, CSF. Central chemoreceptors bathed in CSF can therefore detect blood pH, and signal the appropriate changes in breathing rate. However, the initial response to hypoxia is the stimulation by peripheral chemoreceptors of an increased breathing rate, which blows off CO_2, decreasing blood pH and causing central chemoreceptors to signal a decrease in breathing rate. This 'conflict' between two sets of receptors is eventually resolved, by secretion of CSF containing a lower concentration of HCO_3^-, resetting the pH of CSF so that the hypoxic ventilatory response can maintain an adequate P_{O_2} in alveolar air. Changes in the blood circulatory system induced by low P_{O_2} include rapid release of red blood cells from the spleen followed, within 12 hours, by an increase in the rate of formation of red blood cells stimulated by the hormone erythropoietin. Increased heart rate increases cardiac output and tissue capillary beds open up, enhancing blood supply.

The biochemistry of haemoglobin plays a central role in physiological acclimatization to low P_{O_2}. In particular, the shape of the oxygen dissociation curve means that at the low alveolar P_{O_2} found at high altitude, haemoglobin functions over a steep part of the oxygen dissociation curve. This means that highland residents can deliver enough oxygen to respiring tissues with only a small drop in P_{O_2} between their arteries and veins. The dissociation curve for haemoglobin of acclimatized highland residents is shifted a little to the right by increased levels of 2,3-DPG meaning that more oxygen can be loaded at the lungs. However, for a minority of individuals, the acclimatization process does not work and life-threatening syndromes, high altitude cerebral edema (HACE) and high altitude pulmonary edema (HAPE) develop. For HACE the initial event is an increase in the permeability of the blood–brain barrier in response to hypoxia, mediated by vascular endothelial growth factor (VEGF). For HAPE, constriction of pulmonary small arteries in response to hypoxia results in increased pulmonary artery pressure, causing fluid to leak into the lungs from the blood vessels. The rare syndrome of chronic mountain sickness (CMS) in permanent highland residents is caused by a persistently high pulmonary artery pressure which causes enlargement of the right side of the heart.

Research on acclimatized subjects living sedentary lives in La Paz (at an altitude of about 4000 m) showed that they have a lower capacity for aerobic metabolism in muscle, reduced muscle capillary density and low levels of lipid compared to lowland inhabitants. When those subjects underwent an endurance training programme, the capacity for aerobic metabolism in muscle and muscle capillary density increased to levels comparable to those found in lowland subjects undergoing similar training regimes. In the highland residents, muscle lipid did not increase so much as in lowland subjects, suggesting a shift in oxidative metabolism away from lipid and towards carbohydrate in muscle.

6.3 Genotypic adaptations for life at high altitude in vertebrate species

6.3.1 Genotypic adaptations in human populations

Section 6.2.1 showed that reduced P_{O_2} at high altitude results in a reduced arterial P_{O_2}, which in turn stimulates a compensatory increase in lung ventilation rate, and this is known as the hypoxic ventilatory response (HVR). Prolonged exposure to hypoxia may block the HVR in some high-altitude natives. The physiological response to the resulting hypoventilation can be pathologically high haematocrits. Andean natives have a more pronounced blunting of the HVR than have Tibetans. Reduced HVR and increased haematocrit may contribute to the greater susceptibility to chronic mountain sickness (Section 6.2.5) that is found in Andean natives. Cynthia Beall and her team (1997) carried out a comparative study of high-altitude Tibetan and Aymara (Andean) natives, and reported that the resting ventilatory rate was on average 50% higher in the Tibetans than in the Aymara people. Resting ventilation of Tibetan people showed greater variance than that found in the Aymara people (Figure 6.14). The data in this figure are for females only but males show a similar pattern. There is a brief discussion of heritability in Box 6.1.

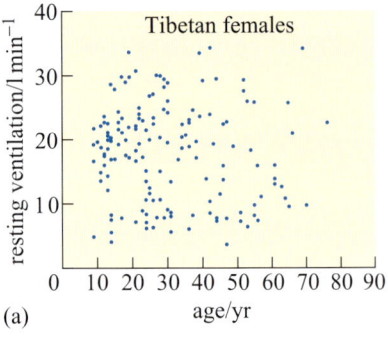

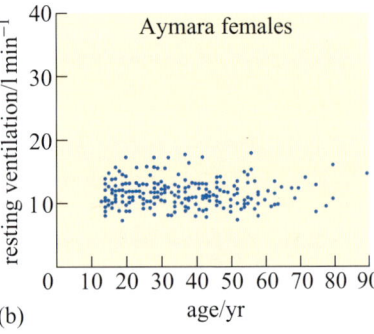

Figure 6.14 Resting ventilation (l min^{-1}) of (a) Tibetan and (b) Aymara females aged from 9 to 90 years.

BOX 6.1 HERITABILITY

Heritability (h^2) is the proportion of the phenotypic variation that is genetically determined and is accessible to selection. The remaining variation is mainly due to phenotypic plasticity arising from environmental factors. For natural selection and consequent evolutionary change to occur, there must be some genetic factors contributing to the selected phenotype. Researchers can estimate the heritability of a trait by comparing the resemblance between relatives, usually parent/offspring pairs. Heritability is defined as:

h^2 = additive genotypic variance/phenotypic variance

and is expressed either as a number from 0–1, or as a percentage.

The heritable contribution to the variance in resting ventilation rate was calculated to be significantly greater in Tibetan than in Aymara people: 35% versus 0%. Heritability of HVR was also calculated to be greater in Tibetan than in Aymara people: 31% versus 21%, respectively.

The absence of heritability of resting ventilation rate in the Aymara people suggests either that selection has eliminated genetic variability, or that there was no genetic variation in that trait within their original founder population. There appears to be greater potential for evolutionary changes in resting ventilation rate and HVR in the Tibetan population.

Kayser and his colleagues have provided further support for genetic adaptations in Tibetan people (Kayser et al., 1996). Working in Kathmandu (Nepal) at 1300 m altitude, the researchers compared muscle ultrastructure and biochemistry, in the vastus lateralis muscle (in the thigh), of eight lowland Tibetans (the sons and daughters of Tibetan refugees, who had lived at > 3500 m altitude in Tibet), and eight native lowland Nepalese people. Table 6.2 summarizes some of the main findings.

Table 6.2 Comparison of rates of oxygen uptake, and features of vastus lateralis muscle in eight lowland Tibetans and eight Nepalese lowlanders. (The values are the means ± SE.) P is the probability of the differences arising by chance.

Measured quantity	Tibetan people	Nepalese people	P
peak oxygen consumption/ $cm^3\ min^{-1}\ kg^{-1}$ body mass	37.9 ± 2.2	40.1 ± 1.4	>0.10
muscle fibre cross-sectional area/μm^2	3413 ± 224	3895 ± 148	<0.05
capillaries per muscle fibre	1.35 ± 0.07	1.46 ± 0.08	>0.10
mitochondrial density/%	3.99 ± 0.17	5.51 ± 0.19	<0.05

■ Describe the data in Table 6.2, drawing comparisons where appropriate.

Peak oxygen consumption is similar in both lowland Tibetans and Nepalese, at around 38–40 $cm^3\ O_2\ min^{-1}\ kg^{-1}$ body mass. The Tibetan people had significantly smaller muscle fibre cross-sectional areas – 3413 μm^2 compared with 3895 μm^2 in the Nepalese ($P<0.05$). The numbers of capillaries per muscle fibre (1.35–1.46) were similar in both groups. Mitochondrial density in muscle tissue was significantly lower at 4% in the Tibetan people, compared to 5.5% in the Nepalese ($P=<-0.05$). Low mitochondrial density in Tibetan people may be interpreted as adaptive, reducing the maximum capacity for aerobic exercise.

The researchers concluded that the low ratios of muscle mitochondrial density : peak O_2 consumption, that are found in lowland-born Tibetan people, are genetically determined.

■ Explain why Kayser's group could discount phenotypic plasticity or flexibility in their explanation of the differences between the Tibetans and the Nepalese.

Unlike their parents, the Tibetan group had *never* lived at high altitude, so it is likely that observed differences between the 'lowland' Tibetans and Nepalese are in fact hereditary.

6.3.2 Genotypic adaptations in physiology and biochemistry of high-altitude animal species

In Section 6.2.4, we learned how pulmonary vasoconstriction, a response in humans and cattle to hypoxia at high altitude, can cause serious health problems – including HAPE and right-side heart failure. Yet the resident fauna at high altitude do not experience such difficulties. Yaks and llamas maintain low pulmonary arterial pressure at low P_{O_2}, and do not have highly muscularized small pulmonary arteries. The pika (*Ochotona curzoniae*; Figure 6.4) lives in Tibet at an altitude of around 4000–4500 m. Pikas are diurnal and spend their days collecting food and basking in the sunshine but retreat to their burrows on cold windy days. They do not hibernate over the winter. Instead, pikas collect huge quantities of selected species of grasses and other plants to build up several haystacks which are food caches for the long winter. They may also burrow under snow where they can eat cushion plants and lichens off the ground. In this way, pikas can obtain sufficient food in the long alpine winters to enable them to remain active and maintain T_b. A Japanese research team measured values for pulmonary arterial pressure (P_{pa}) in 10 pikas captured from a site at 4300 m altitude in China. They compared the data with values for P_{pa} found in 10 Wistar rats, raised at an altitude of 2260 m in Xining, China (Ge et al., 1998). Rats were chosen for comparison because they are not a species associated with high-altitude environments, and their body mass is similar to that of pikas. Table 6.3 shows comparative data of animals in the study with respect to body mass, and the diameter and thickness of the wall of the pulmonary arteriole. Baseline measurements of P_{pa} were made (in Xining) at 2260 m, then the animals were placed in decompression chambers and exposed to *simulated* altitudes of 4300 and 5000 m. Measurements of P_{pa} were carried out by means of miniature catheters connected to pressure transducers. The changes in mean P_{pa} (ΔP_{pa}) are shown in Figure 6.15.

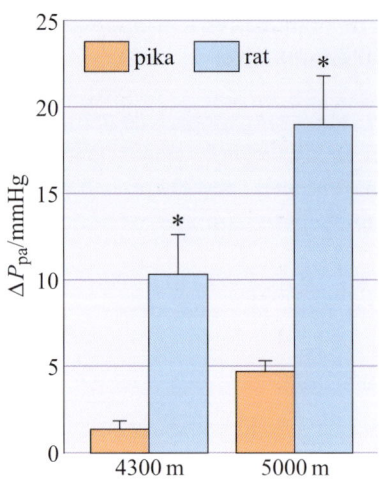

Figure 6.15 Changes in pulmonary artery pressure, P_{pa} (ΔP_{pa}) in pikas and Wistar rats from simulated altitude 2269 m to simulated altitudes 4300 and 5000 m, respectively. (The values are means ± SE. The asterisks indicate those changes that were significant, $P < 0.05$.)

■ Compare the values for ΔP_{pa} in rat and pika. What can you conclude from the data about the pulmonary vasoconstrictor response in the rat and pika?

At simulated altitudes of 4300 m and 5000 m, P_{pa} for pikas was elevated slightly by 1.5 mmHg and 4.8 mmHg, respectively. In contrast, the P_{pa} values for rats were elevated by 10.2 mmHg at 4300 m and by 19.1 mmHg at 5000 m. These findings show that the pulmonary vasoconstrictor response to acute hypoxia was much smaller in the pikas than in the rats.

Histological study of the small pulmonary arteries and arterioles showed that the walls of small pulmonary arteries/arterioles of pikas are very thin and lacking in smooth muscle with only a single elastic lamina, a thin sheet of elastic connective tissue. In contrast the small pulmonary arteries of the rats have a thick wall made of smooth muscle sandwiched between an inner and outer layer of elastic lamina (Figure 6.16a and b). Measurements of the thickness of the pulmonary arteriole walls of pikas and rats showed statistically significant differences (Table 6.3).

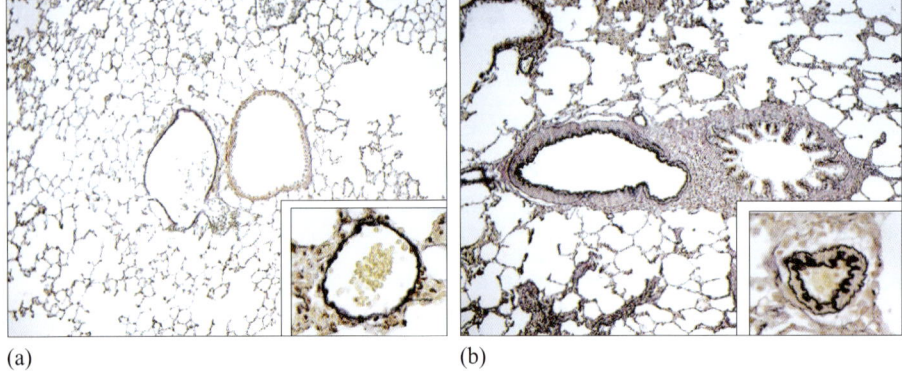

(a) (b)

Figure 6.16 (a) Lung tissue of a pika (magnification ×25). Note how the wall of the pulmonary arteriole is very thin in the pika lung in contrast to the same tissue in rat lung. The inset shows a ×400 view of an arteriole with a wall consisting of just one elastic lamina. (b) Lung tissue of a Wistar rat (magnification ×25). Note the thickness of the pulmonary arteriole wall. The inset shows a ×400 view of a small arteriole with a thick wall comprising thick smooth muscle sandwiched between an inner and outer elastic lamina.

Table 6.3 Tissue measurements taken from sections of pika and rat lungs. (The values are the means ± SE of 10 individuals.)

Anatomical or histological measurement	Pika	Wistar rat
body mass	163.7 ± 5.91	226.2 ± 7.52*
diameter of pulmonary arteriole/μm	113.0 ± 3.92	114.1 ± 3.71
thickness of pulmonary arteriole wall/μm	8.20 ± 0.64	31.30 ± 0.81*
% wall thickness relative to diameter of pulmonary arteriole	9.22 ± 0.70	27.21 ± 0.81*

*$P < 0.01$.

■ Drawing on the data in Table 6.3 and the histology in Figure 6.16, explain why having thin-walled pulmonary arteries is advantageous for high-altitude species such as pikas.

Vasoconstriction in arteries arises from contraction of the thick smooth muscle layer in the vessel wall. In response to hypoxia, the rats in this study showed a sharp increase in P_{pa} (as do humans exposed suddenly to high altitude), mediated by contraction of the smooth muscle in pulmonary arteriole walls. In contrast, pikas show very little increase in P_{pa} in response to hypoxia (Figure 6.15). There is no smooth muscle in the walls of pika pulmonary arterioles, so vasoconstriction in response to hypoxia is not possible. Lack of a vasoconstrictive response prevents pulmonary hypertension and helps to maintain P_{O_2} levels in arterial blood.

■ Study the oxygen dissociation curves in Figure 6.17. Compare the oxygen dissociation curves for vicuna and llama haemoglobins with that for the sea-level species. Explain the implications of your comparison in terms of affinity of the pigment for oxygen at the lungs and in the tissue capillary beds.

The curves for the vicuna and llama are displaced to the left, which means that the vicuna haemoglobin has a higher affinity for oxygen, so it can load up with more oxygen for a particular value of P_{O_2} in the alveoli. However, a high-affinity haemoglobin will release less oxygen at the respiring tissues, where P_{O_2} is low.

Vicunas, llamas and other camelids that live at high altitudes do not have any physiological problems derived from their high-affinity haemoglobin. The disadvantage of reduced oxygen offloading at the respiring tissues must be offset by the advantages of increased loading of oxygen in the lungs. What is most important is the amount of oxygen released to the respiring tissues. It can be suggested that, for humans at high altitude, an *increased* affinity of haemoglobin for oxygen – maintained by 2,3-diphosphoglycerate – is more advantageous for supplying oxygen to the respiring tissues. Vertebrates living at high altitudes have amino acid substitutions in key positions in haemoglobin polypeptide chains, which modulate the oxygen affinity of the pigment by weakening interactions between the sub-units that stabilize the T conformation (Section 6.2.3).

We mentioned in the introduction to this chapter that the migration path of the bar-headed goose (*Anser indicus*) crosses the Himalayas, and thus the geese fly at up to 9000 m altitude. Flapping flight has high energy costs and places heavy demands on the respiratory system. Research in the late 1970s (Black et al., 1980) on this species showed that P_{O_2} at a simulated altitude of 11.6 km is only

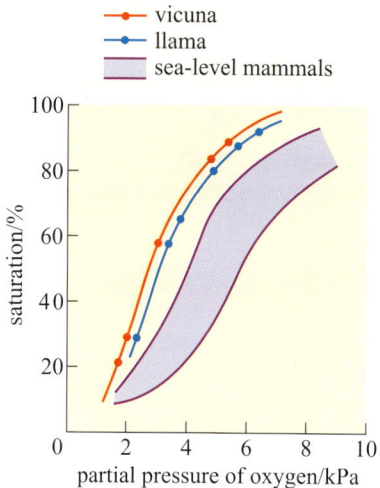

Figure 6.17 Oxygen dissociation curves for vicuna, llama and sea-level mammalian haemoglobin.

0.133 kPa lower than that of the inhaled air, which demonstrates the extraordinarily high efficiency of pulmonary gas exchange in A. indicus. The haemoglobin of this goose has a high affinity for oxygen, with P_{50} of only 3.96 kPa at 37 °C and pH 7.4. In contrast, the greylag goose (Anser anser), has haemoglobin with P_{50} of 5.27 kPa at 37 °C and pH 7.4 (reported in Weber et al., 1993).

Studies on the molecular structure of the haemoglobin of A. indicus, showed that in comparison with haemoglobin of A. anser, there are four amino acid substitutions in the polypeptide chains (Table 6.4). The fourth substitution, at position 119 in the α chains, is unique among birds and mammals, which suggests that it accounts for increased affinity for oxygen in A. indicus.

Table 6.4 Comparison of amino acids in haemoglobin (Hb) of Anser anser and A. indicus.

Position of amino acid in A. anser Hb	Substituted amino acid in A. indicus Hb
α^{18}-glycine	serine
α^{63}-alanine	valine
β^{125}-glucine	aspartamine
α^{119}-proline	alanine

In human and Anser anser haemoglobin, α^{119}-pro participates in a weak interaction (Van der Waals forces), with a methionine residue in the adjacent β chain (β^{55}-met) that stabilizes the T (tense) structure (Figure 6.18). This weak interaction is abolished in A. indicus haemoglobin by replacement of α^{119}-proline with alanine. In haemoglobin from the Andean goose (Chloephaga melanoptera), the same weak interaction is disrupted by substitution of β^{55}-leucine by serine, an amino acid with a shorter side chain than leucine. So the haemoglobins in these high-altitude species both have a similar change in structure associated with a higher affinity for oxygen.

Weber and his team used the technique of site-directed mutagenesis on human haemoglobin, genetically engineered in bacteria, to mimic the amino acid substitution that had been found in Andean and bar-headed geese. This technique involves changing the base sequence of the gene that codes for the protein being studied. The gene is isolated and a specific base or set of bases is altered chemically, in this case the sequence for α^{119}-pro, thereby producing a different codon, α^{119}-ala. The modified gene is then inserted into a bacterial cell, which proliferates producing millions of cells each containing the new gene, which directs synthesis of the altered protein. The protein, haemoglobin α^{119}-ala, is then isolated from the batch of bacterial cells. Weber et al. measured P_{50} for the recombinant human haemoglobin, Hbα^{119}-ala at pH values of 7.0 and 7.4 (Table 6.5). You will recall from Sections 6.2.2 and 6.2.3 that you would expect to see different P_{50} values for haemoglobin at different values of pH, since a change in pH will alter the affinity of haemoglobin for oxygen. Determination of P_{50} at different pH values is a useful check on the procedure. A pH of 7.0 is useful for this procedure but is much lower than would be found in the blood plasma of either humans or birds.

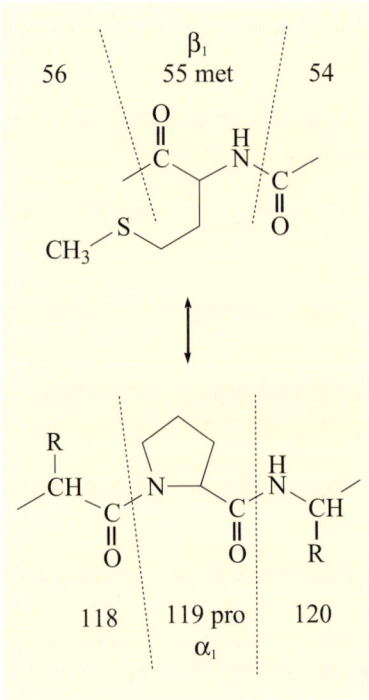

Figure 6.18 Schematic diagram of α^{119}-pro-β^{55}-met contact in human haemoglobin that stabilizes tense (T) structure but is lost when either proline or alanine is substituted by shorter-chain residues, resulting in increased O_2 affinity. The dashed lines indicate the beginning of amino acid residues adjacent to ^{55}met and ^{119}pro.

Table 6.5 Values of P_{50} at pH 7.0 and 7.4 ($T = 37\,°C$) for normal and recombinant human haemoglobin, with and without KCl and 2,3-DPG (modified from Weber et al., 1993).

Haemoglobin	P_{50} at pH 7.0/kPa	P_{50} at pH 7.4/kPa
human Hb, stripped*	1.09	0.82
human Hb, stripped + KCl + 2,3-DPG	5.62	4.00
human Hbα^{119}-ala, stripped	0.77	0.56
human Hbα^{119}-ala, stripped + KCl + 2,3-DPG	2.54	1.72

* The term 'stripped' means that the haemoglobin molecules were separated from other molecules and ions such as Cl⁻ and 2,3-DPG, that normally influence the affinity for oxygen.

■ From the data in Table 6.5 deduce the effect of the replacement of α^{119}-pro in the human Hbα-chain, by alanine, on the affinity of the pigment for oxygen.

The stripped haemoglobin retained sensitivity to KCl and 2,3-DPG, with both factors decreasing the affinity of the pigment for oxygen at pH 7.0 and 7.4 (in Section 6.3.3 you read that increased P_{50} denotes decreased affinity). The recombinant haemoglobin, Hbα^{119}-ala showed increased affinity for oxygen in comparison to normal human Hb at pH 7.0 and 7.4 (P_{50} reduced from 1.09 to 0.77 and from 0.82 to 0.56, respectively). Human Hbα^{119}-ala with KCl and 2,3-DPG showed increased affinity in comparison to stripped human Hbα^{119}-ala, showing that recombinant human Hb retains sensitivity to 2,3-DPG at 37 °C.

So, replacement of α^{119}-pro by alanine in human haemoglobin, results in an increased affinity for oxygen, just as this substitution does in the natural mutant haemoglobin in *A. indicus*.

The sensitivity of *A. indicus* haemoglobin to 2,3-DPG and temperature was not investigated. There are no data available yet on the body temperature of *A. indicus* during its flight at 9000 m, but ambient temperatures at such high altitudes are very low. It is remarkable that *A. indicus* can fly at what must be exceptionally low T_a.

We have seen how small changes in the amino acid sequence of polypeptide chains in haemoglobin can have huge implications for oxygen affinity. The molecular structure of haemoglobin is of crucial importance for maintaining oxygen supply to respiring tissues at high altitudes where P_{O_2} is low. In Section 6.4 we move on from a study of the adaptations at physiological and biochemical levels, to an examination of responses to environmental cues at the molecular level.

Summary of Section 6.3

Genotypic adaptations have been identified in high-altitude human populations. Andean native people have a more pronounced blocking of the hypoxic ventilatory response (HVR) than Tibetan people. On average the resting ventilatory rate was 50% higher in Tibetan people than in Aymara people from the Andes. Calculated heritability values for both resting ventilation rate and HVR are significantly greater in Tibetan people. Kayser et al.'s research provides strong support for genotypic adaptations in people living at high altitudes.

Comparisons between native Nepalese people in Kathmandu, and children of Tibetan refugees who were born and living in Kathmandu (altitude 1300 m) showed that the Tibetan group retained aspects of muscle ultrastructure and biochemistry typical of Tibetans living at high altitudes (>3500 m). Muscle fibre cross-sectional area, and mitochondrial density were significantly lower in the lowland Tibetans than in the Nepalese group. Research on the pika, *Ochotona curzoniae*, a species living at 4000–4300 m in the Qinhai-Tibetan plateau, demonstrated that in this species, pulmonary artery pressure (P_{pa}) increases only slightly at low P_{O_2} associated with simulated altitudes of 4300 m and 5000 m. In contrast, the P_{pa} of rats increases sharply at low P_{O_2}, a similar response to that observed in humans. Lung histology revealed that low P_{pa} at low P_{O_2} in pika derives from the thin walls of small pulmonary arteries and arterioles which lack smooth muscle. In contrast, the thick-walled small pulmonary arteries and arterioles of rats contain thick layers of smooth muscle that constrict at low P_{O_2}. The difference in thickness between rat and pika small pulmonary arteriole walls is statistically significant.

Biochemical genotypic adaptations for low P_{O_2} have been identified in animal species. The haemoglobin of the bar-headed goose has an amino acid substitution, of alanine for proline at position 119, which disrupts a weak interaction that stabilizes the tense structure of haemoglobin. Studies using the technique of site-directed mutagenesis in human haemoglobin to mimic the amino acid substitution of bar-headed goose haemoglobin, confirm that the disrupted weak interaction increases the affinity of haemoglobin for oxygen, which is important for a species subjected at regular intervals to extremely low P_{O_2}.

6.4 Integrating across disciplines

Since the 1980s, there has been an expansion of research into the effects of hypoxia at all levels of structure and function. Molecular biology research at the start of the 21st century has thrown light on the molecular mechanisms of responses to hypoxia at high altitude. It is not possible to cover all aspects here, so the following section concentrates on hypoxia-inducible factor. You should be aware though that there is strong evidence supporting the central role of vascular endothelial growth factor (VEGF) in the physiological response to hypoxia. You will find some references to this work in the Further Reading section at the end of this book.

6.4.1 Hypoxia-inducible factor (HIF)

When subjected to hypoxia, cells in the kidney respond by switching on production of the hormone erythropoietin (Section 6.2.2). There is relatively little information available on either the cellular oxygen sensor that detects hypoxia or the subsequent signal transduction pathway. So in this section, we concentrate on the target of the signalling mechanisms following cellular detection of hypoxia. The response of cells to hypoxia involves the induction of hypoxia-responsive genes. In Chapter 2, *HIF-α* was identified as one of the most important of these genes. Initially, hypoxia-inducible factor (HIF) was identified as the transcription factor that binds to a hypoxia-response element (HRE) in the gene for erythropoietin. The HIF transcriptional complex consists of a heterodimer of one of three α sub-units (HIF-1α, HIF-2α, or HIF-3α), and an α-sub-unit called ARNT (aryl hydrocarbon receptor nuclear translocator). Both HIF-1α and ARNT belong to the helix-turn-helix family of transcription factors (Figure 6.19) which consist of three α-helices and a β-structure.

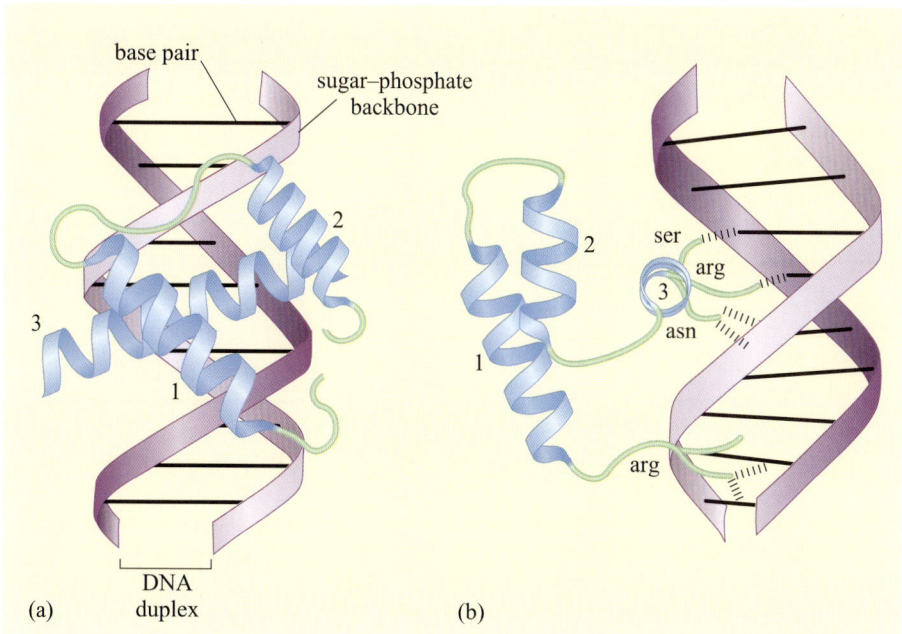

Figure 6.19 Structural motifs of a major type of DNA-binding protein, the helix-turn-helix protein, which consists of three linked α helices. (a) frontal view; (b) lateral view.

We are particularly interested in HIF-1α for our purposes, as this sub-unit controls vascularization. Hypoxia inducible factor α-sub-unit (HIF-1α) is the protein that is a crucial player in the cellular response to hypoxia. When physiological P_{O_2} is normal, i.e. during normoxia, HIF-1α levels are very low and often undetectable. Hypoxia results in a dramatic increase in HIF-1α, brought about by a number of mechanisms acting together. For example, sharp increases in cellular levels of HIF-1α are observed in cell cultures exposed to hypoxia. During normoxia, enzymes mediate the rapid degradation of HIF-1α; these enzymes, prolyl- and asparinyl-hydroxylases, can be regarded as the molecular 'sensors' that detect cellular hypoxia, and mediate the hypoxic response. Degradation of HIF-1α in normoxia involves von Hippel-Lindau tumour suppressor protein, pVHL, which recognizes and links HIF-1α sub-units – via a complex of incompletely characterized proteins, to ubiquitination pathways (Figure 6.20, overleaf). Ubiquitination is simply the addition of ubiquitin molecules (a type of protein) to other molecules, in this case to a transcription factor. We have included this figure so you can see the complexity of the process involved in (a) the role of HIF-1α in gene transcription under hypoxic conditions, and (b) the breakdown of HIF-1α under normoxic conditions – recall of all the detail is not required!

Addition of ubiquitin to a protein molecule is a signal for its degradation. Once HIF-1α has been subjected to ubiquitination, it is transferred to proteasomes where rapid degradation takes place. An interesting finding derived from research into the molecular biology of hypoxia in tumour tissue was that the genetic deficiency of pVHL in cells results in high constitutive levels of HIF-1α and over-expression of oxygen-regulated genes leading to formation of multiple tumours, von Hippel-Lindau syndrome.

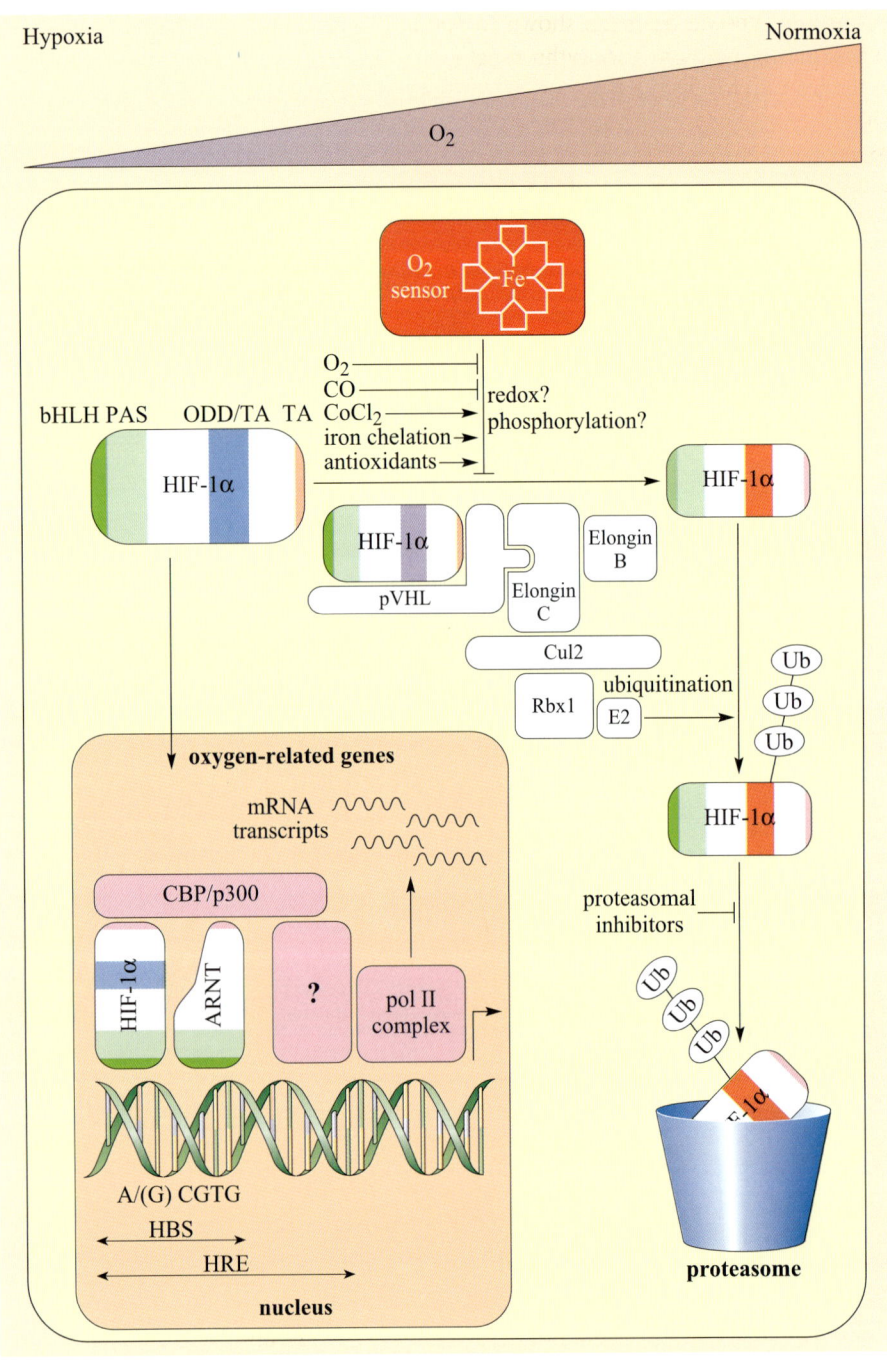

Figure 6.20 Model of oxygen sensing, signalling and gene regulation (Wenger, 2000). Under hypoxic conditions, the ODD/TA domain of HIF-1α (blue) is stable and translocates to the nucleus where it binds to the hypoxia-responsive gene regulatory element (HRE), and initiates gene regulation. Under normoxic conditions, oxygen is detected by an oxygen sensor, here depicted as a haem group. The ODD/TA domain of HIF-1α, shown in orange, receives the signal from the oxygen sensor, which targets HIF-1α for proteolytic breakdown by enzymes within proteasomes (large complexes of proteolytic enzymes). Adding the protein ubiquitin to the HIF-1α molecules targets them for proteolysis; von Hippel-Lindau tumour suppressor protein is involved in this as well as other proteins (elongins, Cul2 and polII). *Abbreviations*: ARNT: aryl hydrocarbon receptor nuclear translocator; bHLH: basic helix-turn-helix; HBS: HIF-1α binding site; HRE: hypoxia-response element; ODD: oxygen-dependent degradation domain; pol II complex: RNA polymerase II; pVHL: von Hippel-Lindau tumour suppressor protein; Ub: ubiquitin.

During hypoxia, the enzymes cease their activity in mediating degradation of HIF-1α, and this transcription factor dimerizes with ARNT, and translocates to the nucleus and binds to the HIF-1α binding site (HBS) within the hypoxia-response element (HRE) of the oxygen-regulated gene, which initiates transcription (see Figure 6.20).

■ Is HIF-1α activity regulated transcriptionally or post-transcriptionally?

As HIF-1α protein is degraded so rapidly when P_{O_2} is at normal physiological levels, HIF-1α must be regulated post-transcriptionally.

So far, the induction of HIF-1α in response to hypoxia has been shown to regulate transcription of more than 40 target genes, including those for erythropoietin, glycolytic enzymes, glucose transporters and VEGF (see Section 6.2.4). Activity of the ARNT sub-unit, which is constitutively expressed, is not affected by hypoxia. HIF-1α is the master transcriptional regulator of the adaptive response to hypoxia. The activation of HIF-1α initiates effects in cells at molecular and biochemical levels, which help to reduce the deleterious effects of chronic exposure to hypoxia.

In order to appreciate the pivotal role of HIF-1α we need to look at the results of research on the role of HIF-1α in vertebrates. Vogt and co-workers (2001) studied molecular adaptations to simulated hypoxic conditions in human skeletal muscle. The researchers explored changes in gene expression resulting from exercise training at two levels of intensity under normoxic and hypoxic conditions (P_{O_2} corresponding to altitude 3850 m, but total atmospheric pressure at sea-level value). Four groups of human subjects trained five times a week for 6 weeks on an exercise bicycle. Muscle biopsies were taken from the vastus lateralis muscle before and after the training period. mRNA samples extracted from the muscle were subjected to the reverse transcriptase polymerase chain reaction (RT-PCR). This technique is used to amplify DNA copies of very small amounts of mRNA, and is therefore useful in studies of gene expression. Comparative measurements of amounts of mRNA in samples can be obtained with the technique. Increased levels of HIF-1α mRNA were found after both high-intensity and low-intensity training under hypoxic conditions (Table 6.6). In contrast, levels of ARNT remained the same.

Under normoxic conditions, HIF-1α mRNA levels were not increased even after 6 weeks of the exercise regime, consistent with the view that the gene for HIF-1α is expressed only in hypoxia. The increased levels of HIF-1α measured after 6 weeks of training in hypoxic conditions, suggest a persistent adaptation to the 6-week training regime under hypoxia, rather than the immediate effects of exposure to hypoxia. The researchers suggest that HIF-1α is a key transcription factor for specific hypoxia-related changes in muscle.

Levels of mRNAs for certain enzymes involved in oxidative pathways – cytochrome oxidase, NADH dehydrogenase, succinate dehydrogenase and phosphofructokinase – increased after high-intensity training under hypoxia ($P < 0.05$, see Table 6.6). Cytochrome oxidase and PFK mRNAs increased significantly after high-intensity training under normoxia. None of the enzyme mRNAs listed in Table 6.6 increased after low-intensity training.

Table 6.6 Measures of mRNA for four metabolic enzymes, HIF-1α and VEGF from muscle biopsies after 6 weeks training, derived from RT–PCR results relative to normalized pre-training values.

mRNA	High-intensity training: normoxia	High-intensity training: hypoxia	Low-intensity training: normoxia	Low-intensity training: hypoxia
COX-4 (cytochrome oxidase)	1.7 ± 0.2*	1.5 ± 0.2‡	0.9 ± 0.1	1.4 ± 0.2‡
NADH6 (NADH dehydrogenase)	1.7 ± 0.3‡	1.4 ± 0.1*	1.0 ± 0.1	1.2 ± 0.2
SDH (succinate dehydrogenase)	1.3 ± 0.1	1.5 ± 0.1*	0.8 ± 0.1	1.4 ± 0.2‡
PFK (phosphofructokinase)	1.5 ± 0.2*	1.9 ± 0.3*	0.8 ± 0.1‡	1.0 ± 0.1
HIF-1α (hypoxia-inducible factor α-sub-unit)	1.3 ± 0.2	1.8 ± 0.1‡	1.2 ± 0.1	1.6 ± 0.2*
VEGF (vascular endothelial growth factor)	1.2 ± 0.1	1.7 ± 0.3*	1.0 ± 0.1	1.2 ± 0.2

Values are means ± SE.

* Significant difference between pre- and post-training values.

‡ Difference between pre- and post-training values $P < 0.10$ but not significant.

Levels of myoglobin mRNA and VEGF mRNA were measured in the same muscle biopsies (Figure 6.21). Statistically significant increases in myoglobin mRNA and VEGF mRNA were observed only after high-intensity training under hypoxic conditions. Note how low intensity exercise under hypoxia does not increase myoglobin mRNA or VEGF mRNA levels in muscle tissue when compared to low-intensity exercise in normoxia. The researchers suggest that as myoglobin and VEGF mRNA were increased in muscle tissue of subjects undergoing intensive training under hypoxia, myoglobin might be regulated, like VEGF, via the oxygen-sensing pathways and the HIF-1α system. (We should add that the difference in myoglobin mRNA is just at the limit of significance.)

Vogt et al. noted that the increase in VEGF was reflected by a parallel increase in capillary density. Capillary length density increased only after high-intensity training under hypoxia, from 701 mm mm^{-3} pre-training to 831 mm mm^{-3} post-training. Their data demonstrated a slight and non-significant decrease in muscle fibre cross-sectional area after training under hypoxia.

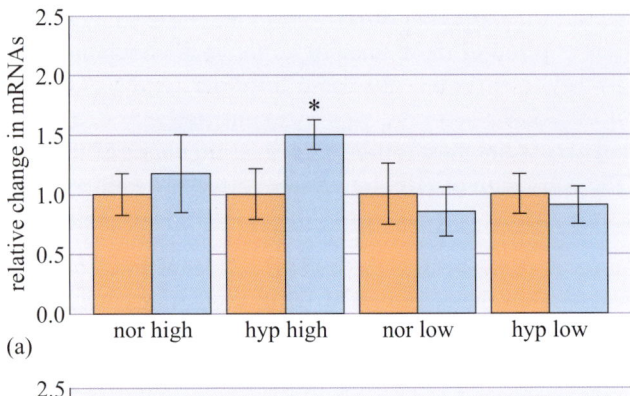

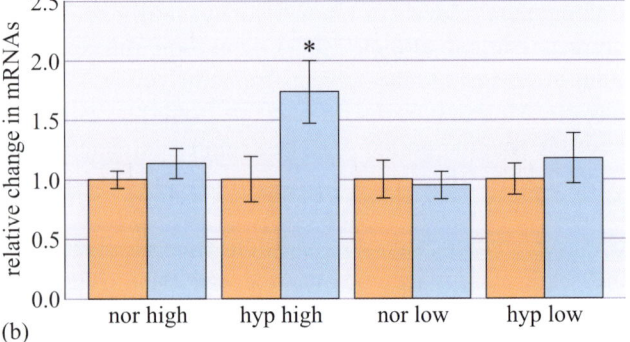

Figure 6.21 Relative changes in mRNAs coding for (a) myoglobin and (b) vascular endothelial growth factor. The subject groups comprise: nor high and nor low: high- and low-intensity level of training under normoxic conditions, respectively and hyp high hyp low. Values are means ± SE. In (a) the difference pre- and post training marked by an asterisk is $P < 0.1$; in (b), $P < 0.05$.

The key point here is that Vogt et al.'s data support the involvement of HIF-1α in the regulation of processes of physiological adaptation in muscle tissue after training under hypoxia. Higher capillarity facilitates supply of oxygen and substrates to cells. Increased myoglobin improves the oxygen storage capacity in muscle cells. Induction of metabolic pathways that promote use of carbohydrates, rather than lipids, as substrate facilitate more efficient use of oxygen. The increased cytochrome oxidase and NADH dehydrogenase mRNA following high intensity exercise, demonstrates that the oxidative capacity of muscle tissue can be increased by training under hypoxia.

Summary of Section 6.4

The transcription factor HIF-1α plays a pivotal role in the response of cells to hypoxia. At normal cellular P_{O_2}, cytoplasmic HIF-1α is degraded via a complex ubiquitination pathway; regulation of cellular HIF-1α is post-transcriptional. When cells are subjected to hypoxia, HIF-1α dimerizes with ARNT and translocates to the nucleus where it binds to a site within the hypoxia-response element (HRE) of a hypoxia-responsive gene and initiates transcription. Important hypoxia-responsive genes include those coding for erythropoietin (VEGF), glucose transporters and glycolytic enzymes. Vogt et al. demonstrated increased levels of HIF-1α mRNA following 6 weeks of low and high-intensity training under hypoxic conditions in human subjects. Levels of mRNA VEGF, NADH dehydrogenase, succinate dehydrogenase and phosphofructokinase were also increased after high-intensity training under hypoxia. Vogt et al.'s research demonstrates the involvement of HIF-1α in the regulation of physiological acclimatization of muscle tissue in subjects undergoing high-intensity training while subjected to hypoxia, mimicking that experienced at 3850 m altitude.

6.5 Integrating across species

So far in this chapter we have looked at particular aspects of animals' adaptations, both phenotypic and genotypic. Within its environment an animal is subjected to many pressures, so for disparate species it is difficult to separate changes linked to the factor of interest, in this case hypoxia, from changes linked to different selective pressures.

The study of adaptation that involves the use of two or more distantly related species for comparisons has drawbacks. Conclusions drawn from the results of such studies assume that the species would be identical physiologically, were it not for the effects of the environmental parameter that is being investigated. With such studies it is not possible to separate adaptations that are linked only to the factor of interest, in this case hypoxia, from those derived from other selective pressures acting on the species. Physiologists assume that in animals living at high altitude, natural selection has led to the evolution of characters that enable the maintenance of ATP synthesis even though the availability of oxygen is always limited. Comparisons of activity of enzymes involved in glucose oxidation in high-altitude species with those in laboratory rats or rabbits are interesting, but the wild species are not closely related to laboratory animals such as rat and rabbit. Therefore, in order to demonstrate adaptation for hypoxia, comparisons between closely related species are required.

Pika and deer mice genera both include species that live at high altitude and species that are found in equivalent habitats at low altitude but at high latitude. Sheafor (2003) compared the activities of muscle enzymes in three species of pika, each living at different altitude. Species living at low latitude inhabit rocky alpine meadows at high altitude, but the altitude at which the alpine habitat is found decreases with increasing latitude. Table 6.7 shows that while the three species live at different altitudes, each lives in an environment with a similar range of temperature.

Sheafor measured activities of oxidative and glycolytic enzymes in muscle tissues of the pikas, including skeletal muscle and diaphragm and cardiac muscle. Wild Alaskan pikas were captured and tissues were removed in a mobile laboratory.

Table 6.7 Environmental, ecological and behavioural data for *Ochotona princeps, Ochotona collaris* and *Ochotona hyperborea* (data taken from Sheafor, 2003).

Environmental, ecological or behavioural parameter	*Ochotona princeps*	*Ochotona collaris*	*Ochotona hyperborea**
country where collected	USA (Alaska)	USA (Alaska)	northern Russia
latitude/°N	40	65	59
altitude/m	3350	1070	sea-level
barometric pressure/kPa	66.7	89.3	101.3
mean annual temperature/ °C	-3.7 ± 0.5	-3.5 ± 0.8	
mean annual snowfall/mm	930 ± 54.2	1764 ± 77.5	
home territory size/m^2	720 ± 15.7	700 ± 12.1	
above ground activity/h day^{-1}	4.1 ± 0.8	4.5 ± 0.6	

* Detailed data are not available on temperature, snowfall, territory or activity for this species. However, the animals used in this study were trapped in a region where the temperature and snowfall were similar to those in the Alaskan habitats.

Heart tissue from *O. hyperborea* was obtained from a frozen tissue bank kept at the University of Alaska. Rabbit data were also collected for use as a control. Table 6.8 and Figure 6.22 summarize the results of the enzyme assays. Sheafor collected data from three types of skeletal muscle (vastus, gastrocnemius and soleus) as they had varying compositions of oxidative and glycolytic muscle fibres. Diaphragm and cardiac muscles were chosen because of their involvement with oxygen delivery and hence give a good indication of the metabolic status of the animal. (S324, Book 3 discusses muscle structure and function in more detail.)

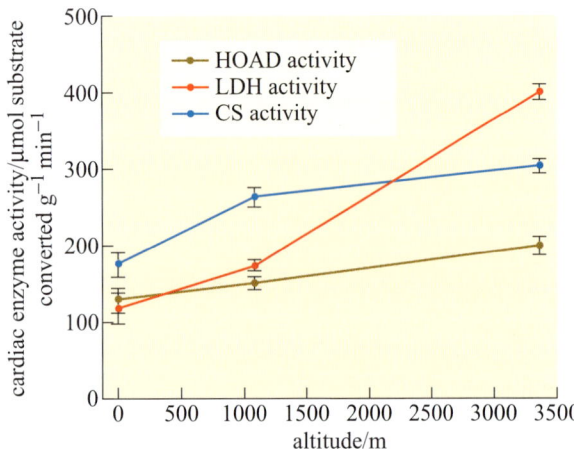

Figure 6.22 Enzyme activities from cardiac muscle of pika collected across an altitudinal gradient. *Ochotona hyperborea* (sea-level), *Ochotona collaris* (1070 m) and *Ochotona princeps* (3350 m). Values are mean ± SE. CS, citrate synthase; HOAD, β-hydroxyacyl CoA dehydrogenase; LDH, lactate dehydrogenase.

Table 6.8 Enzyme activities of muscle in *Ochotona princeps* (3350 m), *Ochotona collaris* (1070 m), *Ochotona hyperborea* (sea-level) and laboratory rabbits (*Oryctolagus cuniculus*) (adapted from Sheafor, 2003).

	Heart	Diaphragm	Vastus	Gastrocnemius	Soleus
Citrate synthase activity/μmol substrate converted g^{-1} wet tissue min^{-1}					
O. princeps	304 ± 9.2[a]	160 ± 7.2[a]	50 ± 2.7[a]	52 ± 4.6[a]	61 ± 6.2[a]
O. collaris	262 ± 13.3[b]	123 ± 2.0[b]	47 ± 3.4[a]	47 ± 2.3[a]	56 ± 2.1[a]
O. hyperborea	173 ± 16.6[c]	–	–	–	–
rabbit	131 ± 3.9[c]	53 ± 2.9[c]	22 ± 1.9[b]	24 ± 3.5[b]	42 ± 2.6[b]
β-hydroxyacyl CoA dehydrogenase activity/μmol substrate converted g^{-1} wet tissue min^{-1}					
O. princeps	200 ± 12.0[a]	98 ± 4.0[a]	23 ± 1.0[a]	28 ± 1.2[a]	42 ± 2.8[a]
O. collaris	149 ± 8.0[b]	71 ± 3.2[b]	31 ± 1.1[b]	31 ± 2.5[a]	36 ± 2.0[a]
O. hyperborea	126 ± 16.6[ab]	–	–	–	–
rabbit	217 ± 9.9[a]	63 ± 3.5[b]	24 ± 1.9[ab]	30 ± 3.6[a]	53 ± 3.4[b]
Lactate dehydrogenase activity/μmol substrate converted g^{-1} wet tissue min^{-1}					
O. princeps	401 ± 9.7[a]	592 ± 15.6[a]	1011 ± 6.1[a]	1216 ± 64.5[a]	1145 ± 85.8[a]
O. collaris	172 ± 6.4[b]	289 ± 13.6[b]	590 ± 13.3[b]	586 ± 20.6[b]	506 ± 21.8[b]
O. hyperborea	116 ± 19.4[b]	–	–	–	–
rabbit	151 ± 8.1[b]	524 ± 6.8[c]	1124 ± 31.2[c]	963 ± 14.2[c]	178 ± 9.2[c]

Values are mean ± SE. For *O. princeps* and for rabbit, $N = 8$; *O. collaris*, $N = 9$; for *O. hyperborea*, $N = 4$. Statistical analyses were carried out to obtain comparisons of the activities of each of the enzymes between the four species, within each tissue. Note that for *each* enzyme those values within each column which do not have the same superscript letter, differ significantly ($P < 0.05$). The vastus, gastrocnemius and soleus muscles are all locomotory muscles.

- What are the respective roles of the enzymes (a) citrate synthase, (b) β-hydroxyacyl CoA dehydrogenase, and (c) lactate dehydrogenase? What measure is provided by assaying the actvity of each enzyme?

(a) Citrate synthase catalyses the synthesis of citrate (from oxaloacetate and acetyl CoA), the first reaction in the TCA cycle. Therefore measuring citrate synthase activity provides a measure of relative overall oxidative capacity.

(b) β-hydroxyacyl CoA dehydrogenase indicates relative capacity for fatty acid oxidation.

(c) Lactate dehydrogenase converts lactate into pyruvate, the first stage of a set of reactions that allows lactate to be metabolized.

Important comparisons can be drawn from the data in Table 6.8. Both citrate synthase and β-hydroxyacyl CoA dehydrogenase (HOAD) activities correlate positively with altitude in heart and diaphragm tissues. The level of cardiac citrate synthase (CS) activity in *O. princeps*, the high-altitude species, was significantly higher than that in the low-altitude species, *O. collaris* and the sea-level species, *O. hyperborea*. There is no significant difference in cardiac CS activity in rabbit and *O. hyperborea*. *Ochotona princeps* showed greater CS activity in diaphragm muscle than did *O. collaris*, both species having higher CS activity than rabbit. Cardiac HOAD activity correlated positively to altitude in the three species of pika – HOAD activity in cardiac and diaphragm muscle of *O. princeps* was significantly higher than that of *O. collaris*. Skeletal muscle CS activity was not significantly different in low and high altitude pikas, but the rabbit values were lower than those for all pikas. HOAD activity in skeletal muscle showed little correlation with altitude.

We can conclude from Sheafor's data that oxidative metabolic rate is increased in both heart and diaphragm tissues, but not in skeletal muscle.

- Why is it important for oxidative metabolic rate to be scaled up in heart and diaphragm tissues? What does the increased activity of CS in cardiac tissue from the high-altitude pika tell us?

Heart and diaphragm muscle are continuously active and require a constant supply of ATP. It is therefore important that the respiring cells can utilize all of the available oxygen. Increased activity of the enzymes involved in aerobic glucose breakdown indicates that the tissue has a high capacity for aerobic glucose breakdown.

As citrate synthase is located inside the mitochondria, an increase in CS activity generally indicates an increase in mitochondrial density and/or mitochondrial size.

- Why would a greater density of mitochondria in cells be advantageous for a pika at high altitude?

The greater the mitochondrial density, the greater the chance that an oxygen molecule meets a cytochrome *c* oxidase molecule, the oxidative enzyme site for the electron transport chain. Therefore increased mitochondrial density increases the ability of the cell to utilize the available oxygen.

The increase in CS activity and the inferred increase in mitochondrial density in cardiac and diaphragm muscle enables high-altitude pikas to maintain aerobic metabolism. Activity of β-hydroxyl CoA dehydrogenase provides a measure of oxidative capacity derived from fatty acid oxidation. Data in Table 6.8 show that HOAD activity is low in pika compared to the rabbit. Low HOAD activity in pikas may relate to their diet of alpine and tundra plants, with which has a very low fat content. Nevertheless, HOAD activity does correlate with altitude in heart and diaphragm muscle in pika species. The highest HOAD activity in cardiac and diaphragm tissue is seen in *O. princeps*, and values are significantly higher than those measured in *O. collaris*, and *O. hyperborea*. Sheafor interprets the data as a reflection of the adaptive upward scaling of the activity of oxidative enzymes in tissues that rely heavily on oxidative catabolism in animals at altitude.

For the most part, increases in anaerobic capacity have not been demonstrated in mammals acclimated to hypoxic conditions. This contrast has been called the lactate paradox.

- Why would it be disadvantageous for a mammal acclimatized to high altitude to rely on anaerobic ATP synthesis?

Build-up of lactate, especially in hypoxic conditions, where oxygen is in short supply, would affect cell function adversely because the product of anaerobic metabolism is lactate. Increased concentrations of lactate cause a decrease in cell function through decreased pH and inhibition of glycolysis.

Summary of Section 6.5

Conclusions from comparisons of adaptations of two or more distantly related species are difficult to interpret because it is impossible to separate adaptations linked to the factor of interest, in this case hypoxia, from other selective pressures acting on unrelated species. Sheafor's study of enzyme activities in muscle of three pika species, each living at a different altitude, provides more useful comparisons. The species studied were *Ochotona hyperborea*, *O. collaris* and *O. princeps*. Sheafor used rabbit data as an outgroup control. Sheafor's data show that citrate synthase (CS) activity is increased in both heart and diaphragm tissues in high-altitude species in comparison to *O. hyperborea* and rabbit, but not in skeletal muscle. Heart and diaphragm muscle are continuously active and require ATP constantly. As CS is located in mitochondria, Sheafor suggests that increased CS may indicate an increased mitochondrial density in heart and diaphragm muscle in the high altitude species, which would increase the capacity of cells to utilize available oxygen. Levels of heart β-hydroxyacyl CoA dehydrogenase (HOAD) activity in the *Ochotona* species correlate positively with increasing altitude; HOAD activity in *O. princeps* was similar to that of rabbits. The activity of oxidative enzymes in locomotory muscle showed no consistent pattern. Lactate dehydrogenase activity was higher in locomotory muscles, diaphragm and heart of *O. princeps* than in *O. collaris* and and also higher than in heart muscle of *O. hyperborea*.

Summary of Chapter 6

Our brief survey of the environmental physiology of animals and humans at high altitude has taken us from the anatomical level of structure right down to the molecular mechanisms of responses to hypoxia. We have learned about the difficulty of separating true genetic adaptations from genotypic plasticity and flexibility. Yet all three mechanisms of 'adaptation' contribute to survival of animals and humans at high altitude. Most people moving from sea-level to high altitude can adapt physiologically. This is phenotypic flexibility or acclimatization. There is limited evidence that Tibetan and Andean human populations have true genetic adaptations, including anatomy and physiology, for life at high altitude. Estimates of heritability support the view of genetic adaptations in populations living in the Andes and in Tibet. Compelling evidence that such adaptations cannot be explained by phenotypic plasticity includes the observations on the adult offspring born to native Tibetans and people from the high Andes who have moved to sea-level. These second-generation adult Tibetan people living in Nepal had *never* visited or lived at high altitude. There is more available evidence that animal species living at high altitude have genetic adaptations that enable the animals to cope with low P_{O_2}. Pika and yak have thin-walled pulmonary small arteries and thereby avoid the problem of pulmonary hypertension and HAPE. Adaptations in the molecular structure of haemoglobin in high-altitude species such as the bar-headed goose, provide increased affinity of the pigment for oxygen, maintaining adequate provision of oxygen to respiring tissues.

Studies on the molecular responses to hypoxia in cells have provided insights into observed biochemical and physiological effects of hypoxia. We should not be surprised that a transcription factor, hypoxia-inducible factor (HIF-1α), is a central player in cellular responses to hypoxia. Transcription factors initiate the transcription of genes, and intuitively there should be a molecule that is activated in response to hypoxia and initiates a complex sequence of appropriate regulatory responses. For example HIF-1α switches on the gene coding for VEGF, a polypeptide cytokine that stimulates increased growth and branching of capillaries. Unfortunately, the first stage in the process is increased permeability of capillary endothelial cells, which can lead to HACE in vulnerable individuals exposed to high altitude.

Genetic adaptations in animal species may be supported or refuted by comparisons across related species. 'Genetic adaptations' in animals may in fact derive from phenotypic plasticity and/or flexibility, both of which are prominent in most animal species. Sheafor's work, a comparative study of enzyme activities in *Ochotona princeps* – a pika living at altitude 3350 m – and low-altitude and sea-level species demonstrated increased levels of enzymes associated with both aerobic *and* anaerobic glucose breakdown in cardiac muscle of the high-altitude species.

Learning Outcomes for Chapter 6

When you have completed this chapter you should be able to:

6.1 Define and use, or recognize definitions and applications of, each of the **bold** terms.

6.2 Describe the variety of altitude environments, and describe some examples of vertebrate species that live at high altitude.

6.3 Explain why mammalian respiration is more sensitive to changes in P_{CO_2} than in P_{O_2}, and describe the functioning of the central and peripheral chemoreceptors.

6.4 Explain why acute exposure to high altitude is more dangerous for humans than a gradual ascent, and describe the various physiological and biochemical changes associated with acclimatization to low P_{O_2}.

6.5 Describe the aetiology of HAPE and HACE, and explain the link to increased permeability of endothelial cells in brain and lung capillary networks.

6.6 Interpret data and tissue histology, and derive information about phenotypic and genotypic adaptations of life at high altitudes from your interpretation.

6.7 Explain the crucial role of variations in molecular structure of haemoglobin in genotypic adaptation for life at high altitudes, and interpret oxygen dissociation curves for different haemoglobins.

6.8 Explain the central role of hypoxia-inducible factor, HIF-1α, in the response of cells to hypoxia, and interpret data that demonstrates the molecular response to hypoxia, in particular those relating to induction of VEGF.

6.9 Interpret data from related species living at a range of altitudes to draw conclusions about the adaptive significance of physiological features.

Questions for Chapter 6

(Answers to questions are at the end of the book.)

Question 6.1 (LOs 6.1 and 6.9)

Describe the data for lactate dehydrogenase activity in three species of *Ochotona* (Table 6.8) and how this relates to altitude.

Question 6.2 (LOs 6.1, 6.4 and 6.6)

Each of (a) to (f) gives 2 measurements for lowlanders; one who recently moved from sea-level to live at high altitude, and one who has remained at sea-level. Both individuals lead normal sedentary lives. For each option, select the measurement that is most likely to refer to the individual living at high altitude. Give a brief explanation of your answer.

(a) The haemoglobin concentration is (i) 16 g 100 cm^{-3} blood, or (ii) 20 g 100 cm^{-3} blood.

(b) The number of red blood cells is (i) 45 × 10^{12} l^{-1} blood, or (ii) 18 × 10^{12} l^{-1} blood.

(c) The difference in the P_{O_2} between arteries and veins is (i) 7.33 kPa, or (ii) 1.86 kPa.

(d) The level of 2,3-DPG is (i) 15 µmol g^{-1} haemoglobin, or (ii) 25 µmol g^{-1} haemoglobin.

(e) The P_{50} in the blood is (i) 3.56 kPa, or (ii) 3.97 kPa.

(f) The volume density of mitochondria in muscle tissue is (i) 3.94%, or (ii) 5.12%.

Question 6.3 (LOs 6.1, 6.4 and 6.5)

(a) A healthy 38-year-old woman who normally lives at sea-level was transported rapidly to an altitude of 3500 m. She was keen to begin hiking in the alpine meadows as soon as possible after her ascent, and went out just 1 hour after arrival, climbing up to 4000 m altitude. However, she found that exercise produced rapid fatigue. She also developed headache and nausea. Outline the physiological, biochemical and molecular processes that account for the signs observed in the patient.

(b) The following day, the woman went hiking again having taken pills for the headache, but by the second evening, she was short of breath and bubbling sounds could be heard as she struggled for breath. Suggest a possible diagnosis.

Question 6.4 (LOs 6.1, 6.3 and 6.4)

Which of the biochemical and physiological responses of humans to hypoxia at high altitude:

(a) is associated with the development of respiratory alkalosis?

(b) results in elevation of P_{O_2} (alveolar P_{O_2}) relative to inspired P_{O_2}.

(c) is the result of increased sensitivity of central chemoreceptors to carbon dioxide.

(d) results in improved unloading of oxygen at the tissues but an impaired uptake of oxygen at the lung.

(e) is responsible for an improvement in the rate of delivery of oxygen from the blood to the mitochondria.

(f) results in an increased capacity of the animal to metabolize aerobically.

Question 6.5 (LOs 6.7 and 6.9)

Figure 6.23a shows oxygen dissociation curves for *Telmatobius peruvianus*, an aquatic frog that lives at 3800 m altitude in the Andes. Figure 6.23b shows oxygen dissociation curves for *Xenopus*, a lowland aquatic frog. Both pairs of curves include one curve for haemoglobin alone, and one curve for haemoglobin in the presence of added Cl^- ions. The rationale for testing the effect of Cl^- was that in most amphibians, Cl^- functions as a modulator of oxygen affinity of haemoglobin.

(a) Recall from Section 6.2.3 that the P_{50} of haemoglobin is the P_{O_2} at which the pigment is 50% saturated with oxygen. Calculate P_{50} for haemoglobin of *Telmatobius* and *Xenopus* in (i) the absence of Cl^- and (ii) the presence of Cl^-, and present your results as a table. (Make a direct comparison with the oxygen dissociation curve in Figure 6.11b, by converting values for P_{50} in mmHg, into kPa, by multiplying each by 0.1332.)

(b) Describe the effects of added chloride ions, Cl^- on the oxygen dissociation curve for each species in terms of affinity of haemoglobin for oxygen, using the calculated P_{50} values as reference points.

(c) Explain the adaptive advantages for *Telmatobius*, the high-altitude species, of the response of its haemoglobin to Cl^- as demonstrated in Figure 6.23a.

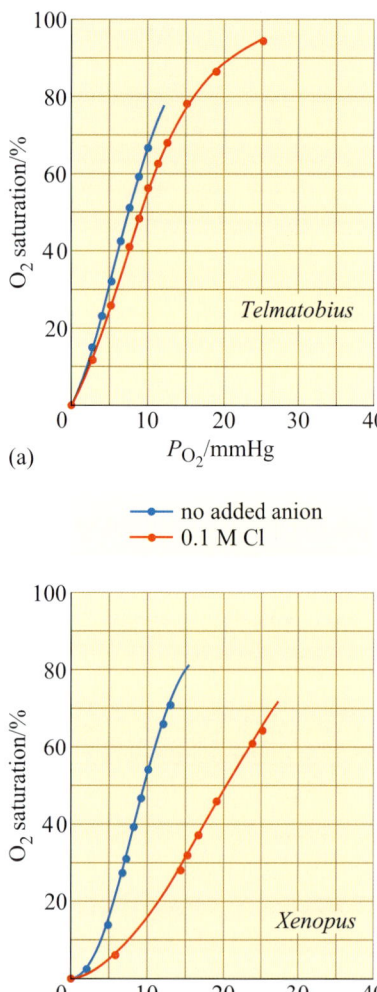

Figure 6.23 Oxygen dissociation curves for haemoglobin of (a) *Telmatobius peruvianus* and (b) *Xenopus laevis* at pH 7.0 and temperature 20 °C in the absence of added Cl^- (blue line) and the presence of added Cl^- (red line).

REFERENCES AND FURTHER READING

Chapter 1

Block, B. A., Dewar, H., Blackwell, S. B., Williams, T. D., Prince, E. D., Farwell, C. J., Boustany, A., Teo, S. L., Seitz, A., Walli, A. and Fudge, D. (2001) Migratory movements, depth preferences, and thermal biology of Atlantic bluefin tuna. *Science*, **293**, 1310–1314.

Block, B. A., Finnerty, J. R., Stewart, A. F. and Kidd, J. (1993) Evolution of endothermy in fish: mapping physiological traits on a molecular phylogeny. *Science*, **260**, 210–214.

Buck, M. J., Squire, T. L. and Andrews, M. T. (2002) Coordinate expression of the PDK4 gene: a means of regulating fuel selection in a hibernating mammal. *Physiological Genomics*, **8**, 5–13.

Carey, F. G. (1990) Further observations on the biology of the swordfish. In R. H. Stroud (ed.), *Planning the Future of Billfishes*, pp. 103–122. Savannah, Georgia: National Coalition for Marine Conservation.

Clarke, A. and Johnston, N. M. (1999) Scaling of metabolic rate with body mass and temperature in teleost fish. *Journal of Animal Ecology*, **68**, 893–905.

Fields, P. A. and Somero, G. N. (1998) Hot spots in cold adaptation: localized increases in conformational flexibility in lactate dehydrogenase A4 orthologs of Antarctic notothenioid fishes. *Proceedings of the National Academy of Sciences*, **95**, 11 476–11 481.

Gorman, M. L., Mills, M. G., Raath, J. P. and Speakman, J. R. (1998) High hunting costs make African wild dogs vulnerable to kleptoparasitism by hyenas. *Nature*, **391**, 479–480.

Heldmaier, G., Klingenspor, M., Werneyer, M., Lampi, B. J., Brookes, S. P. and Storey, K. B. (1999) Metabolic adjustments during daily torpor in the Djungarian hamster. *American Journal of Physiology – Endocrinology and Metabolism*, **276**, E896–906.

Li, Q., Sun, R., Huang, C., Wang, Z., Liu, X., Hou, J., Liu, J., Cai, L., Li, N., Zhang, S. and Wang, Y. (2001) Cold adaptive thermogenesis in small mammals from different geographical zones of China. *Comparative Biochemistry and Physiology Part A: Molecular and Integrative Physiology*, **129**, 949–961.

O'Brien, J. and Block, B. A. (1996) Effect of Ca^{2+} on oxidative phosphorylation in mitochondria from the thermogenic organ of marlin. *Journal of Experimental Biology*, **199**, 2679–2687.

Sidell, B. D. (2000) Life at body temperatures below 0 degrees C: the physiology and biochemistry of Antarctic fishes. *Gravitational and Space Biology Bulletin*, **13**, 25–34.

Somero, G. N. (1995) Proteins and temperature. *Annual Review of Physiology*, **57**, 43–68.

Chapter 2
References

Barros, R. C. H., Zimmer, M. E., Branco, G. S. L. and Milsom, W. K. (2001) Hypoxic metabolic response of the golden-mantled ground squirrel. *Journal of Applied Physiology*, **91**, 603–612.

Buck, C. M. and Barnes, B. M. (2000) Effects of ambient temperature on metabolic rate, respiratory quotient, and torpor in an arctic hibernator. *American Journal of Physiology – Regulatory, Integrative and Comparative Physiology*, **279**, R255–R262.

Butler, P. J., Woakes, A. J., Smale, K., Hillidge, C. J. and Martin, D. J. (1993) Respiratory and cardiovascular adjustments during exercise of increasing intensity and during recovery in thoroughbred racehorses. *Journal of Experimental Biology*, **179**, 159–180.

Covell, D. F., Miller, D. S. and Karosov, W. H. (1996) Cost of locomotion and daily energy expenditure by free-living swift foxes (*Vulpes velox*): a seasonal comparison. *Canadian Journal of Zoology*, **74**, 283–290.

Geffen, E., Degen, M., Kam, M., Hefner, R. and Nagy, K. A. (1992a) Daily expenditure and water flux of free-living Blanford's foxes (*Vulpes cana*), a small desert carnivore. *Journal of Animal Ecology*, **61**, 611–617.

Geffen, E., Hefner, R., Macdonald, D. W. and Ucko, M. (1992b) Diet and foraging behaviour of the Blanford's fox, *Vulpes cana*, in Israel. *Journal of Mammology*, **73**, 395–402.

Girard, I. (1998) The physiological ecology of a small canid, the kit fox (*Vulpes macrotis*) in the Mojave desert. Unpublished PhD dissertation, University of California, Los Angeles.

Marquet, P. A., Ortiz, J. C., Bozinovic, F. and Jaksic, F. M. (1989) Ecological aspects of thermoregulation at high altitudes: the case of Andean *Liolaemus* lizards in northern Chile. *Oecologia*, **81**, 16–20.

Nagy, K. A., Girard, I. A. and Brown, T. K. (1999) Energetics of free-ranging mammals, reptiles and birds. *Annual Review of Nutrition*, **19**, 247–277.

Southwood, A. L., Andrews, R. D., Lutcavage, M. E., Paladino, F. V., George, R. H. and Jones, D. R. (1999) Heart rates and diving behaviour of leatherback turtles in the Eastern Pacific Ocean. *Journal of Experimental Biology*, **202**, 1115–1125.

St-Pierre J., Charest, P. and Guderley, H. (1998) Relative contribution of quantitative and qualitative changes in mitochondria to metabolic compensation during seasonal acclimatisation of rainbow trout *Oncorhynchus mykiss*. *Journal of Experimental Biology*, **201**, 2961–2970.

Stebbins, R. C. and Barwick, R. E. (1968) Radiotelemetric study of thermoregulation in a lace monitor. *Copeia*, **3,** 541–547.

Tattersall, G. J. and Milsom, W. K. (2003) Transient peripheral warming accompanies the hypoxic metabolic response in the golden-mantled ground squirrel. *Journal of Experimental Biology*, **206**, 33–42.

Tieleman, B. I., Williams, J. B. and Bloomer, P. (2003) Adaptation of metabolism and evaporative water loss along an aridity gradient. *Proceedings of the Royal Society, London, Series B*, **270**, 207–214.

Williams, J. B., Ostrowski, S., Bedin, E. and Ismail, K. (2001) Seasonal variation in energy expenditure water flux and food consumption of Arabian oryx, *Oryx leucoryx*. *Journal of Experimental Biology*, **204**, 2301–2320.

Williams, J. B., Lenain, D., Ostrowski, S., Tieleman, B. I. and Seddon, P. J. (2002) Energy expenditure and water flux of Rüppell's foxes in Saudi Arabia. *Physiological and Biochemical Zoology*, **75**, 479–488.

Wilson, R. P. and Gremillet, D. (1996) Body temperatures of free-living African penguins (*Spheniscus demersus*) and bank cormorants (*Phalacrocorax neglectus*). *Journal of Experimental Biology*, **201**, 2215–2223.

Further reading

Cooper, J. (1986) Diving patterns of cormorants Phalacrocoracidae. *Ibis*, **128**, 562–570.

Willmer, P., Stone, G. and Johnstone, I. (2000) *Environmental Physiology of Animals*. Oxford: Blackwell Science.

Chapter 3
References

Bozinovic, F., Gallardo, P. A., Visser, G. H., Ortes, A. (2003) Seasonal acclimatization in water flux rate, urine osmolality and kidney water channels in free-living degus: molecular mechanisms, physiological processes and ecological implications. *Journal of Experimental Biology*, **206**, 2959–2966.

Bulova, S. (2002) How temperature, humidity, and burrow selection affect evaporative water loss in desert tortoises. *Journal of Thermal Biology*, **27**, 175–189.

Meigs, P. (1953) World distribution of arid and semi-arid homoclimates. *Reviews of Research on Arid Zone Hydrology*. Paris: UNESCO.

Mueller, P. and Diamond, J. (2001) Metabolic rate and environmental productivity: well provisioned animals evolved to run and idle fast. *Proceedings of the National Academy of Sciences*, **98**, 12 550–12 554.

Schmidt-Nielsen, K., Crawford, E. C., Newsome, A. E., Rawson, K. S. and Hammel, H. T. (1967) *American Journal of Physiology*, **212**, 341–346.

Tieleman, B. I. and Williams, J. B. (2000) The adjustment of avian metabolic rates and water fluxes to desert environments. *Physiological and Biochemical Zoology*, **73**, 461–479.

Tieleman, B. I., Williams, J. B. and Bloomer, P. (2003) Adaptation of metabolism and evaporative water loss along an aridity gradient. *Proceedings of the Royal Society, London, Series B*, **270**, 207–214.

UNEP (1992) *World Atlas of Desertification*. Sevenoaks: Edward Arnold.

Williams, J. B. (2001) Energy expenditure and water flux of free-living dune larks in the Namib: a test of the re-allocation hypothesis on a desert bird. *Functional Ecology*, **15**, 175–185.

Williams, J. B. and Tieleman, B. I. (2000) Flexibility in basal metabolic rate and evaporative water loss among hoopoe larks exposed to different environmental temperatures. *Journal of Experimental Biology*, **203**, 3153–3159.

Williams, J. B., Ostrowski, S., Bedlin, E. and Ismail, K. (2001) Seasonal variation in energy expenditure, water flux and food consumption of Arabian oryx, *Oryx leucoryx*. *Journal of Experimental Biology*, **204**, 2301–2320.

Williams, J. B., Lenain, D., Ostrowski, S., Tieleman, B. I. and Seddon, P. J. (2002) Energy expenditure and water flux of Rüppell's foxes in Saudi Arabia. *Physiological and Biochemical Zoology*, **75**, 479–488.

Willmer, P., Stone, G. and Johnston, I. (2000) *Environmental Physiology of Animals*. Oxford: Blackwell Science Ltd.

Zatsepina, O. G., Ulmasov, K. A., Beresten, S. F., Molodtsov, V. B., Rybtsov, S. A. and Evgen'ev, M. B. (2000) Thermotolerant desert lizards characteristically differ in terms of heat-shock system regulation. *Journal of Experimental Biology*, **203**, 1017–1025.

Further reading

Crane, C. M. and Kishore, B. K. (2003) Aquaporins: the membrane water channels of the biological world. *Biologist*, **50**, 81–86.

Huang, Y., Tracey, R., Walsberg, G. E., Makkinje, A., Fang, P., Brown, D. and van Hoek, A. N. (2001) Absence of aquaporin-4 water channels from kidneys of the desert rodent *Dipodomys merriami merriami*. *American Journal of Physiology – Renal Physiology*, **280**, 794–802.

Louw, G. (1993) *Physiological Animal Ecology*. Harlow: Longman Ltd.

van Hoek, A. N., Tonghui, M., Baoxue Y., Verkman, A. S. and Brown, D. (2000) Aquaporin-4 is expressed in basolateral membranes of proximal tubule S3 segments in mouse kidney. *American Journal of Physiology – Renal Physiology*, **278**, 310–316.

Chapter 4
References

Azzam, N. A., Hallenbeck, J. M. and Kachar, B. (2000) Membrane changes during hibernation. *Nature*, **407**, 317–318.

Boutilier, R. G. and St-Pierre, J. (2002) Adaptive plasticity of skeletal muscle energetics in hibernating frogs: mitochondrial proton leak during metabolic depression. *Journal of Experimental Biology*, **205**, 2287–2296.

Buck, C. L. and Barnes, B. M. (2000) Effects of ambient temperature on metabolic rate, respiratory quotient, and torpor in an arctic hibernator. *American Journal of Physiology – Regulatory, Integrative and Comparative Physiology*, **279**, R255–R262.

Florant, G. L. and Heller, H. C. (1977) CNS regulation of body temperature in euthermic and hibernating marmots (*Marmota flaviventris*). *American Journal of Physiology – Regulatory, Integrative and Comparative Physiology*, **232**, R203–R208.

Frerichs, K. U. (1998) Suppression of protein synthesis in brain during hibernation involves inhibition of protein initiation and elongation. *Proceedings of the National Academy of Sciences USA*, **95**, 14 511–14 516.

Hashimoto, M., Gao, B., Kikuchi-Utsumi, K., Ohinata, H. and Osborne, P. G. (2002) Arousal from hibernation and BAT thermogenesis against cold: central mechanism and molecular basis. *Journal of Thermal Biology*, **27**, 503–515.

Hayward, J. and Lyman, C. P. (1967) Nonshivering heat production during arousal from hibernation and evidence for the contribution of brown fat. In K. C. Fisher, A. R. Dawe, C. P. Lyman, E. Schonbaum and F. E. South Jr. (eds), *Mammalian Hibernation III*, pp. 346–355. New York: Elsevier.

Heller, H. C., Colliver, G. W. and Beard, J. (1977) Thermoregulation during entrance to hibernation. *Pflügers Archive*, **369**, 55–59.

Lyman, C. P. (1948) The oxygen consumption and temperature regulation of hibernating hamsters, *Journal of Experimental Zoology*, **109**, 55–78.

Lyman, C. P. and O'Brien, R. C. (1960) Circulatory changes in the thirteen-lined ground squirrel during the hibernating cycle. *Bulletin of the Museum of Comparative Zoology*, **124**, 353–372.

Körtner, G. and Geiser, F. (2000) The temporal organization of daily torpor and hibernation: Circadian and circannual rhythms. *Chronobiology International*, **17**, 103–128.

Koteja, P., Jurczyszyn, M. and Woloszyn, B. W. (2001) Energy balance of hibernating mouse-eared bat *Myotis myotis*: a study with a TOBEC instrument. *Acta Theriologica*, **46,** 1–12.

Malatesta, M., Zancanaro, C., Baldelli, B. and Gazzanelli, G. (2002) Quantitative ultrastructural changes of hepatocytes constituents in euthermic, hibernating and arousing dormice (*Muscardinus avellanarius*). *Tissue and Cell*, **34**, 397–405.

Mayer, W. V. (1960) Histological changes during the hibernating cycle in the artic ground squirrel. *Bulletin of the Museum of Comparative Zoology*, **124**, 131–154.

Mussachia, X. J. and Volkert, W. A. (1971) Blood gases in hibernating and active ground squirrels: HbO_2 affinity at 6 and 38 °C. *American Journal of Physiology*, **221**, 128–130.

Ortmann, S. and Heldmaier, G. (2000) Regulation of body temperature and energy requirements of hibernating Alpine marmots (*Marmota marmota*). *American Journal of Physiology – Regulatory, Integrative and Comparative Physiology*, **278**, R698–R704.

Ouezzani, S. E., Tramu, G. and Rabia, M. (1999) Neuronal activity in the mediobasal hypothalamus of hibernating jerboas (*Jaculus orientalis*). *Neuroscience Letters*, **260**, 13–16.

Panula, P., Karlstedt, K., Sallmen, T., Peitsaro, N., Kaslin, J., Michelsen, K. A., Anichtchik, O., Kukko-Lukjanov, T. and Lintunen, M. (2000) The histaminergic system in the brain: structural characteristics and changes in hibernation. *Journal of Neuroanatomy*, **8**, 65–74.

Pitrosky, B., Delagrange, P., Rettori, M. C. and Pevet, P. (2003) S22153, a melatonin antagonist, dissociates different aspects of photoperiodic responses in Syrian hamsters. *Behavioural Brain Research*, **138**, 145–152.

Popov, V. I., Bocharova, L. S. and Bragin, A. G. (1992) Repeated changes of dendritic morphology in the hippocampus of ground squirrels in the course of hibernation. *Neuroscience*, **48**, 45–51.

Sallmen, T., Lozada, A. F., Beckman, A. L. and Panula, P. (2003) Intrahippocampal histamine delays arousal from hibernation. *Brain Research*, **966**, 317–320.

Schleucher, E. (2001) Heterothermia in pigeons and doves reduces energetic costs. *Journal of Thermal Biology*, **26**, 287–293.

Strumwasser, F. (1960) Some physiological principles governing hibernation in *Citellus beecheyi*. *Bulletin of the Museum of Comparative Zoology*, **124**, 285–320.

Wang, L. (1987) Mammalian hibernation. In B. W. W. Groot and G. J. Morris (eds), *Effects of Low Temperatures on Biological Systems*. London: Edward Arnold.

Wang, L. C. H. and Hudson, J. W. (1971) Temperature regulation in normothermic and hibernating eastern chipmunk, *Tamias striatus*. *Comparative Biochemistry and Physiology (Part A)*, **38**, 59–90.

Yu, E. Z., Hallenbeck, J. M., Cai, D. and McCarron, R. M. (2002) Elevated arylalkylamine-*N*-acetyltransferase (*AA-NAT*) gene expression in medial habenular and suprachiasmatic nuclei of hibernating ground squirrels. *Molecular Brain Research*, **102**, 9–17.

Zimmer, M. B. and Milson, R. (2002) Ventilatory pattern and chemosensitivity in unanesthetized, hypothermic ground squirrels (*Spermophilus lateralis*). *Respiratory Physiology and Neurology*, **133**, 49–63.

Further reading

Boyer, B. B. and Barnes, B. M. (1999) Molecular and metabolic aspects of mammalian hibernation – expression of the hibernation phenotype results from the coordinated regulation of multiple physiological and molecular events during preparation for and entry into torpor. *Bioscience*, **49**, 713–724.

Van Breukelen, F. and Martin, S. L. (2002) Molecular Biology of Thermoregulation. Invited review. Molecular adaptations in mammalian hibernators: unique adaptations or generalised responses? *Journal of Applied Physiology*, **92**, 2640–2647.

Chapter 5
References

Donahue, S. W., Vaughan, M. R., Demers, L. M. and Donahue, H. J. (2003) Serum markers of bone metabolism show bone loss in hibernating bears. *Clinical Orthopaedics and Related Research*, **408**, 295–301.

Groscolas, R. (1982) Changes in plasma lipids during breeding, molting and starvation in male and female emperor penguins (*Aptenodytes fosteri*). *Physiological Zoology*, **55**, 45–55.

Groscolas, R. (1986) Changes in body mass, body temperature and plasma fuel levels during the natural breeding fast in male and female emperor penguins *Aptenodytes forsteri*. *Journal of Comparative Physiology*, **156B**, 521–527.

Johnston, I. A. (1989) Antarctic fish muscles – structure, function and physiology. *Antarctic Science*, **1**, 97–108.

Larsen, T. S., Nilsson, N. Ö. and Blix, A. S. (1985) Seasonal changes in lipogenesis and lipolysis in isolated adipocytes from Svalbard and Norwegian reindeer. *Acta Physiologica Scandinavica*, **123**, 97–104.

Lindgård, K. and Stokkan, K.-A. (1989) Daylength control of food intake and body weight in Svalbard ptarmigan *Lagopus mutus hyperboreus*. *Ornis Scandinavica*, **20**, 176–180.

Nelson, R. A., Beck, T. D. I. and Steiger, D. L. (1984) Ratio of serum urea to serum creatinine in wild black bears. *Science*, **226**, 841–842.

Pond, C. M. and Mattacks, C. A. (1985) Body mass and natural diet as determinants of the number and volume of adipocytes in eutherian mammals. *Journal of Morphology*, **185**, 183–193.

Pond, C. M. and Ramsay, M. A. (1992) Allometry of the distribution of adipose tissue in Carnivora. *Canadian Journal of Zoology*, **70**, 342–347.

Ramsay, M. A., Nelson, R. A. and Stirling, I. (1991) Seasonal changes in the ratio of serum urea to creatine in feeding and fasting polar bears. *Canadian Journal of Zoology*, **69**, 298–302.

Schekkerman, H., Tulp, I., Piersma, T. and Visser, G. H. (2003) Mechanisms promoting higher growth rate in arctic than in temperate shorebirds. *Oecologia*, **134**, 332–342.

Scholander, P. F., Van Dam, L., Kanwisher, J. W., Hammel, H. T. and Gordon, M. S. (1957) Supercooling and osmoregulation in arctic fish. *Journal of Cellular and Comparative Physiology*, **49**, 5–24.

Stokkan, K.-A., Sharp, P. J. and Unander, S. (1986) The annual breeding cycle of the high-arctic Svalbard ptarmigan (*Lagopus mutus hyperboreus*). *General and Comparative Endocrinology*, **61**, 446–451.

Chapter 6
References

Beall, C. M., Strohl, K. P., Blangero, J., Williams-Blangero, S., Almasy, L. A., Decker, M. J., Worthman, C. M., Goldstein, M. C., Vargas, E., Villena, M., Soria, R., Alarcon, A. M. and Gonzales, C. (1997) Ventilation and hypoxic ventilatory response of Tibetan and Aymara high altitude natives. *American Journal of Physical Anthropology*, **104**, 427–447.

Black, C. P. and Tenney, S. M. (1980) Oxygen transport during progressive hypoxia in high altitude and sea-level waterfowl. *Respiratory Physiology*, **39**, 217–239.

Ge, Ri-Li, Kubo, K., Kobayashi, T., Sekiguchi, M. and Honda, T. (1998) Blunted hypoxic pulmonary vasoconstrictive response in the rodent *Ochotona curzoniae* (pika) at high altitude. *American Journal of Physiology – Heart and Circulatory Physiology*, **274**, H1792–H1799.

Groves, B. M., Droma T., Sutton, J. R., McCullough, R. G., McCullough, R. E., Zhuang, J., Rapmund, S., Sun, C., Janes, C. and Moore, L. G. (1993) Minimal hypoxic pulmonary hypertension in normal Tibetans at 3658 m. *Journal of Applied Physiology*, **74**, 312–318.

Hoppeler, H., and Vogt, M. (2001) Muscle tissue adaptations to hypoxia. *Journal of Experimental Biology*, **204**, 3133–3139.

Houston, C. (1998) *Going Higher: Oxygen, Man and Mountains*. Seattle, USA: Mountaineer Books.

Kayser, B., Hoppeler, H., Desplanches, D., Marconi, C., Broers, B. and Cerretelli, P. (1996) Muscle ultratructure and biochemistry of lowland Tibetans. *Journal of Applied Physiology*, **81**, 419–425.

Sheafor, B. A. (2003) Metabolic enzyme activities across an altitudinal gradient: an examination of pikas (genus *Ochotona*). *Journal Experimental Biology*, **206**, 1241–1249.

Vogt, M., Puntschart, A., Geiser, J., Zuleger, C., Billeter, R. and Hoppeler, H. (2001) Molecular adaptations in human skeletal muscle to endurance training under hypoxic conditions. *Journal of Applied Physiology*, **91**, 173–182.

Weber, R. E., Jessen, T., Malte, H. and Tame, J. (1993) Mutant haemoglobins $\alpha\text{-}^{119}$-Ala and $\beta\text{-}^{55}$-Ser: functions related to high-altitude respiration in geese. *Journal of Applied Physiology*, **75**, 2646–2655.

Wenger, R. H. (2000) Mammalian oxygen sensing, signalling and gene regulation. *Journal of Experimental Biology*, **203**, 1253–1263.

Further reading

Fischer, S., Causs, M., Wiesnet, M., Renz, D., Schaper, W. and Karliczek (1999) Hypoxia induces permeability in brain microvessel endothelial cells via VEGF and NO. *American Journal of Physiology – Cell Physiology*, **276**, C812–C820.

Heminger, G., Endo, M., Ferrara, N., Hlatky, L. and Jain, R. K. (2000) Growth factors: formation of endothelial cell networks. *Nature*, **405**, 139–141.

Rupert, J. L. and Hochachka, P. W. (2001) Genetic approaches to understanding human adaptation to altitude in the Andes. *Journal of Experimental Biology*, **204**, 3151–3160.

Willmer, P., Stone, G. and Johnston, I. (2000) *Environmental Physiology of Animals*. Oxford: Blackwell Science Ltd.

ANSWERS TO QUESTIONS

Question 1.1

The first design consideration is the mass that the seal can carry without affecting its behaviour and the position in which it is placed so that it both collects the data that you want and is not removed by the animal. Another factor to consider is where to attach the device to limit the effects on streamlining. Since you would not expect to be able to retrieve an implanted tag in the normal course of events, an external tag would be required, and a pop-up type would probably be the best option. You would need to set the period over which you wanted to collect data and pre-programme the pop-up time. You would also have to bear in mind that as seals haul themselves out onto land at certain seasons, the tag might be released onto land instead of into water, which might affect the transmission of signals.

Question 1.2

The doubly-labelled water technique uses two isotopes because the best measure of metabolism is the rate of carbon dioxide production. The rate of loss of carbon dioxide can be measured by monitoring the rate of disappearance of labelled oxygen from the blood, providing it is possible to separate out the loss of oxygen in water from the loss in carbon dioxide. By labelling hydrogen, the rate of loss of hydrogen in water from the blood can be monitored. The difference in loss of the two labels allows the carbon dioxide loss to be calculated and hence the rate of metabolism.

Question 1.3

Phylogenetic studies contribute to an understanding of physiology because they provide the evolutionary relationships between groups of organisms, in particular between those that show similar physiological adaptations. By knowing the evolutionary relationships, the significance of the adaptations becomes clearer. For example, the loss of haemoglobin and myoglobin in certain species of antarctic fish could be interpreted as a single event in the evolutionary line with associated physiological adaptations to cope with the loss. However, phylogenetic analysis shows that the loss of myoglobin probably occurred on four separate occasions in the evolutionary history of these fish, which argues for a strong selection pressure for myoglobin loss in species that live continuously in very cold water. That selection pressure undoubtedly has an interesting physiological basis and changes concepts developed for studying the fish. Instead of asking what physiological adaptations were necessary to cope with a loss of myoglobin one would ask what physiological *advantage* would accrue from the loss.

Question 1.4

There are many ways in which this diagram could be drawn. Figure 1.7 shows one example.

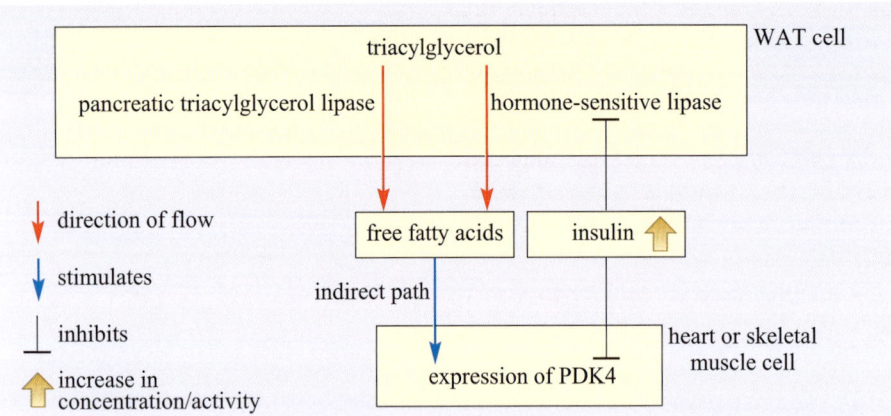

Figure 1.7 Diagram illustrating the possible mechanism by which *PDK4* gene expression is regulated in the 13-lined ground squirrel.

Question 1.5

Tuna are fast-moving carnivorous fish of economic importance. Pop-up tags and implanted tags would show the distances travelled by individual tuna and if their movements are seasonal, information that could be used to set fishing levels, for example. (In fact, such tags have shown that tuna migrate over much larger distances than was previously thought and that the population in US waters is linked to the population in the Mediterranean. So, over-fishing in Europe depletes stocks on the other side of the Atlantic.) Temperature information from the tags would show whether tuna were confined to ranges of particular water temperature, enabling conservation areas to be set up where conditions were optimal for the species and there was little or no fishing.

Question 2.1

(a) As the horses without the face mask progressed from rest to trotting, P_{CO_2} of arterial blood fell from about 6.1 kPa to 5.5 kPa. As running speeds increased, arterial P_{CO_2} rose almost linearly to about 6.4 kPa. From rest to trotting, the pattern of P_{CO_2} for arterial blood in horses wearing the face mask was similar to that for horses without the face mask. However, from trotting to maximum canter speed, arterial blood P_{CO_2} was consistently higher than when measured in horses without the face mask. From cantering speed of 6 m s^{-1} to 12 m s^{-1}, arterial blood P_{CO_2} increased from about 6.1 kPa to 7.1 kPa. Note that where standard error bars do not overlap it is likely that the difference between two data sets is statistically significant.

(b) Both sets of horses were exercising over the same range of intensity. However, the horses wearing the face masks had higher P_{CO_2} levels in arterial blood during cantering than did the horses without the face masks. The higher arterial P_{CO_2} in the horses wearing the masks suggests that the face mask was hindering air flow in the lungs, causing a reduction in the rate of loss of CO_2 from arterial blood into alveolar air. So measurements of arterial P_{CO_2} in horses subjected to high workloads, and wearing a face mask, would provide an overestimate of CO_2 production in respiring tissues. In contrast, the P_{CO_2} measured in exhaled air in the horses wearing face masks may be at a lower level than in the equivalent measurements made in horses not wearing face masks. (The latter may be exercising within a metabolic chamber.)

Question 2.2

(a) Mean values for field metabolic rate of male Rüppell's foxes expressed as kJ day^{-1} (mass$^{0.869}$)$^{-1}$ are higher than the equivalent value in female foxes. The mean FMR of male foxes in winter 1998 was 1306.5 kJ day^{-1}, in contrast to the mean FMR for females, 722.8 kJ day^{-1}.

(b) Figure 2.24 shows that the higher the percentage of night-time that foxes are active, the higher the measured metabolic rate. The males have a higher metabolic rate which suggests that they are active for a higher proportion of the night.

(In fact Williams et al.'s field observations of the foxes at night confirmed that in winter, males travel longer distances at night than do females. Males spend much time patrolling their territories and defending them against intruders at night during the winter. This interpretation of Figure 2.24 demonstrates the importance of field observations. The results of physiological measurements are only useful when considered in the context of the animal's normal daily activities and behaviour patterns.)

Question 2.3

(a) Figure 2.27 shows the graphs you should have drawn as your answer to Question 2.3a.

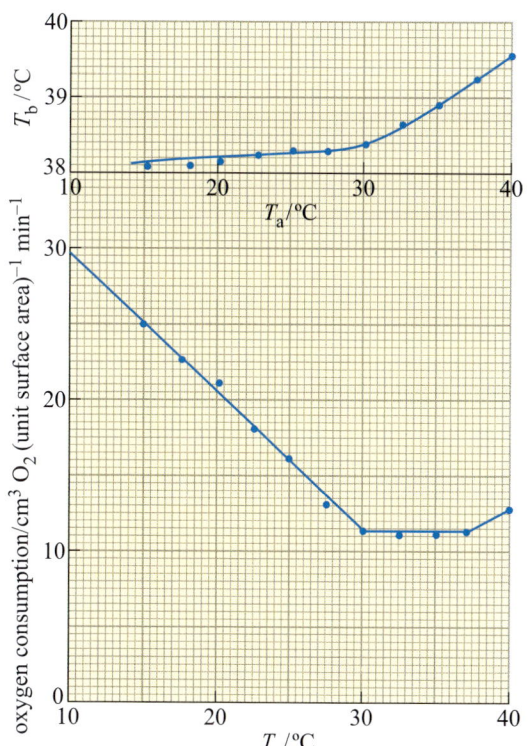

Figure 2.27

(b) (i) T_b remains almost constant for T_a 15 °C–37.5 °C, so the cat is maintaining T_b within a fairly narrow range. The data suggest that the cat is a homeotherm.

(ii) For T_a 10 °C–30 °C, the slope of the line is:

$$\frac{-19.0}{30-10} = -0.95$$

For T_a 30 °C–37.5 °C the slope of the line is zero, i.e. the line is flat. There is no change in metabolic rate within this range of T_a.

For T_a 37.5 °C–40 °C, the slope of the line is:

$$\frac{1.2}{40-37.5} = 0.48$$

(iii) For T_a 10 °C–30 °C, metabolic rate of the cat decreases, with slope -0.95 min kg cm^{-3} O$_2$ °C^{-1}. For T_a 30 °C–37.5 °C, the slope is zero: there is no change in metabolic rate over this range of T_a. For T_a 37.5 °C–40 °C, metabolic rate increases, the slope of the line being 0.48 min kg cm^{-3} O$_2$ °C^{-1}.

(iv) The thermoneutral zone must be T_a 30 °C–37.5 °C, as there is no change in metabolic rate over this range.

(v) The data indicate that the lower critical temperature, T_{lc} for the cat, is 30 °C, because as T_a decreases from 30 °C, metabolic rate increases, suggesting production of extra metabolic heat by thermogenesis. The data indicate that the upper critical temperature, T_{uc} for the cat, is 37.5 °C, because as T_a increases from 37.5 °C, metabolic rate increases, suggesting initiation of physiological mechanisms that enhance heat loss.

Question 2.4

(a) The signal from the radiotransmitter attached to each turtle could be used to locate the turtles as they left the sea to walk to their nesting sites at night. The signal could also be used to locate the turtles when they surfaced at sea, if a receiver was within range of the signal.

The advantage of using a data logger rather than radiotelemetry is that the data logger stores all data collected while the animal is both on the surface and underwater. When the data logger is retrieved the data can be downloaded into a laptop computer. If the data have to be transmitted to a receiver via an aerial attached to the animal the data could not be transmitted while the animal was submerged. Radiowaves are not transmitted through water, so data generated while the turtles were submerged would be lost.

(b) (i) Turtle 7610's dive lasted for a total of 13 minutes. Maximum depth reached was 50 m.

(ii) Before the dive the turtle's heart rate was about 30 beats min^{-1}. As the turtle dived, heart rate decreased progressively, reaching a nadir of about 10–14 beats min^{-1} at 30 m depth within 1.5 min of submergence. When the turtle began swimming towards the surface, heart rate increased sharply from 30 m depth rising from 18 beats min^{-1} to 30 beats min^{-1} at zero depth over a period of just 1.5 min.

(c) A data logger attached to a diving animal provides insight into diving physiology. Data on body temperature, heart rate, depth of submergence, and other variables can be collected and stored in the memory of the data logger.

However, the process of attaching and detaching the equipment might cause the animal stress, which in turn affects heart rate and T_b.

However, the fact that the female turtles behaved normally after attachment of the equipment to their shells suggests that they were not stressed. The turtles returned to the beach to lay more clutches of eggs after spending time at sea.

Question 2.5

(a) This scenario is a classic example of *acclimatization*. Acclimatization is possible because of *phenotypic flexibility*, differential gene expression in response to new environmental signals. As the mountaineer ascends the slopes at walking pace, he meets progressively lower ambient temperatures and decreasing P_{O_2}. Phenotypic flexibility enables physiological adjustments in response to low T_a and P_{O_2}. The result is acclimatization. As these processes take time to complete, a slow rate of ascent is essential. Phenotypic flexibility and the consequent ability to acclimatize have enabled a number of climbers to reach the summit of Everest without needing to use supplementary oxygen.

(b) The lizard is a *poikilotherm*; over 24 hours the T_b of the lizard has varied from 15 °C to 40 °C. Basking in the sun and absorbing solar radiation and thermal radiation from the rock demonstrates *ectothermy*, whereby an animal uses external sources of heat to increase T_b.

(c) The fox, like nearly all mammals, is an *endotherm*. At low T_a the T_b of the fox is maintained at 38 °C by heat generated from catabolism of fuel. The fox is also a *homeotherm* and maintains T_b within narrow limits. A rise in T_b of just 0.5 °C initiates a physiological cooling mechanism, panting.

(d) The two groups of birds are undergoing *acclimation* in the laboratory, each to a different T_a. All other conditions, day/night regime, diet, water are the same for the two groups of birds.

Question 3.1

(a) During the night *Ammospermophilus* remains inside the burrow. At about 06.00 h, the squirrel leaves the burrow and is active, presumably collecting seeds. A short period of activity is followed by a return to the burrow. From 06.00 h to 17.00 h the squirrel left the burrow on nine occasions, and was active for a short period before returning to the burrow. Peaks of activity were recorded up to about 17.00 h when the squirrel retired to the burrow for the night.

(b) During the night T_b was fairly stable at 36 °C. During bouts of daytime activity, T_b rose sharply peaking at 43 °C. Returning to the burrow enabled the squirrel to cool off, as T_b dipped to 38–39 °C after only short periods in the burrow.

(c) Use of the burrow suggests *Ammospermophilus* is an evader, but tolerance of T_b as high as 43 °C indicates a degree of endurance too.

Question 3.2

(a) *False*. Most small animals avoid rather than endure heat stress, but a few species, notably ground squirrels, can tolerate T_b up to 42–43 °C for short periods, so for them relaxed homeothermy is part of their strategy for maintaining activity during the day. Burrow-dwelling rodents can readily unload small amounts of stored heat by returning to cool burrows periodically.

(b) *True.* Exposure of bare skin to the heat of solar radiation would promote evaporative water loss. The coat insulates the skin from this intake of heat by reflecting solar radiation and in its absence, water loss is likely to increase. Note that the coat must not be so thick that it impedes the effective vaporization of water on the skin and a relevant point here is that many camels have a thin uneven coat (Figure 3.37) with thicker fur on the back and sparse fur on the sides.

(c) *False.* Measured BMR for hoopoe larks acclimated to T_a 15 °C in the laboratory was at 46.8 kJ day^{-1}, close to that measured for a temperate species, the woodlark, 49.4 kJ day^{-1}. This example illustrates the importance of awareness of phenotypic flexibility and the need to draw comparisons between more than two species in order to prove that a feature of a desert species is adaptive.

(d) *False.* Larger animals tend to rely more on sweating than on panting but panting is observed in some reptiles and is important in most birds, even those species that live in deserts.

(e) *False.* Although behavioural responses form the major thermoregulatory strategies of reptiles, these animals have autonomic responses too, including vasoconstriction, vasodilation and panting. Birds and mammals also respond to heat stresses by behaviour, thereby making savings of evaporative water loss and energy. In all three groups, reptiles, birds and mammals, both behavioural and autonomic responses have a role.

Question 3.3

1 Jack rabbits use (a), (f) and possibly (b). Insulation (c) would not be effective in a relatively small animal, and the heat storage effect (e) would be useful for a short time only. Sweating (d) is not an option for small species because they have a large surface area to volume ratio and would lose too much water.

2 The oryx uses (a) moving into shade during hot summer days, and (e) heat storage. No information is available about whether the species produces concentrated urine, but this is likely. Oryx do not have a thick pelage so (c) is unlikely and lack of bare skin means (f) is not possible.

3 The hoopoe lark uses (a) moving into burrows on hot summer days, (b) production of uric acid which involves very little water loss, and also (f) losing heat by conduction from bare skin surfaces. Insulation (c) and heat storage (e) are not effective for such a small bird. Birds do not sweat (d), although there is some evaporative cooling by evaporation of water from the skin.

4 Reptiles use (a) and excrete uric acid (b). Options (c) and (d) are not possible in reptiles; (e) is unlikely although T_b of desert reptiles, e.g. *Dipsosaurus*, can reach high levels that would not be tolerated by most mammals. (f) is used but all reptilian skin is bare!

5 Kangaroo rats use (a) and (b). A small animal is unlikely to rely on (c) and (e) and kangaroo rats cannot sweat (d). Kangaroo rats are nocturnal and remain in burrows during the day so (f) is not used.

6 The camel uses (b), (c) (d) and (e). Use of shade when possible (a) is probably important too. The coat of camels makes (f) unlikely although short-coated species have sparse fur on their sides through which bare skin can be seen.

Question 3.4

(a) If BMR is comparatively low, the amount of heat generated internally by metabolism is decreased which lessens the total heat load in a stressful environment. Water loss is reduced as less is evaporated to cool the body. In a hot, arid environment where food may be in short supply, low BMR would reduce daily energy demand.

(b) Phenotypic plasticity, the process by which individual features of physiology are shaped during development, can cause variation in BMR between individuals. Variations resulting from phenotypic flexibility are usually fixed for life. Phenotypic flexibility or acclimatization, the capacity of an animal to vary physiological parameters in response to environmental stresses, may alter BMR. Williams and Tieleman's work on hoopoe larks showed that at thermoneutral T_a, larks acclimated to T_a of 15 °C had a significantly lower BMR than individuals acclimated to 36 °C (Table 3.5).

Question 3.5

The problem with the experimental design is that the researchers are comparing solar heat gain in two completely different situations. An isolated skin preparation has a huge surface area to volume ratio and would inevitably warm up very quickly in the sun (and dry up). In the intact live animal, the skin is part of a whole larger body. Therefore, the physical properties of the isolated skin are not the same as those for skin in a live animal. It is inevitable that isolated skin would not have any physiological properties; in contrast, in the whole animal, a number of physiological processes could come into play to reduce solar heat gain. Vasoconstriction would slow transfer of solar heat to the rest of the body. The skin may become paler in colour, which also slows down heat gain.

A better design would involve using either a model of the animal or a mounted preserved specimen of the same size as the live specimen. Thermistors just below the 'skin' and in the deep body of the model or preserved specimen, and the intact animal, can be used to monitor heat gain just below the skin and in the body core.

Question 3.6

BMR increases linearly with body mass for the 15 species. An artiodactyl weighing 100 kg has a BMR of about 25 litres O_2 h^{-1}. The values of BMR and body mass of the oryx are almost on the plotted line suggesting that oryx do not have an unusually low BMR. The camel is a little below the plotted line, suggesting a trend for reduced BMR in this species.

Question 4.1

The three measures are (a) induction of thermal dormancy, (b) the suppression of behavioural activity and (c) the depression of metabolic activity. Although the Svalbard reindeer becomes behaviourally inactive and ceases feeding in the winter months, its BMR and T_b are not depressed as in true hibernators.

Question 4.2

Adaptive hypothermia appears to have evolved in birds and in several mammalian orders that are only distantly related. Within single families, the occurrence of torpor is not universal, suggesting adoption of the strategy at species level based on energy cost-benefit considerations. Furthermore, internal and environmental triggers for entry to torpor vary considerably between species.

Question 4.3

(a) False. Many non-hibernating (e.g. migrating) animals lay down fat on a seasonal basis. Some hibernators store food as well as fat.

(b) True.

(c) False. Although in some species the heart may be the warmest tissue during arousal, thermographic imaging shows that for most mammals, deposits of BAT, which are sources of thermogenesis, are the warmest.

(d) True.

(e) False. Unless BMR increases, the animal will not warm and thus there would be no effective response to the fall in temperature.

Question 4.4

(a) The main signals that initiate entry into hibernation are environmental, i.e. food supply, day-length and ambient temperature.

(b) Periodic arousal from torpor is sometimes linked to the time of day, although mechanisms for initiating periodic arousal may be diverse. Early arousal may also be initiated by alarm mechanisms, such as mechanical stimuli or a fall in T_a to below 0 °C. Final arousal is normally independent of T_a and may be triggered by endogenous seasonal changes or reduction in fat stores below a critical level.

Question 4.5

Your flow chart should show that most or all of the following steps were followed:

- Perform subtractive gene expression analysis in an experimental mammalian model using tissue from hibernating and euthermic liver.
- Identify genes that are expressed at a higher level in hibernating liver.
- Obtain a partial or complete DNA base sequence of one or more of these genes.
- Predict an amino-acid sequence for all or part of the encoded protein.
- Delete the gene to determine the impact of its absence on liver cell metabolism.
- If you have identified a new gene with no existing sequence homologies, you may want to design experiments to determine the function of the protein.

Question 4.6

Protein synthesis is dramatically reduced; cell cross-sectional area is reduced; changes occur in the structure of the Golgi apparatus; there are changes in the numbers of glycogen granules and lipid droplets; membrane-associated proteins undergo redistribution; metabolic processes designed to neutralize reactive oxygen species are activated; there are metabolic changes relating to use of alternative respiratory fuel sources; there is reduced electron transport and protein-leak in mitochondria.

Question 4.7

Heart rate is reduced to less than 10 min^{-1}. Breathing is episodic, punctuated by periods of apnoea. Blood P_{CO_2} is increased up to four-fold. BMR drops rapidly at the onset of hibernation. BMR may increase periodically during hibernation despite a sustained low T_b. Heart and respiratory rates may increase if T_a falls below a critical level.

Question 4.8

Lipid is the preferred energy source for long-term hibernation. Glucose is the preferred source of energy in the short-term for animals in which torpor episodes last a day or less. Glucose is also used in species that hibernate at ambient temperatures below freezing. Protein is used to supply amino acids for gluconeogenesis when carbohydrate fuel sources are exhausted.

Question 4.9

The factors include the comparative energy cost of maintaining euthermia over a comparable period and the energy savings made possible for costly physiological processes such as gestation. The savings must take into account: animal size; the energy cost of re-warming during arousal; and the nature of the habitat. In Richardson's ground squirrel, an obligative hibernator (Table 4.5), energy is saved through periods of torpor both in July and November, but is greatest in November when the difference between T_a and T_b is larger and the number of arousal episodes is reduced.

Question 4.10

There are a number of studies that support the rheostasis theory. Two key ones are the experiments on marmots and ground squirrels in Section 4.6. Experiments on marmots show that there is a steady shift down in the set point when the hypothalamus is cooled. The work on golden-mantled ground squirrels (Figure 4.36) shows progressive and continuous resetting of the thermostat.

Studies that do not support the rheostasis theory would be that a gradual, controlled decline in T_s is not universal, either in different species or within individuals of the same species entering torpor. The theory does not explain the metabolic response to temperature changes applied to the hypothalamus of Eastern chipmunks (Figure 4.37).

Question 4.11

Rapid-response genes controlling electrical firing activity of neurons are activated, neuronal structure is altered, the critical temperature of firing is depressed, and there is an increase in the number of cold-sensitive neurons. Biosynthesis of histamine and serotonin is also increased. Rapid-response genes are characterized by their activation very soon after the onset of hibernation – they initiate many of the other changes which hibernating neurons undergo, and can thus be viewed as a preparative signal.

Question 4.12

Slow-wave sleep activity (SWA) is depressed in the brain of hibernating mammals. This is characteristic not of sleep, but of sleep deprivation. SWA activity cycles do not follow normal sleep activity cycles in torpor but become shortened.

Question 5.1

Only (e) is correct. In general, the climate is harsher in Antarctica than in the Arctic. There are plenty of fish in the Arctic Ocean (although they are of different species from those in the Southern Ocean). Polar bears eat seals, not birds (although they might eat a penguin if the species ever met). Leopard seals are common in coastal waters of the Southern Ocean and penguins are their principal prey, but the two species have coexisted successfully for millions of years. Many migratory birds breed in the Arctic and a few, notably ptarmigan, are resident. With bears, reindeer, arctic foxes, birds, walruses and numerous fish, as well as seals, the Arctic has more species of vertebrates than Antarctica, which is without any fully terrestrial vertebrates.

Question 5.2

Svalbard and Norwegian reindeer ate different amounts even when they had unlimited access to food and were exposed to the same environmental conditions, so the appetites of the two subspecies must be at least partially under endogenous control (see Figure 5.3).

The effects of exposure to continuous light on food intake and body mass depends upon the season at which the treatment is started. At very high latitudes, there is continuous light or continuous darkness for several months (see Figure 5.1), so daylength would not, by itself, be a sufficiently precise clue to seasonal breeding that must be accurately timed to favourable weather and food supply, e.g. for ptarmigan (see Figure 5.6).

Question 5.3

Only (b) is generally true. It would be impossible to sustain permanently more animals than the food supply could support. At the beginning of fasting, bears are both fat and lethargic, but one does not cause the other: male penguins are least active during incubation, and walk briskly back to the open sea at the end of the fast when they are thin. Bears and penguins, like large animals everywhere, have few predators, but smaller birds, lemmings and fish have plenty of predators. Since many fish and terrestrial animals are fat while in arctic regions, it cannot be true that the available food is less nutritious. There is no clear evidence for (g): BMR actually falls below normal for weeks at a time in bears and penguins and is less in winter- than summer-acclimatized foxes (Figure 5.4).

Question 5.4

The marker for bone breakdown, ICTP, was more abundant during dormancy, but the differences between denning and activity for the same bears were not significant for PICP, the marker for bone formation (Figure 5.13). These observations explain how the skeleton is weakened during dormancy, but not how it is reformed during activity. PICP has a brief peak at the start of the active period (Figure 5.14), probably indicating rapid strengthening of the skeleton that was not detected by this sampling procedure.

Question 5.5

(a) Body temperature is a crude and somewhat indirect measure of activity. In both bears and penguins, body temperature falls slightly at the start of fasting (see Figure 5.11). Since there is no evidence that the body insulation has changed, the rate of production of heat (i.e. BMR) must have decreased. Body temperature rises abruptly during egg-laying in female penguins; the synthesis of components of the relatively large egg and the effort of laying would raise BMR.

(b) RER provides information about the chemical composition of the fuel broken down by respiration. An RER close to 1 indicates that carbohydrates are being used; a lower RER means that proteins and/or lipids are being used.

(c) Chemical analysis of blood plasma provides detailed information about the concentrations of different fuels available to cells (e.g. fatty acids, glucose, β-hydroxybutyrate) and about the rates of formation and accumulation of breakdown products of metabolism (e.g. urea, creatinine, β-hydroxybutyrate).

Question 5.6

Only (d), (g) and (h) are true. Many arctic mammals have subcutaneous adipose tissue for part of the year, because many are large and they naturally become obese because the food supply is seasonal. However, in mammals, having subcutaneous adipose tissue is not an essential or unique feature of living in an arctic climate. In Carnivora, including bears, the partitioning of adipose tissue is related to body size, not to habits or habitat (see Figure 5.16). In any one animal, the adipocytes enlarge with increasing fatness, but the total adipocyte complement, and hence the relationship between mean adipocyte size and fatness, differs from individual to individual.

Question 5.7

(a) Without haemoglobin in red blood cells, oxygen is carried in solution in the blood, but the solubility of oxygen in water increases with decreasing temperature. The enormous reduction in density of cells in the blood offsets the increase in its viscosity at low temperature, so the blood flows faster for the same work of pumping by the heart. These factors mean that at $-1.5\,°C$, the muscles of an icefish receive nearly as much oxygen as those of a fish with haemoglobin, but such fish would be unable to swim efficiently in warm water.

(b) The salt concentration of the body fluids of fish is less than that of seawater so their supercooled body fluids are at risk from freezing if they come into contact with ice crystals (Figure 5.24). Because ice is less dense than water, it floats at the surface. The salt concentration of the body fluids of most invertebrates is similar to that of seawater, so they are not vulnerable to freezing until the seawater itself freezes. There are also more diving birds such as penguins and skuas, and marine mammals, all of which eat fish, in surface waters.

Question 5.8

(a) The abundance of marine plankton, and hence of all the animals that eat it, changes greatly with the season, and with currents and ice movements. The climate, and hence the presence and availability of prey species, change erratically and may differ greatly between apparently similar sites. So large predatory fish and mammals (and seabirds such as albatross) range over a wide area and have a varied diet.

(b) The composition of fatty acids in the storage lipids is determined by the metabolism of the microbes in the rumen and by that of the reindeer tissues, more than by the composition of fatty acids in the diet.

Question 6.1

There is a correlation between lactate dehydrogenase (LDH) activity and altitude in all of the tissues studied. *O. princeps* has the highest LDH activity – for most tissues, LDH activity is at least twice as high in *O. princeps* as it is in *O. collaris*. Cardiac muscle of *O. princeps* was 401 $\mu mol\,g^{-1}$ wet mass tissue, in comparison to 172 $\mu mol\,g^{-1}$ for *O. collaris*. Only cardiac muscle was assayed for LDH activity in *O. hyperborea*, and although LDH activity was lower than that of *O. collaris*, the difference was not significant. Within each of the two pika species, LDH activities were similar across all skeletal muscles. Rabbits, in contrast, showed the highest LDH activity in vastus lateralis muscle, moderate activity in gastrocnemius muscle, and lowest activity in soleus muscle.

Sheafor suggests that the high anaerobic potential in pika tissues may be linked to ecological factors. Pika attempt escape from predators by sprinting from their foraging grounds to hide in rock piles. Therefore, pika depend on the ability to generate high-speed burst activity for short periods of time. Rapid ATP synthesis is required to power burst activity; thus, it appears that pika need to maintain anaerobic capacity in order to power their escape from predators.

Question 6.2

(a) (ii) The increase is largely due to an increase in the number of red blood cells.

(b) (i) Some reserve red blood cells, e.g. from the spleen, are added to the circulating blood within hours of exposure to high altitude. However, in the longer term, the increased haematocrit level is due to an increased rate of production of red blood cells.

(c) (ii) Recall that the same amount of oxygen can be delivered to the tissues of highlanders for a smaller fall in arterio-venous P_{O_2}.

(d) (ii) The concentration of 2,3-DPG increases during acclimatization.

(e) (ii) At high altitude, the affinity of haemoglobin for oxygen falls, so there is an increase in P_{50}.

(f) (i) In their study of a group of students resident at high altitude (La Paz), Hoppeler and Vogt (2001) found that the capacity of muscle for aerobic metabolism (measured by density of mitochondria, 3.94%), was 30% less than would be expected for untrained young lowlanders. So, although the intuitive choice might be (ii), for individuals living a *sedentary* life at high altitude, mitochondrial density in muscle was measured as being less than in individuals living at sea-level (Section 6.2).

Question 6.3

(a) At 4000 m, atmospheric pressure is about 65% of that at sea-level (Figure 6.1), and P_{O_2} is about 12.6 kPa. Figure 6.6 demonstrates that at an atmospheric P_{O_2} of 12 kPa, alveolar P_{O_2} is about 7 kPa, which gives a maximum of 17 volume percent of oxygen in arterial blood (Figure 6.11b). In theory there should be sufficient oxygen reaching respiring tissues, as a slight drop in P_{O_2} would result in haemoglobin giving up much of its oxygen to the tissues. However, before acclimatization, the conflict between the receptors for oxygen and CO_2 means that, initially, the rate of breathing is not increased, which depresses alveolar P_{O_2} and hence arterial P_{O_2}. Eventually breathing rate increases, but this results in a depression of plasma P_{CO_2}. This in turn increases blood pH and depresses oxygen delivery to the muscles (low P_{CO_2} shifts the oxygen dissociation curve of haemoglobin to the left, so the pigment gives up less of its oxygen to the respiring tissues, Figure 6.11). Resulting hypoxia in tissues may initiate activity of HIF-1α and the expression of VEGF, resulting in increased permeability of capillaries in the brain with consequent slight edema causing headache, dizziness and nausea.

(b) The description suggests that the woman was suffering from HAPE. Bubbling sounds during breathing suggest accumulation of fluid in the lungs which is the salient feature of HAPE. The fluid is forced out of the capillaries into the tissues because pulmonary vasoconstriction in response to hypoxia raises blood pressure in the lungs.

Question 6.4

(a) Within a few hours of exposure of humans to hypoxia at high altitudes, sustained hyperventilation develops, which 'blows-off' carbon dioxide, and results in an increased pH of the blood (see Henderson–Hasselbalch equation, Equation 6.2). Normal [H$^+$] in blood is restored eventually by excretion of excess HCO_3^- by the kidneys.

(b) Hyperventilation – this replaces a higher proportion of alveolar air with freshly inspired air.

(c) Hyperventilation – the initial increased ventilation rate on exposure to high altitude is probably a response to detection of low blood P_{O_2} by the peripheral chemoreceptors (Section 6.2.1). Sustained hyperventilation is achieved because the the central chemoreceptors develop an increased sensitivity to carbon dioxide, a response that develops within about 15 hours of exposure to high altitude.

(d) A rightward shift of the oxygen dissociation curve. A decrease in the affinity of haemoglobin for oxygen probably increases unloading of oxygen in the tissues and maintains high issue P_{O_2}, but at the expense of oxygen loading in the lungs. Increased levels of 2,3-DPG cause the rightward shift in the oxygen dissociation curve (Section 6.2.3).

(e) Possibly an increase in capillary density, an elevation of tissue myoglobin and increased numbers of mitochondria (Sections 6.3.1 and 6.4.1).

(f) An increase in the number of mitochondria, and in the activities of key mitochondrial enzymes, that may contribute towards an increased capacity of the animal to metabolize aerobically (Section 6.5).

Question 6.5

Table 6.9 P_{50} values for haemoglobin of *Telmatobius* and *Xenopus*, with and without, added Cl$^-$ ions.

	P_{50} without added ATP or Cl$^-$/mmHg (kPa)	P_{50} with added Cl$^-$/mmHg (kPa)
Telmatobius	7 (0.93*)	8 (1.1)
Xenopus	9 (1.2)	20 (2.7)

* The values for P_{50} in mmHg obtained directly from Figure 6.23 were converted to kPa by multiplying by the conversion factor, 0.1332, as suggested in Question 6.5.

(a) (i) P_{50} without added ATP or Cl$^-$/mmHg (kPa). (ii) P_{50} with added Cl$^-$/mmHg (kPa).

(b) The stripped haemoglobins of both *Telmatobius* and *Xenopus* display high affinity for oxygen. Haemoglobin of *Telmatobius* has P_{50} of about 0.93 kPa, and that of *Xenopus* has P_{50} of 1.2 kPa (compare with P_{50} of about 4.8 kPa that can be calculated for human blood from data in Figure 6.13b).

Haemoglobin of *Telmatobius* has virtually no response to Cl^-, in contrast to many amphibians in which Cl^- functions as a modulator of the oxygen affinity of haemoglobin. Values for P_{50}, and therefore the oxygen affinity of *Telmatobius* haemoglobin, are very close in the presence and absence of Cl^-. In contrast, *Xenopus* haemoglobin shows a marked response to Cl^-, with P_{50} increasing to a P_{O_2} of 2.7 kPa from a value of just 1.2 kPa in the absence of Cl^-. In *Xenopus*, added Cl^- reduces the affinity of haemoglobin for oxygen.

(c) *Telmatobius* is an aquatic frog living at high altitude. Low atmospheric P_{O_2} results in an even lower P_{O_2} in water, so the environment of this species is markedly hypoxic. Having a haemoglobin with a very high affinity for oxygen is advantageous for *Telmatobius* as this means the animal can extract sufficient oxygen from water, even at low P_{O_2}. Loss of sensitivity to Cl^- is advantageous too, as this means that Cl^- ions, ubiquitous in living tissue cannot affect the affinity of haemoglobin for oxygen.

ACKNOWLEDGEMENTS

The present Course Team gratefully acknowledges the work of those involved in the chapters who are not also listed as authors in this book.

Grateful acknowledgement is made to the following sources for permission to reproduce material in this book:

Cover

Courtesy of Ardea Picture Library.

Figures

Figure 1.1: Andrews, M. T. Department of Biochemistry and Molecular Biology, University of Minnesota, USA; *Figure 1.2*: Masa Ushioda/Image Quest Marine; *Figure 1.3*: Cathy Walters, Company of Biologists Ltd, Cambridge; *Figure 1.4*: Sidell, B. D. (2000) Life at body temperatures below 0 degrees C: The physiology and biochemistry of antarctic fishes. *Gravitational and Space Biology Bulletin*, **13** (2), June 2000. American Society for Gravitational and Space Biology; *Figure 1.5*: William Ervin/Science Photo Library; *Figure 1.6*: Peter Chadwick/Science Photo Library; *Figure 2.5*: Scholander, Walters, Hock, Johnson and Irving (1950) Body insulation of some arctic and tropical mammals and birds. *Biological Bulletin*; *Figure 2.6*: Ingram, D. L. and Mount, L. E. (1975) *Man and Animals in Hot Environments*, Springer-Verlag GmbH & Co KG; *Figure 2.7*: Adapted from Withers, P. C. (1992) *Comparative Animal Physiology*, Copyright © 1992 Saunders College Publishing, reproduced by permission of the publisher; *Figure 2.8*: Hull, D. and Smales, O. R. C. (1978) *Temperature Regulation and Energy Metabolism in the Newborn*, Grune & Stratton Inc; *Figure 2.15*: Cossins, A. R. and Bowler, K. (1987) *Temperature Biology of Animals*. Kluwer Academic Publishers, B. V.; *Figures 2.17a, b, 2.18*: Tattersall, G. J. and Milsom, W. K. (2003) Transient peripheral warming accompanies the hypoxic metabolic response in the golden-mantled ground squirrel. *Journal of Experimental Biology*, **206**. Company of Biologists Ltd; *Figure 2.19*: Wilson, R. P. and Gremillet, D. (1996) Body temperatures of free-living African penguins (*Spheniscus demersus*) and bank cormorants (*Phalacrocorax neglectus*), *Journal of Experimental Biology*, **199**, 2215-2223. Copyright © Company of Biologists Ltd; *Figure 2.21*: Gabai, V. L. and Sherman, M. Y. (2002) Molecular biology of thermoregulation. Invited review. *Journal of Applied Physiology*, **92**, 1744–1745. © American Physiological Association; *Figure 2.22*: Willmer, P., Stone, G. and Johnston, I. (2000) *Environmental Physiology of Animals*. Blackwell Publishers Ltd; *Figure 2.23*: Butler, P. J. et al. (1993) Respiratory and cardiovascular adjustments during exercise of increasing intensity and during recovery in thoroughbred racehorses. *Journal of Experimental Biology*, **179**. Copyright © Company of Biologists Ltd; *Figure 2.24*: Williams J. B. et al. (2002) Energy expenditure and water flux of Rüppell's foxes in Saudi Arabia, *Physiological and Biochemical Zoology*, **75**, 479–488. University of Chicago Press; *Figure 2.26*: Southwood, A. L. et al. (1999) Heart rates and diving behaviour of leatherback sea turtles in the Eastern Pacific Ocean. *Journal of Experimental Biology*, **202**. Copyright © Company of Biologists Ltd; *Figure 3.1*: Willmer et al. (2000) The occurrence of deserts on a worldwide basis…, *Environmental Physiology of Animals*, Chapter 14. Blackwell Science Limited; *Figure 3.2*: Marion Hall, Open University; *Figure 3.3*: Science Photo Library; *Figures 3.4–3.7*: David Robinson, Open University;

Figure 3.8: Willmer, P., Stone, G. and Johnston, I. (2000) *Environmental Physiology of Animals*. Blackwell Science Limited; *Figure 3.10*: Robinson, M. (1999) Water-holding frog…, *A Field Guide to Frogs of Australia*, p. 76. New Holland Publishers; *Figure 3.11*: Robinson, M. (1999) Family Myobatrachidae, *A Field Guide to Frogs of Australia*. New Holland Publishers; *Figure 3.13, 3.15*: Dr Peter Davies; *Figure 3.14*: Wardene Weisser/Ardea; *Figure 3.16*: Brad Alexander; *Figure 3.17*: Louw, G. N. (1993) Temperature and thermoregulation, *Physiological Animal Ecology*, Longman Group UK Ltd; *Figure 3.18*: NHPA/Rod Planck; *Figure 3.19*: Bulova, S. J. (2002) How temperature, humidity…, *Journal of Thermal Biology*, **27**. Elsevier Science; *Figure 3.20*: Dr Lloyd Glenn Ingles, California Academy of Sciences; *Figure 3.21*: Courtesy of Texas Parks & Wildlife Dept Copyright © 2003 Glen Mills; *Figure 3.22*: Based on Folk, G. E. (1974) *Textbook of Environmental Physiology* (2nd edn), Lea and Febiger; *Figure 3.23*: Kenneth W. Fink/Ardea London Ltd; *Figure 3.24*: Dr Joseph B. Williams; *Figure 3.25*: Williams, J. B. (2001) Energy expenditure and water flux…, *Functional Ecology*, **15**. British Ecological Society; *Figure 3.31*: Withers, P. C. (1992) *Comparative Animal Physiology*. Saunders College Publishing, Fort Worth; *Figures 3.32, 3.33*: Taylor, C. (1977) Exercise and environmental heat loads. *International Review of Physiology: Environmental Physiology II*, **15**. University Park Press; *Figure 3.42*: Pockley, G. (2001) Heat shock proteins in health and disease…, *Expert Reviews in Molecular Medicine*. Cambridge University Press; *Figure 3.34*: Irene Tieleman; *Figure 3.43, 3.44*: Zatsepina et al. (2000) Thermotolerant desert lizards…, *Journal of Experimental Biology*, **203**. Copyright © Company of Biologists Ltd; *Figure 3.35*: Courtesy of NHPA/Hellio & Van Ingen; *Figure 3.36*: Redrawn from Gordon, M. S. (1977) *Animal Physiology: Principles and adaptations* (3rd edn). Macmillan Publishing, New York; *Figure 3.45*: Mueller, P. and Diamond, J. (2001) Metabolic rate and environmental productivity: well-provisioned animals evolved to run and isle fast. *Proceedings of the National Academy of Sciences*. **98** (22), National Academy of Sciences; *Figures 3.46, 3.47*: Tieleman, I. et al. Adaptation of metabolism and evaporative water loss…, *Proceedings of the Royal Society of London*, **270**. The Royal Society; *Figure 3.48*: Willmer, P. Stone, G. and Johnston, I. (2000) *Environmental Physiology of Animals*. Blackwell Science Limited; *Figure 3.49*: Williams, J. B. et al. (2001) Seasonal variation in energy expenditure, water flux and flood consumption of Arabian oryx, *Oryx leucoryx, Journal of Experimental Biology*, **204**. Copyright © Company of Biologists Ltd; *Figure 4.1*: Michael and Diane Porter, American Goldfinch, Ideaform Inc.; *Figure 4.2*: Tom and Cathy Saxton, Hummingbird, Saxton.org.; *Figure 4.3*: John Franklin, birdinfo@mirrorpole.com; *Figure 4.4*: Art Wolfe/Science Photo Library; *Figure 4.5*: Roger W. Barbour/Morehead State University; *Figure 4.6*: Peter Menzel/Science Photo Library; *Figure 4.7*: Leonard Lee Rue/Science Photo Library; *Figure 4.8*: Roger W. Barbour/Morehead State University; *Figures 4.9, 4.13*: Strumwasser, F. (1960) Some physiological principles governing hibernation. *Bulletin of the Museum of Comparative Zoology*, **124**, Harvard University; *Figures 4.10, 4.18*: Lyman, C. P. and O'Brien, R. C. (1960) Circulatory changes in the thirteen-lined ground squirrel during the hibernating cycle. *Bulletin of the Museum of Comparative Zoology*, **124**, Harvard University; *Figures 4.14, 4.16*: Mussacchia, X. J. and Volkert, W. A. (1971) *American Journal of Physiology*, **221**. American Physiological Society; *Figure 4.20a, b*: Hayward, J. and Lyman, C. P. (1967) Nonshivering heat production during arousal from hibernation and evidence for the contribution of brown fat, Fisher, K. et al. (eds), *Mammalian Hibernation III*.

1967 Oliver and Boyd; *Figure 4.21*: Leming Shi, Ph.D., Principal Investigator at the U.S. FDA's National Center for Toxicological Research (NCTR), Jefferson, Arkansas; *Figure 4.23*: Erik Z. Yu and John M. Hallenbeck (2002) Elevated arylalkylamine-*N* acetyltranserase (AA-NAT)…, *Molecular Brain Research*, **102**. Elsevier Science; *Figure 4.24a, b*: Frerichs, K. U. and Smith, C. B. et al. (1998) Suppression of protein synthesis in brain…, *Proceedings of the National Academy of Sciences*, **95**. National Academy of Sciences; *Figure 4.25a*: Malatesta, M. et al. (2002) Quantitative ultrastructural changes of hepatocyte constituents…, *Tissue and Cell*, **34**. Elsevier Science; *Figure 4.25b*: Azzam, N. A., Hallenbeck, J. M. and Kachar, B. (2000) Membrane changes during hibernation, *Nature*, **407**. Nature Publishing Group; *Figure 4.26a and b*: Ortmann, S. and Heldmaier, G. (2000) Regulation of body temperature and energy requirements…, *American Journal of Physiology*, **278**. Copyright © American Physiological Society; *Figure 4.27*: Dawn Sadler, Open University; *Figure 4.28*: Buck, C. L. and Barnes, B. M. (2000) Effects of ambient temperature on metabolic rate…, *American Journal of Physiology – Regulatory Integrative Comparative Physiology*, **279**. Copyright © American Physiological Society; *Figure 4.29*: Boutilier, R. G. and St-Pierre, J. (2002) Adaptive plasticity of skeletal muscle energetics in hibernating frogs: mitochondrial proton leak during metabolic depression, *Journal of Experimental Biology*, **205**. Copyright © Company of Biologists Ltd; *Figure 4.30a and b*: Zimmer, M. B. and Milsom, W. K. (2002) Ventilatory pattern and chemosensitivity…, *Respiratory Physiology and Neurobiology*, **113**. Elsevier Science; *Figure 4.31*: Schleucher, E. (2001) Heterothermia in pigeons and doves reduces energetic costs, *Journal of Thermal Biology*, **26**. Elsevier Science; *Figure 4.32*: Koteja, P. et al. (2001) Energy balance of hibernating mouse-eared bat *Myotis myotis*… *Acta Theriologica*, **46**. *Polska Akademia Nauk, Zaklad Badania Ssakow*; *Figure 4.33*: Diana Weedman Molavi, PhD, Washington University School of Medicine; *Figure 4.36*: Heller, H. C. (1977) *Pflugers Archiv*, **369**, Springer Verlag GmbH & Co KG; *Figure 4.38*: Masaaki Hashimoto et al. (2002) Arousal from hibernation and BAT thermogenesis against cold…, *Journal of Thermal Biology*, **27**. Elsevier Science; *Figure 4.39*: Quezzani, S. E. et al. (1999) Neuronal activity in the mediobasal hypothalamus of hibernating jerboas, *Neuroscience Letters*, **260**. Elsevier Science; *Figure 4.40a*: Popov, V. I. (1992) Repeated changes of dendritic morphology in the hippocampus of ground squirrels…, *Neuroscience*, **48**. Elsevier Science; *Figure 4.40b*: Panula, P. et al. (2002) The histaminergic system in the brain…, *Journal of Chemical Neuranatomy*, **18**. Elsevier Science; *Figure 4.40c*: Sallmen, T. et al. (2003) Intrahippocampal histamine delays arousal from hibernation, *Brain Research*, **966**. Elsevier Science; *Figure 4.41*: Pitrosky, B. et al. (2003) Research report – S22153, a melatonin antagonist dissociates…, *Behavioural Brain Research*, **138**. Elsevier Science; *Figures 5.2a, 5.5*: Dr Alison Ames; *Figures 5.2b, 5.17, 5.18b, c, 5.22a and b*: Caroline Pond, Open University; *Figure 5.7*: Mark Eaton; *Figure 5.10a*: Bryan and Cherry Alexander Photography; *Figure 5.10b*: © Kim Crosbie; *Figure 5.13 and 5.14*: Donahue, S. W. et al. (2003) Serum markers of bone…, *Clinical Orthopaedics and Related Research*, **406**. Copyright © Dr S. W. Donahue; *Figure 5.18a*: Dill, P. B. and Irving, L. (1964) Polar biology, *Handbook of Physiology*, Chapter 5, American Physiological Society; *Figure 5.18b, c*: Dr Malcolm A. Ramsay, University of Saskatchewan, Canada; *Figure 5.20*: Captain Budd Christman, Office of NOAA Photo Library; *Figure 5.21*: Office of NOAA Photo Library; *Figure 6.1*: Dejours, P. (1966) *Respiration*, p. 187, Oxford University Press; *Figure 6.2*: Francois

Gohier/Ardea London Ltd; *Figure 6.3*: rex.morrey@thalesgroup.com; *Figure 6.4*: Pat and Tom Leeson/Science Photo Library; *Figure 6.5*: Glyn Ryland Photographic; *Figure 6.6*: Heath, D. and Williams, D. R. (1981) *Man at High Altitude*, Churchill Livingstone; *Figure 6.8*: Houston, C. (1998) HACE: high altitude cerebral edema, *The Mountaineers* (4th edn); *Figure 6.10*: Vander, A., Sherman, J. and Luciano, D. (2001) Circulation, *Human Physiology*, 8th edn. McGraw-Hill Companies Inc; *Figure 6.13*: Hannon, J. P., Shields, J. L. and Harris, C. W. (1969) Effects of altitude acclimatization on blood composition of women. *Journal of Applied Physiology*, **26** (5). The American Physiological Society; *Figure 6.14*: Szathmary, E. J. E. (1997) Ventilation and hypoxic ventilatory response…, *American Journal of Physical Anthropology*, **104**. Wiley-Less, Inc.; *Figures 6.15, 6.16*: Ri-Li Ge et al. (1998) Blunted hypoxic pulmonary vasoconstrictive response…, *American Journal of Physiology – Heart Circulatory Physiology*, **274**. American Physiological Society; *Figure 6.17*: Hall, F. G., Dill, D. B. and Barron, E. S. G. (1936) Comparative physiology in high altitudes, *Journal of Cellular and Comparative Physiology*, **8**; *Figure 6.18*: Weber, R. E. et al. (2002) Mutant hemoglobins: functions related to high-altitude respiration in geese, *American Journal of Physiology*, American Physiological Society; *Figure 6.19*: Alberts, B. et al. (1998) *Essential Cell Biology: An Introduction to the Molecular Biology of the Cell*, Chapter 8, Reproduced by permission of Taylor and Francis, Inc./Routledge Inc., http://www.routledge-ny.com; *Figure 6.20*: Wenger, R. H. (2000) Mammalian oxygen sensing…, *Journal of Experimental Biology*, **203**. Company of Biologists Ltd; *Figure 6.21*: Vogt, M. et al. (2001) Molecular adaptations in human skeletal muscle endurance…, *Journal of Applied Physiology*, **91**. American Physiological Society; *Figure 6.22*: Sheafor, B. A. (2000) Mammalian oxygen sensing…, *Journal of Experimental Biology*, **203**. Company of Biologists Limited; *Figure 6.23*: Weber, R. E. (2002) Novel mechanism for high-altitude adaptation…, *American Journal of Physiology – Regulatory Integrative Comparative Physiology*, **283**. American Physiology Society.

Tables

Table 2.5: Marquet, P. A. (ed.) (1989), Ecological aspects of thermoregulation at high altitudes, *Oecologia*, **81** (1). Springer Verlag GmbH & Co. KG; *Table 2.6*: St-Pierre, J., Charest, P. M. and Guderley, H. (1998) Relative contribution of quantitative and qualitative changes in mitochondria to metabolic compensation during seasonal acclimatisation of rainbow trout, *Journal of Experimental Biology*, **201**. Copyright © Company of Biologists Ltd; *Table 2.7*: Tattersall, G. J. and Milsom, W. K. (2003) Transient peripheral warming accompanies the hypoxic metabolic response in the golden-mantled ground squirrel, *Journal of Experimental Biology*, **206**. Copyright © Company of Biologists Ltd; *Table 3.6*: Williams et al. (2001) Seasonal variation in energy expenditure…', *Journal of Experimental Biology*, **204**. Copyright © Company of Biologists Ltd; *Table 3.8*: Mueller, P. and Diamond, J. (2001) Metabolic rate and environmental productivity: well-provisioned animals evolved to run and idle fast. *Proceedings of the National Academy of Sciences*, **98** (22). National Academy of Sciences; *Table 6.8*: Sheafor, B. A. (2000) Mammalian oxygen sensing…, *Journal of Experimental Biology*, **203**. Copyright © Company of Biologists Ltd.

Every effort has been made to contact copyright holders. If any have been inadvertently overlooked the publishers will be pleased to make the necessary arrangements at the first opportunity.

INDEX

Glossary terms are in bold. Page numbers in italics indicate items mainly, or wholly, in a figure or table.

A

AA-NAT enzyme, in hibernation, 143
absolute humidity, **28**
   in the desert, 79–80
absorption coefficient, 47
acclimatization, **41**
   of desert birds, 97–98
   to high altitude, 228
acetyl CoA, 21, 34, 143, *144*
adaptive hypothermia, **123**, **124**–**129**
   behaviour characteristics, 130–140
   in hibernation, 152
adipocytes, 191–193
adipose tissue,
   in hibernation, 146–147
   regulatory enzymes, *144*
   in seals, 199–200
   structure of, 191–193
   thermal insulation, 196–197
   see also brown adipose tissue; white adipose tissue
aestivation, 75, 80, 125
African Namaqua dove (*Oena capensis*), hibernation, 155–156
African wild dog (*Lycaon pictus*), energy expenditure, 16–18
air,
   pressure, 45
   properties, 28
Alaemon alaudipes, *see* hoopoe lark
alleles, in genetic drift, 59
Alopex lagopus (arctic fox), insulation, 27
alpine marmot (*Marmota marmota*), hibernation, 149
altitude,
   effect on respiration, 218–221
   effect on the blood, 221–224
   effects of high, 24, 45, 62, 215–216
   genotypic adaptations, 230–236
   illness from, 224–227
   integration across disciplines, 236–240
   integration across species, 241–244
   phenotypic adaptation, 227–228
alveoli, exchange of gases, 46–49, 225–226
ambient temperature (T_a), **9**
   in cats, 64–65
   in the desert, 78–80, 84
   in hibernation, 127–129, 134–135
   in a range of animals, *38*, *39*, *40*–*41*

American goldfinch (*Carduelis tristis*), torpor, *127*
Ammospermophilus leucurus (antelope ground squirrel), 120
amphibians,
   temperature compensation, 43
   see also frogs
Andean goose (*Chloephaga melanoptera*), 234
Andes region, 216, 228, 229, 230–231
Anser indicus (bar-headed goose), *216*, 233–235
antarctic silver fish (*Pleuragramma antarcticum*), *202*, 204, 208
Antarctica, 171, 174
antelope ground squirrel (*Ammospermophilus leucurus*), 120
antidiuretic hormone, 91
   in sweating, 101–102
apnoea, during hibernation, 134
apoptosis, **56**–57
Aptenodytes forsteri (emperor penguin), 182–184
aquatic frogs, oxygen dissociation curve, 248
Arabian desert, 104
Arabian oryx (*Oryx leucoryx*), *70*, 84
   adaptations, 104–106
archive tag, 9
Arctic, 171–173
arctic fox (*Alopex lagopus*), *173*
   fecundity, 180
   insulation, 27, 194
arctic ground squirrel (*Spermophilus parryii*),
   hibernation, 134, 150
   metabolic rate, *29*
arctic marmot (*Marmota caligata*), hibernation, 133, 136
aridity, **69**
   of deserts, 69–72
   role in phylogeny, 115–118
arousal, during hibernation, 135–140
ATP,
   generation, 10–11, 21–22, 138
   in lizards, 99
   synthesis, 30, 31, 34
atrichial sweat glands, 100, *101*
Australian hopping mouse (*Notomys*), adaptations, 87, 91

autoradiogram, showing protein synthesis, *145*
Awassi sheep, thermoregulation, 100
Aymara people, 216, 230–231
Azzam, N. A., 147

B

bank cormorants (*Phalacrocorax neglectus*), hypoxia, 54–55
bar-headed goose (*Anser indicus*), *216*, 233–235
Barents, Willem, 201
Barnes, B. M., *29*, 150
Barros, R. C. H., 50
basal metabolic rate (BMR), **31**–32
   in the Arabian oryx, 105
   in birds, 96–98
   in fish, 14
   in hibernation, 149–151
   in lark species, 115–118
   in polar animals, 176
   in polar fish, 205–206, 208
bats, 156, *157*
   arousal, 139
   renal function, *92*
   torpor, 128
Beall, C. M., 230
bears, 184–193
   see also black bear
behavioural suppression, **125**
behavioural thermoregulation, 75–81
Bering, Vitus, 201
birds,
   adaptations to desert conditions, 96–98
   in the desert, 82–83, 84
   exhaled air temperatures, *86*
   migration for breeding, 177–179
   regulation of breeding, 179–180
   showing torpor, 127–128
   thermal insulation, 200
   thermogenesis, 139–140
Black, C. P., 233–234
black bear (*Ursus americanus*),
   dormancy, 184–188
   hibernation, 155
black-tailed prairie dog (*Cynomys ludovicianus*), *129*
Blanford's fox (*Vulpes cana*), *36*, 37
Block, B. A., 9, 10, 11
blood,
   changes at high altitude, 221
   pigments, 204–205

blood-brain barrier, and
    chemoreceptors, *219*, 220–221
blood-gas exchange, in the lungs,
    48–49
blubber, 210
blue marlin (*Makaira nigricans*), 9
    thermogenesis, *11*
bluefin tuna, tachymetabolism, 39
BMR, *see* basal metabolic rate
body mass,
    adaptations in birds, 97–98
    fluctuations in penguins, 183–184
    in hibernation, *131*, *157*
    in lark species, 115–118
body size,
    and evaporation, 74–75
    in hibernation, 155
    in torpor, 127
body temperature (T_b), 9–10
    adaptation to cold, *124*, 125
    in arctic animals, 193–197
    catalytic rate constant, 15
    in cats, 64–65
    in desert reptiles, 76–78
    during hibernation, 129, 137
    in hypoxia, 51–53
    metabolic effects, 31
    in a range of animals, *38–39*, 40
    in reptiles, 99
    in wild animals, 41–43
bone formation, during dormancy,
    187–188
Boreogadus saida (fjord cod), *203*
Bos gruniens (yak), *216*, 231
Bos taurus (cow), at high altitude, 216
Boutilier, R. G., 152
Bowman's capsule, 89
Bozinovic, F., 93
bradymetabolic changes, **125**
bradymetabolism, 31–32, **33**, 38
brain, control of hibernation, 158–166
breeding,
    environmental regulation, 179–180
    in migrating birds, 177–179
    in penguins, 182–183
    in polar bears, 190–191
brown adipose tissue (BAT),
    in hibernation, 138–140
    thermogenicity, 10
brown bats (*Myotis lucifugus*), *129*
    hibernation, 156
brown bear (*Ursus arctos*), dormancy,
    184–188
Buck, C. L., 150
Buck, M. J., *29*
Bulova, S., 79–80

burrows,
    in the desert, 79–80
    in hibernation, 134
Butler, P. J., 64

C

c-fos genes, 162
Calidris canutus (red knot), 178–179
Calidris mauri (sandpiper), 179
Californian ground squirrel,
    hibernation, 132–133
calorimetry, 31
camels,
    thermoregulation, 100
    water balance, 103–104
Camelus dromedarius (dromedary
    camel), *70*
carbohydrates,
    in hibernation, 150–151
    oxidation, 5–6
carbon dioxide,
    in catabolism, 34–35
    in the environment, 45
    in the lungs, 46–47
    in respiration, 218–221
    see also respiratory exchange ratio
cardiac output, 221
Carduelis flammea (redpoll), torpor,
    127
Carduelis tristis (American goldfinch),
    torpor, *127*
Carey, F. G., 10
carnitine palmitoyl transferase, 43–44
caspases, in cell death, 56–57
catabolism, 21–22, 34
    in acclimatization, 43–44
cats, temperature effects, 64–65
cell death pathway, 56–57
cells, effects of temperature, 55–58
cerebral edema, at high altitude,
    224–225
Chaenocephalus aceratus, *see* icefish
Chalcides bedriagai (Bedriagai's
    skink), *76*
chaperone proteins, 56, 108
chemoreceptors, and the blood-brain
    barrier, *219*, 220–221
chipmunks, hibernation, 130–131,
    160–161
Chloephaga melanoptera (Andean
    goose), 234
chuckwalla (*Sauromalus ater*), body
    temperature, 41
circadian cycle, 164
    in hibernation, 158

citrate synthase, 43–44, 242–243
clades, 60
cladistics, 59–61
    and phylogeny, 115–118
cladogram, 60, *117*, 118
Clarke, A., 14
cloven-feathered dove (*Drepanoptila
    holosericea*), hibernation, 155–156
Colombian ground squirrel
    (*Spermophilus columbianus*),
    hibernation, 165
conduction, 26–27
conservation, 16–18
convection, 27–28
convective conductance, 217–218
Cook, James, 201
cormorants, *see* bank cormorants
Covell, D. F., 36
cow (*Bos taurus*), at high altitude, 216
creatinine, 185, 189
Cricetus cricetus (European hamster),
    brain activity, 162
Crocuta crocuta (spotted hyena), 16
cryoprotectants, 203
Cyclorana platycephala (water-
    holding frog), 75
Cynomys ludovicianus, (black-tailed
    prairie dog), *129*
cytochrome *c* oxidase, 43–44
    escape from mitochondria, 56–57
    at high altitude, *239*, 242–244

D

data loggers, 54, 66
daylength,
    initiating hibernation, 130, *131*
    in polar regions, 172, *173*, 175
De Martonne's aridity index, 71
deer mouse (*Peromyscus*),
    at different altitudes, 241
    metabolic rate, 113–114
degu (*Octodon degus*), adaptations,
    92–93
dehydration,
    in camels, 103–104
    in humans, 102
Dermochelys coriacea (leatherback sea
    turtle), 66
desert frogs, 75
desert grouse (*Pterocles*), cooling, 96
desert iguana (*Dipsosaurus dorsalis*),
    adaptations, 98–99
    body temperature, 15, 41
desert tortoise (*Xerobates agassizii*),
    thermoregulation, 79–80

desert wood rat (*Neotoma lepida*), thermoregulation, 80–81
deserts, 69–73
    animals in, 73–74
    environment, 73–85
    integration across disciplines, 108–112
    integration across levels of analysis, 85–107
    integration across species, 112–115
    vertebrates in, 74–85
Diamond, J., 113
diffusion, of gases, 47–49
diffusive conductance, **217**–218
Dipodomys, *see* kangaroo rat
Dipsosaurus dorsalis, *see* desert iguana
Dissostichus mawsoni (toothfish), *202*
Djungarian hamster (*Phodopus campbelli*),
    sleep, 166
    torpor, 150
DNA microarray, 12, **142**
dogs,
    energy expenditure, 18
    heat conservation, *94*, 95
    see also African wild dog; black-tailed prairie dog; husky dog
Donahue, S. W., 187
Dorcas gazelle (*Gazella dorcas*), *70*, 84
'**dormancy**,' **184**–188
dormouse (*Muscardinus avellanarius*),
    arousal, 135
    hibernating cells, *146*
    in hibernation, 134, 159
doubly-labelled water technique, **16**–17, 34–37, 93
    bird migration, 178
doves, hibernation, 155–156
dragonfly, tachymetabolism, 39
Drepanoptila holosericea (cloven-feathered dove), hibernation, 155–156
dromedary camel (*Camelus dromedarius*), *70*
Drosophila (fruit fly), at high temperatures, 108–109
dune lark (*Mirafra erythroclamys*), 82–83, 96

E

Eastern chipmunk (*Tamias striatus*), hibernation, 160, *161*
eccritic body temperature, **43**
ectotherms, **8**, 38
    polar, 14, 202–212

electromagnetic spectrum, *23*
elephant shrews (*Elephantulus myurus*), torpor, 130
Elephantulus myurus, (elephant shrews), torpor, 130
emerald rockcod (*Trematomus bernacchii*), *202*, 204, 207
emperor penguin (*Aptenodytes forsteri*), 182–184
endocrine glands, during hibernation, 135
endotherms,
    aquatic, 197–200
    hibernation and torpor, 126–129
    terrestrial, 193–197
endothermy, **8**–9
endurers,
    behaviour, 84–85
    integration of anatomy and behaviour, 98–107
energy budgeting,
    in hibernation, 150–151, 153–156
    in polar regions, 175–177, 179
enzyme activity, temperature effects, 29–30, 43–44
Eothenomys miletus (vole), thermogenesis, 13
epitrichial sweat glands, 100, *101*
erythropoietin, 221, 236
European hamster (*Cricetus cricetus*), brain activity, 162
eutherms, hibernation in, 128–129
evaders,
    behaviour, 75–81
    integration of anatomy and behaviour, 85–93
evaporation, 40
    and body size, 74–75
    and heat loss, 95, 96
    thermoregulation, 81–83
evaporators, integration of anatomy and behaviour, 93–98
evolutionary adaptation, 58–61
extra-terrestrial life, 211–212

F

facultative hibernators, **131**
fasting,
    in penguins, 181–184
    in polar bears, 189–191
fatty acids,
    indicators of diet, 209–210
    levels in fasting, 184, 186
    see also triacylglycerols
feathers,
    insulation properties, 26–27, 194
    thermoregulation, 84

fecundity, of polar animals, 180–181
fennec fox (*Fenecus zerda*), 95
Fick's equation, **47**–49
field metabolic rate, 34–37
    in birds, 96–98
    in the oryx, 105
field water flux, 95–96
field water influx rates, 105–106
Fields, P. A., 15
Fischer, S., 225
fish,
    basal metabolic rate, 14
    body temperature, 8–10
    eyes, 204
    muscles in, 206–209
    in polar regions, 202, 203
    temperature compensation, 43
fjord cod (*Boreogadus saida*), *203*
Florant, G. L., 159–160
foxes,
    body temperatures, 195–196
    see also arctic fox; kit fox; Rüppell's fox
Franklin, John, 201
freeze tolerance, 203
Frerichs, K. U., 145
frogs,
    aquatic, 248
    mitochondria, 151–152
    see also desert frogs; water-holding frog
fruit fly (*Drosophila*), at high temperatures, 108–109
fur, insulation properties, 26–27, 194, 197–198

G

gas exchange, physics of, 46–49
Gazella dorcas (Dorcas gazelle), *70*
Ge, R. L., 232
gecko, Moorish, 77
Geffen, E., 36, 37
Geiser, F., 124
gene expression,
    in heat-shock proteins, 108–112
    during hibernation, 141–146
    in hypoxia, 236–240
genetic drift, **59**
genotypic adaptations, to high altitudes, 230–236
gerbil, Mongolian, 13
Gillichthys mirabilis (mudsucker), hypoxia, 12–13
Girard, I., 36
glycerol, production in dormancy, 185–186

golden hamster (*Mesocricetus auratus*),
    circadian cycle, 158
    hibernation, 136–137, 154
golden-mantled ground squirrel (*Spermophilus lateralis*),
    hibernation, *128*, *131*, 152–154, 160
    hypoxia, 50–53
Gorman, M. L., 16–18
Gremillet, D., 54
grizzly bear, *see* brown bear
Groscolas, R., 182
ground squirrels,
    brain activity, 162–165
    energy budgeting, 153–154
    genetic drift, 59
    hibernation, 6–7, 134, 136–137, 143
    protein synthesis, *145*
    see also named ground squirrels
Groves, B. M., 226
gular flutter, **96**
Gymnodactylus caspius, 110–111

H

haemoglobin,
    biochemistry of, 221–224
    at high altitudes, 227, 233–235
    oxygenation, 49
hair, role in insulation, 99–100, 175
hamsters,
    brain activity, 162
    hibernation, 129, 130–131, 136
    see also Djungarian hamster; golden hamster
Hayward, J., 139
heat energy,
    on Earth's surface, 22–24
    methods of meaurement, 31–32
heat-shock proteins (Hsps), **55**–57, 108–112
    in desert lizards, 59
Heldmaier, G., 149
heliothermy, **43**
Heller, H. C., 159–160, *160*
Henderson-Hasselbalch equation, **218**–219
heritability, *230*
heterothermy, **38**
hibernation, **124**–126
    arousal, 135–140
    cellular changes, 146–148
    control systems, 158–167
    energy budgeting, 153–156
    energy sources, 150–151
    genetic basis, 141–146
    and hormones, 164–166
    integrative physiology, 5–8
    maintenance of, 133–135
    mitochondrial adaptations, 151–152
    onset, 130–133, 160
    respiration, 152–153
    see also dormancy; torpor
high altitude cerebral edema (HACE), 225
high altitude pulmonary edema (HAPE), 225–227
hippocampus, 164
histamine, 163–164
homeostasis, 158
homeotherms, **9**
    adaptation to cold, 123–126
homeothermy, **38**–39, *40*
homology, **143**
hoopoe lark (*Alaemon alaudipes*),
    acclimatization, 97–98
    in phylogenetic tree, 118
Hoppeler, H., 227–228
hormones, and hibernation, 164–166
horses, respirometer measurements, 64
Houston, C., 225
Hsp gene, 108–112
Hsps, *see* heat-shock proteins
humans,
    adaptation to high altitude, 230–231, 234–235
    at high altitudes, 216, 218, 220–228, 239–240
    obesity, 192
    in polar regions, 200–202
humming birds, torpor, 127
husky dog, surface temperature, *196*
β-hydroxybutyrates, levels in fasting, 183–184, 186
Hydrurga leptonyx (leopard seal), 182
hyena, spotted, 16
hypercapnia, **218**
hyperthermia, in deserts, 72–73, 106
hypocapnia, **218**
hypothalamus,
    control of hibernation, 158–161
    neurotransmission, 163–164
hypothermia, *see* adaptive hypothermia; torpor
hypoxia, **218**, 220, 221
    adaptations to, 60–61
    in cells, 55–58
    effects on metabolism, 50–55
    at high altitude, 45, 227–228
    in the mudsucker, 12–13
hypoxia-inducible factor, 236–240
hypoxic ventilatory response, 230

I

ice crystal formation, 174, 203
icefish (*Chaenocephalus aceratus*), *202*
    muscles, *207*
    phylogeny, 16
    skin, 204–205
iguanas,
    metabolic rate, 32, *33*
    see also desert iguana
infrared thermography, 50–53
insulation, *see* thermal insulation
integration, 5–8
    across disciplines, 11–13
        in the desert, 108–112
        at high altitude, 236–240
    across levels of analysis, 8–11
        in the desert, 85–107
    across species, 13
        in the desert, 112–115
        at high altitude, 241–244
Inuit people, 200–201

J

jack rabbit (*Lepus californicus*),
    evaporation, 81–82
Johnston, I. A., 206
Johnston, N. M., 14

K

kangaroo rat (*Dipodomys*), 80
    adaptations, 85–87
    hypoxia, 50
Kayser, B., 230–231
ketone bodies, 186
kidneys, function, 87–93
kit fox (*Vulpes macrotis*), evaporation, 82
Körtner, G., 124
Koteja, P., 156

L

lactate dehydrogenase,
    in fish, 15
    at high altitude, 242–243
lactation, in polar bears, 190
Lagopus lagopus (willow ptarmigan),
    metabolic rate, 176–177
Lagopus mutus hyperboreus (Svalbard ptarmigan), 176–177
    breeding, 179–180
larks,
    acclimitization, 97–98
    cladistic analysis, 115–118
    see also dune lark

Larsen, T. S., 175
leatherback sea turtle (*Dermochelys coriacea*), 66
leopard seal (*Hydrurga leptonyx*), 182
Lepus californicus (jack rabbit), evaporation, 81–82
Li, Q., 13
Lindgård, K., 176
Liolaemus, see lizards
lipids,
    in hibernation, 149–150
    in polar fish, 208–210
    see also adipose tissue
lizards,
    adaptations in, 59
    ATPase activity, 99
    body temperature, 15, 41, *42*
    role of Hsp70, 109–112
    thermoregulation, 76–78
    see also Mojave fringe-toed lizard; monitor lizard
llama, 231, 233
loop of Henle, 87–92
lower critical temperature (T_{lc}), 40–41, *83*
lungs,
    function at high altitude, 225–226, 233
    gas diffusion, 46–49
Lycaon pictus (African wild dog), energy expenditure, 16–18
Lyman, C. P., 132, 137, 139

M

Makaira nigricans (blue marlin), *9*
    thermogenesis, *11*
Maletesta, M., 146
mammals,
    aquatic, 197–199
    in desert regions, 87–107
    at high altitude, 231–236
    kidney function, 87–93
    in polar regions, 173, 175–177, 180–181, 193–197
    torpor in, *126*, 127
Marmota caligata (arctic marmot), hibernation, 133, 136
Marmota marmota (alpine marmot), hibernation, 149
Marmota monax, see woodchuck
marmots,
    hibernation, *129*, 133, 136
    metabolic rate, 158–160
Marquet, P. A., 41–42
marsupial mouse (*Sminthopsis crassicaudata*), *30*
Mattacks, C. A., 191
Mayer, W. V., 134, 137

meadow jumping mouse (*Zapus hudsonius preblei*), *129*
medulla, in the mammalian kidney, 91–92
Meigs, P., 71
melatonin, 143, 164, *165*
Merino sheep, thermoregulation, 100
Meriones unguiculatus (Mongolian gerbil), thermogenesis, 13
Merriam's kangaroo rat (*Dipodomys merriami*), thermoregulation, 80
mesic environments, **35**
Mesocricetus auratus, see golden hamster
metabolic chambers, 31, 50–51
metabolic inhibition, 125
metabolic rate,
    effects of hypoxia, 50–55
    in hibernation, 134, 137, 144–145, 160–161
    in polar fish, 205–209
    within species, 113–114
    and temperature, 29–37
    see also basal metabolic rate (BMR); field metabolic rate
micro-array analysis, 11–13
microclimates, 24
midbrain, role in hibernation, 161–162
migration, for breeding, 177–179
Milsom, W. K., 50, 52–53
Milson, R., 152
Mirafra erythroclamys (dune lark), 82–83, 96
mitochondria,
    in acclimatization, *44*
    in catabolism, 34
    escape of cytochrome *c*, 56–57
    fluid properties, 30
    in hibernation, 151–152
    at high altitude, 228, 243–244
    temperature compensation, 207
Mojave desert, 72, 78–79
Mojave fringe-toed lizard (*Uma scoparia*), thermoregulation, 77–78
Mongolian gerbil (*Meriones unguiculatus*), thermogenesis, 13
monitor lizard (*Varanus*), body temperature, 42–43
Moorish gecko (*Tarentola mauretanica*), 77
mountain sickness, 216, 224
mouse, see deer mouse; dormouse; marsupial mouse
mouse-eared bats (*Myotis myotis*), hibernation, 156, *157*

mRNA,
    analysis of, 142, 145
    in heat-shock proteins, 108–111
    in hypoxia, 239–240
    PDK4 expression, 6–7
mudsucker (*Gillichthys mirabilis*), hypoxia, 12–13
Mueller, P., 113
Musacchia, X. J., 134
Muscardinus avellanarius, see dormouse
muscles,
    heat conduction, 198
    at high altitude, 241–244
    in polar fish, 206–209
Myotis lucifugus (brown bats), *129*
    hibernation, 156
Myotis myotis (mouse-eared bats), hibernation, 156, *157*

N

Nagy, K. A., 36
nasal counter-current heat exchanger, *86*, 94
Nelson, R. A., 186
Neobatrachus sudelli (painted burrowing frog), 75
Neotoma lepida (desert wood rat), thermoregulation, 80–81
nephrons, *88*, 89
neurons, in hibernation, 162–164
neurotransmission, in the hypothalamus, 163–164
'niche,' 73
node, 60
non-shivering thermogenesis (NST), 13, 140
normoxia, 50–53
    at high altitude, 237, 239
northern fur seal (*Callorhinus ursinus*), insulation, 198
Notomys (Australian hopping mouse), adaptations, 87, 91
Notothenia neglecta, 202, 204
    muscles, *207*
nutrient budgeting, 175–177

O

obesity, 191–192
obligative hibernators, 131
O'Brien, J., 10, 132
ocean, freezing, 174
Ochotona curzoniae, see plateau pika
Octodon degus (degu), adaptations, 92–93

Odobenus rosmarus (walrus), insulation, 198–199
Oena capensis (African Namaqua dove), hibernation, 155–156
operative environmental temperature, T_e, **82**
opsins, 204
Ortmann, S., 149
Oryx leucoryx, see Arabian oryx
osmolarity, 87, 90–91, 93
    during sweating, 101–102
osmosis, 87
ostrich (*Struthio camelus*), adaptations, 84, 107
Ouezzani, S. E., 162
outgroup, 60, *61*
oxygen,
    in catabolism, 34–35
    in the environment, 45–46
    at high altitudes, *215*, 222–224, 231
    in hypoxia, 50–51
    in the lungs, 46–49
    partial pressures, 217–218
    uptake in metabolism, 31–33
    see also hypoxia; reactive oxygen species
oxygen cascade, 217–218
oxygen dissociation curve, 222–224, 233
    in aquatic frogs, 248

P

painted burrowing frog (*Neobatrachus sudelli*), 75
panting, 96, 99
    in evaporative cooling, 95
Panula, P., 163
parsimony, 60, *61*
partial pressures,
    of carbon dioxide, 218–219
    of gases, 45–47
    of oxygen, 217–218, 221–224
penguins, 181–184
Peromyscus, see deer mouse
Phalacrocorax neglectus (bank cormorants), hypoxia, 54–55
Phalaenoptilus nuttalli (poor will), torpor, 127
phenotypic adaptation, 58
    in desert birds, 97–98
    in desert mammals, 93
    at high altitude, 227–228
phenotypic flexibility, 62
phenotypic plasticity, 62
Phoca hispida, see ringed seal

Phodopus campbelli, see Djungarian hamster
Phodopus sungorus, see Siberian hamster
phylogenetic autocorrelation, 116, 118
phylogenetic constraint, 115
phylogenetic inertia, 61
phylogenetic tree, 59–61, *117*, 118
phylogeny, 14–16
    and cladistic analysis, 115–118
pineal gland, 143, 164
Pitrosky, B., 164
plateau pika (*Ochotona curzoniae*), 216, 231–233
    at different altitudes, 241–244
    phylogeny, 60–61
    thermogenesis, 13
Pleuragramma antarcticum (antarctic silver fish), *202*, 204, 208
poikilothermy, 38
polar animals, 173–174
    adaptations, 14–15
    fecundity, 180–181
    insulation, 193–197
    nutrient budgeting, 175–177
polar bears (*Ursus maritimus*),
    adipose tissue, 196–197
    fasting, 189–191
    temperature, 195
polar biology, 171–212
polar ectotherms, 202–212
polar environment, 171–174
    for humans, 200–202
polar fish,
    lipids in, 208–210
    metabolic rate, 205–209
Pond, C. M., 191, 196
poor will (*Phalaenoptilus nuttalli*), torpor, 127
Popov, V. I., 163
primary productivity, metabolic rate, 113–114
protective measures, 125
proteins,
    metabolism in dormancy, 185–186
    metabolism in fasting, 184
    metabolism in hibernation, 143–146
    see also heat-shock proteins
proximal convoluted tubule, 88–91
psychrophiles, 174
ptarmigan (*Lagopus*),
    breeding, 179–180
    metabolic rate, 176–177
Pterocles (desert grouse), cooling, 96
pulmonary arteries, adaptations, 61–62, 225–226, 232

pulmonary edema, at high altitude, 225–227
pyruvate dehydrogenase kinase, 6

Q

Q_{10}, **29**, 205
Quechuan people, 216

R

rabbits,
    enzyme activities, *242*
    metabolic rate, 33
    see also jack rabbit
radiation, electromagnetic, 23, 25
radiotelemetry, 42, 54, 66
radiotelethermometry, 42, 54
rainbow trout, acclimatization, 43–44
Ramsay, M. A., 185, 196
Rangifer tarandus platyrhynchus, 175, *see* reindeer
rapid eye movement (REM) sleep, 166
rapid-response genes, 162
rats, *see* desert wood rat; kangaroo rat; Wistar rat
reactive oxygen species (ROS), 147–148
red fox (*Vulpes vulpes*), metabolic rate, 176
red knot (*Calidris canutus*), 178–179
redpoll (*Carduelis flammea*), torpor, *127*
reindeer (*Rangifer tarandus platyrhynchus*), *173*, 175
    body shape, 193–194
    fecundity, 180
    foraging, 175–176
relative humidity, 28
    in the desert, 79–80
'relaxed homeothermy,' 38
    in the camel, 104
reptiles,
    body temperature, 41–42
    temperature compensation, 43
    thermoregulation, 76–78
    see also lizards
respiration,
    control of, 217–221
    in hibernation, 152–153
respiratory exchange ratio, in dormancy, 186
respiratory quotient, in hibernation, *144*, 150–151
rete mirabile, in desert animals, 106
Richardson's ground squirrel (*Spermophilus richardsonii*), hibernation, *136*, *154*
    sleep, 166

ringed seal (*Phoca hispida*), 189
  insulation, 26, *27*
RNA, *see* mRNA
rock ptarmigan (*Lagopus mutus*),
  metabolic rate, 176–177
rodents,
  hibernation, *128*, 129
  renal function, 92
  see also ground squirrels
rufous humming bird (*Selasphorus rufus*), torpor, 127
Rüppell's fox (*Vulpes rueppelli*),
  adaptations, 95–96
  metabolic rate, 35–37
  thermoregulation, 41

S

Sallmen, T., 163
sandpiper (*Calidris mauri*), 179
saturated vapour pressure, 46
Sauromalus ater (chuckwalla), body temperature, 41
Sauromalus obesus, body temperature, 99
Schekkerman, H., 178
Schleucher, E., 156
Schmidt-Nielsen, K., 81, 103
Scholander, P. F., 203
Schultheis thermometer, 41
sealions, thermal insulation, 197–198
seals, *see* leopard seal; ringed seal
serotonin, 163, 164
Sheafor, B. A., 241–243
sheep, thermoregulation, 100
shivering, 51, 139–140
shuttling, 43, 76
Siberian hamster (*Phodopus sungorus*)
side-blotched lizard (*Uta stansburiana*), 77
Sidell, B. D., *15*
skin,
  pigmentation in fish, 204
  variable colour, 77
sleep, 166
slow-wave sleep, 166
Sminthopsis crassicaudata (marsupial mouse), *30*
solar radiation, 22–24, 43
Somero, G. N., *15*
Sonoran desert, 71–72, 80
Southwood, A. L., 66
Spermophilus columbianus (Colombian ground squirrel), hibernation, 165

Spermophilus lateralis, *see* golden-mantled ground squirrel
Spermophilus parryii, *see* arctic ground squirrel
Spermophilus richardsonii, *see* Richardson's ground squirrel
Spermophilus tridecemlineatus, *see* thirteen-lined ground squirrel
spotted hyena (*Crocuta crocuta*), 16
squirrels, *see* ground squirrels
St-Pierre, J., 43, *44*, 152
standard temperature and pressure, 47
Stokkan, K. A., 179
stress responses, 22
Strumwasser, F., 131, 132
Struthio camelus (ostrich), adaptations, 84, 107
succinate-cytochrome *c* reductase, 30
summit metabolic rate, 31–32
supercooling, 203
superoxides, 147–148
surface area, to volume ratio, 74–75, 84, 98, 99
Svalbard, environment, 172–173, 175
Svalbard ptarmigan (*Lagopus mutus hyperboreus*), 176–177
  breeding, 179–180
Svalbard reindeer (*Rangifer tarandus tarandus*), 175
  fecundity, 180
sweating, 100–102, 100–104

T

tachymetabolism, 31–32, **33**, 38–39
Tamias striatus (Eastern chipmunk), hibernation, 160, *161*
Tarentola mauretanica (Moorish gecko), 77
Tattersall, G. J., 50, 52–53
taxidermic mounts, 82–83
TCA cycle, 21, 34
Telmatobius peruvianus (aquatic frog), 248
temperature,
  in deserts, *71*
  on Earth, 23, 24
  effect on cells, 55–58
  and metabolic rate, 29–37
temperature compensation, 43–44
thermal adaptation, *see* adaptive hypothermia
thermal dormancy, 125
thermal generalists, 76

thermal inertia, 43
thermal insulation, 26–27, 175, 193–202
thermal specialists, 76
thermoconformers, 8, 38
thermogenesis,
  in hibernation, 161–162
  mechanism, 10–11
  role of BAT, 138
  see also non-shivering thermogenesis
thermogenic tissue, 8
thermoregulation, 26–27, **38–45**
  behavioural, 75–81
  in desert endurers, 99–101
  in desert mammals, 84–85
  evaporation, 81–83
thermoregulator, 38
thirteen-lined ground squirrel (*Spermophilus tridecemlineatus*), hibernation, *137*, 143, 151
Tibetan people, 230–231
Tieleman, B. I., 61, 96–97, 115–116
toothfish (*Dissostichus mawsoni*), *202*
torpor, 124
  degrees of, 126–127
  energy budgeting, 153–156
  energy sources, 150–151
  length of bouts, 140
  protective measures, 125
  see also hibernation
torpor metabolic rate, *29*
tortoise, desert, 79–80
total evaporative water loss, 95–98
  in Arabian oryx, 105
  in lark species, 115–118
transcription,
  factor, 57–58, 62
  in heat-shock proteins, 108–112
tree shrew (*Tupaia belangeri*), thermogenesis, 13
Trematomus bernacchii (emerald rockcod), *202*, 204, 207
triacylglycerols,
  in dormancy, 186
  in hibernation, 5–7
trout, *see* rainbow trout
tuna,
  bluefin, 39
  body temperature, 9
Tupaia belangeri, *see* tree shrew
turbinates, 86
turtles, leatherback, 66

U

ubiquitin, in hypoxia, 237–238
U:C ratios, 185, 189–190

Uma scoparia (Mojave fringe-toed lizard), thermoregulation, 77–78
uncoupling protein, 138–139
upper critical temperature (T_{uc}), 40–41, 83
urea, production in dormancy, 185, 189
uricotelic, 96, 99
urine,
    concentration, 87, 90–93
    reduction of, 104
Ursus americanus (black bear), dormancy, 184–188
    hibernation, 155
Ursus arctos (brown bear), dormancy, 184–188
Ursus maritimus, *see* polar bears
Uta stansburiana (side-blotched lizard), 77

V

vapour pressure, 28
    saturated, 46
Varanus (monitor lizard), body temperature, 42–43
vascular endothelial growth factor, 225, *239*, 240
vasoconstriction, 226, 232
vasodilation, 226
ventilation, 217–218

vicuna (*Vicugna vicugna*), *216*, 233
Vogt, M., 227–228, 239–240
vole (*Eothenomys miletus*), thermogenesis, 13
Volkert, W.A., 134
Vulpes cana (Blanford's fox), *36*, 37
Vulpes macrotis (kit fox), evaporation, 82
Vulpes rueppelli, *see* Rüppell's fox
Vulpes vulpes (red fox), metabolic rate, 176

W

walrus (*Odobenus rosmarus*), insulation, 198–199
Wang, L. C. H., 153, 154
water,
    in catabolism, 34–35
    content in plants, 92–93
    diuretics, 90–91
    intake rate, 105–106
    loss during sweating, 102–104
    in polar regions, 174, 202–204
    in respiration, 46
    see also doubly-labelled water technique
water-holding frog (*Cyclorana platycephala*), 75
Weber, R. E., 234–235
Wenger, R. H., *238*

white adipose tissue (WAT), 138
    PD4 expression, 6–7
Williams, J. B., 31, 35–37, 82, 95, 96–97, 104
Willmer, P., 74, 81
willow ptarmigan (*Lagopus lagopus*), metabolic rate, 176–177
Wilson, R. P., 54
Wistar rat, pulmonary arteries, 232–233
wolves, insulation, 194
wood rats, *see* desert wood rat
woodchuck (*Marmota monax*), *129*
    entering torpor, 132
    hibernation, 165

X

Xenopus laevis (aquatic frog), 248
Xerobates agassizii (desert tortoise), thermoregulation, 79–80

Y

yak (*Bos gruniens*), *216*, 231

Z

Zapus hudsonius preblei (meadow jumping mouse), *129*
Zatsepina, O. G., 109–111
Zimmer, M. B., 152